CONSCIOUSNESS & REALITY!

Final and Definitive Conclusions

Jerome Iglowitz

Copyright
January 1, 2012
Jerome Iglowitz

ISBN 978-0-9850593-0-9
Library of Congress Control Number:
2012900864

JERRYSPLACE Publishing
Granite Bay, California

Dedication

For Chen

Who has taught me more about courage
than I had ever known.
Chenin-Blanc Yic-Mun-Fuung Iglowitz
(Gentle Phoenix)
March 18, 1974 – May 2, 2010
I love you kid.

Baba[I]

"... Every attempt to transform logic must concentrate above all upon this one point: all criticism of formal logic is comprised in criticism of the general doctrine of the construction of concepts."

(Ernst Cassirer)

Table of Contents

A NOTE FOR IMPATIENT OR SKEPTICAL MINDS 13
   A KANTIAN CAUTION FOR YOUR READING: 15

PREFACE: 19

PREAMBLE: (ON REALISM!) 23

CHAPTER 1. INTRODUCTION AND NEW PRÉCIS: "IN A NUTSHELL" 27
   EMERGENCE 28
   A VERY BASIC ARGUMENT – AN ARGUMENT FROM FUNDAMENTALS 29
   THE "HARD CORE": 34
   ERNST CASSIRER: 34
   DAVID HILBERT: 35
   MATURANA AND FREEMAN: 37
   BUT HOW CAN "AXIOMS" BE *PHYSICALLY* CONCEIVED? 38
   KANT 41
   THE BRAIN: A MATERIALIST PERSPECTIVE: 42
   THE "MENTAL" PERSPECTIVE: 47
   THE CONCORDANCE: 48
   "SYMBOLIC FORMS" 48
   "THE INTERFACE" 49
   AN ARGUMENT FOR VALIDITY 50
   CONCLUSION: 50

CHAPTER 2: EXOTIC MATHEMATICS: WHAT IS IT, AND HOW IS IT RELEVANT TO THE MIND-BRAIN PROBLEM? 55
   MATHEMATICAL STRUCTURALISM AND CATEGORY THEORY: 63
   BACK TO THE SOURCES OF MATHEMATICAL STRUCTURALISM AND CATEGORY THEORY: 67
   "REAL SPACE" 74
   BACK TO THE MATHEMATICAL PROBLEM: 77
   BACK TO THE SOURCES OF CATEGORY THEORY AND STRUCTURALISM: 78
   SHAPIRO PART ONE 78

CHAPTER 3: ADVENTURES OF THE MIND: A CRITICAL TURNING POINT AND THE ORIGINS OF MY CONCEPTION . 87
   CASSIRER AND LOGIC: 88
   ON CANTOR'S DIAGONAL ARGUMENT – WRITTEN 50 YEARS AGO! 97
   A REITERATION OF MY LATER REFLECTIONS ON CANTOR'S ARGUMENT: 99
   A POWERFUL ARGUMENT FOR THE STRUCTURALIST PERSPECTIVE 101
   MY CONCLUSION: 102

THE CONCEPT OF IMPLICIT DEFINITION 105
BUT HOW CAN WE CONCEIVE OF PURELY OPERATIONAL OBJECTS AS
CORRELATING WITH THE REAL WORLD? 108
THE ANTHROPIC PRINCIPLE 110
BACK TO MAC LANE AGAIN: 110
A FINAL COMMENT BY CASSIRER RELEVANT TO THIS CURRENT PROBLEM
.... 112
MAC LANE AND CATEGORY THEORY 114
BACK TO SHAPIRO: 117
THE ROSEN LETTER: (A REFLECTION ON SHAPIRO'S POSITION) 118
MODERN PTOLEMEAN PHYSICS 120
AESTHETICS 122
CASSIRER'S SYMBOLIC FORMS 123
ANOTHER LOOK AT HILBERT 125
LOGIC AS BIOLOGY: 128
RAICHLE: 130
ADDRESSING SHAPIRO AGAIN: 133

STRUCTURAL DIVIDER: MOST OF THE REMAINDER OF THIS BOOK, (CHAPTERS 4 THROUGH 13), CONSTITUTES MY PROPOSED SPECIFIC SOLUTION TO THE MIND-BRAIN PROBLEM. THE INITIAL PART WAS TO ENABLE YOU TO UNDERSTAND IT. 135

CHAPTER 4: MY FIRST HYPOTHESIS IN DETAIL: (BIOLOGY PART ONE) 137

1. ON REPRESENTATION: THE PERSPECTIVE FROM BIOLOGY 137
2. "THE SCHEMATIC MODEL": DEFINITION AND EXAMPLES. 138
   2.1 The Simplest Case: A Definition by Example 138
      2.1.1 Reversing our perspective: 142
   2.2 A Case for Schematism More Specific to Our Special Problem: Narrowing the Focus 145
   2.3 The "G.U.I.", the Most Pertinent and Sophisticated Example of a Schematic Model: the Special Case! 146
   2.4 Towards a Better Biological Model 151
      2.4.1 Biology, The Real Thing: Freeman's Model 151
      Walter J. Freeman 152
         The Peripheral Code: 156
         Cortical Mapping is Very Different, However: 157
      2.4.2 An Explicit Model of the Mind: 162
         GOD'S EYE? 163
   ON P.S.CHURCHLAND: 165
3. THE FORMAL AND ABSTRACT PROBLEM: 165
   3.1 The formal argument 165
   3.2 The Specific Case of Biology 167
      Turning our Perspective Around –a Model of Process! 167
   But How can these Brain "Axioms" be Physically Conceived? 169
   "AXIOMS" 169

REVISITING CASSIRER: 169
A UNICELLULAR PERSPECTIVE.. 170
   3.3 Retrodictive Confirmation *171*
   A Profound Teleological Consequence.. *171*
   3.4 Conclusion, (section 3)... *171*
   4. The Concordance: Biology's Proper Conclusion *173*
5. PLAIN TALK: 176
BOUNDS AND LIMITS 179
APPENDIX, (FREEMAN & AUTOMORPHISM) 181
GOD'S EYE? 184
D'ESPAGNAT CONTRIBUTES ANOTHER PERSPECTIVE: 186
   The Realism of the Accidents.. *186*
CHAPTER CONCLUSIONS AND A FORESHADOWING: 187

CHAPTER 5: MY SECOND HYPOTHESIS –A SHORT SKETCH 191

AN ASIDE FOR CLARIFICATION:... 192
THE CONCEPT OF IMPLICIT DEFINITION 197
FROM APPENDIX C: CASSIRER AGAIN 199
(AN EXTRACT FROM IGLOWITZ 1998: CHAPTER 2 TO EXPAND THE
CASSIRER AND CONCEPT OF IMPLICIT DEFINITION DIALOGUES
INCORPORATED HERETOFORE.) 199
HOW? THE LOGICAL PROBLEM OF CONSCIOUSNESS 199
CASSIRER AND CLASSICAL LOGIC: 201
CONCEPT VS. PRESENTATION: 201
CONTRA THE THEORY OF ATTENTION:... 203
MAJOR CONSEQUENCES:.. 204
   Re Presentation:.. *205*
THE CONCEPT OF IMPLICIT DEFINITION: 207
IMPLICIT DEFINITION VIS A VIS PRESENTATION: 211
WHY IS THIS RELEVANT TO MIND?... 213
CONTRA CASSIRER: 215
THE CRUX OF THE ISSUE: PRESENTATION 217

CHAPTER 6: MATURANA & VARELA & KANT. BIOLOGY-PART II: TOWARDS THE WHERE AND THE WHAT? 219

BIOLOGY & EPISTEMOLOGY, (MATURANA AND VARELA AND KANT) ... 219
CLOSURE: 221
MATURANA AND VARELA:... 226
   "Triggering" vs "Causation": *230*
   The Conservation of Autopoiesis:... *231*
   Behavior as an Aspect of Structural Coupling: *232*
   Operational Closure: *234*
   The Structural Present: *235*
   Maturana's Paradox *237*
THE AXIOM OF EXTERNALITY 243
RELATIVIZED MATERIALISM 252
AN ANSWER TO THE NEW DILEMMA: 253

  Kant's "Critical Idealism": What it *really* means! 253
  A New and More Recent Perspective on Maturana: 256
  The Parallel Postulate 260

CHAPTER 7: COGNITION AND EXPERIENCE 265

  Quine and Cassirer 265
  A Fantasy: 267
  The Axiom of Experience: 271
  The Epistemological Problem: 271
  Cassirer Revisited: 276
  The Axiom of Experience 278
  Cassirer's Theory of Symbolic Forms, an Analysis: 279

CHAPTER 8: CASSIRER'S "SYMBOLIC FORMS" 291

  Contra Cassirer: (What are the real parameters?) 298
  (1) The Unicellular Perspective – a New Cognitive Symbolic
  Form 310
  The Biochemical Symbolic Form in Cognition 313
  The Power of Naturalism: 315
  The Base Issue 323
  Conclusions: 329
  A Remarkable Parallelism 330
  Chapter Conclusions: 335

CHAPTER 9: A SIMPLER ALTERNATIVE APPROACH TO CASSIRER'S SYMBOLIC FORMS: "MATHEMATICAL IDEALS": 339

  An Alternative Approach to Cassirer's and My Ideas:
  "Mathematical Ideals": 339
  Cassirer's Theory of Symbolic Forms: 346
  The Substance of Mind: 347

CHAPTER 10: "THE INTERFACE" 351

  Here follows my Personal Metaphysical Assertion! 351

CHAPTER 11: THE LAST HURDLE 353

  The Third Hypothesis: 353
  A formal statement of my third hypothesis: 355
  The Axiom of the Interface: 355
  The Strategic Brain: a crude graphic overview 357

CHAPTER 12: TWO (RELATIVELY) CONTEMPORARY REALIST CRITICISMS OF MY CONCLUSIONS 359

  Durant on Kant: 359
  Durant Critiques Kant: 366
  The JCS Review 369
  Where Cassirer and I Fundamentally Differ: 377
  The Anthropic Principle 380

CASSIRER AND GOD'S EYE: 381
IN DEFENSE OF KANT: 384

CHAPTER 13: DISCOVERING BERNARD D'ESPAGNAT AND MANJIT KUMAR: ON QUANTUM PHYSICS AND REALITY. (BOHR, PENROSE, AND EINSTEIN, OF COURSE, WERE ALREADY PRESENT IN THE ROOM!) 387

WHY INSERT PHYSICS INTO A DISCUSSION OF THE MIND/BRAIN? 388
PREAMBLE: 389
PREFACE: 390
BUT *HOW* CAN WE CHANGE THE VERY WAY WE VIEW OUR EXTERNAL WORLD? 394
INTRODUCTION: ABOUT BERNARD D'ESPAGNAT: 396
D'ESPAGNAT PROVIDES POSSIBLE SOLUTIONS TO SOME OF MY OWN CONCEPTUAL DIFFICULTIES: 401
AN EXCERPT FROM APPENDIX C 404
TO RETURN THE FAVOR -A COUNTERSUGGESTION ON THE ISSUE OF "DECOHERENCE": 408
A PROFOUND PARALLELISM 409
BUT WHAT *IS* "MEASUREMENT" FOR THE BRAIN? 411
THE CRUCIAL TEST OF REALISM: D'ESPAGNAT ON BELL'S THEOREM: .. 413
THOUGHTS ABOUT EINSTEINIAN REALISM: 415
IS MINE A *SIMPLE* SOLUTION? 417
ABOUT "OTHER MINDS" 418
D'ESPAGNAT CONTRIBUTES ANOTHER PERSPECTIVE: 421
HOW THEN DO WE GET TO THE ASPECT RESULTS? 422
ANOTHER IDEA INSPIRED FROM READING D'ESPAGNAT 425
QUANTUM PHYSICS AND "NORMAL" COGNITION: 428
I THINK I HAVE ESTABLISHED MY CASE VIS A VIS MODERN PHYSICS ... 430
AN IMPORTANT AND VERY DEEP IMPLICATION FOR MY THESIS: 432
SO WHERE DO WE GO FROM HERE? 433
BACK TO EINSTEINIAN REALISM: 434
D'ESPAGNAT REPLIES: 436
CONCLUSIONS AND A SERIOUS PROPOSAL FOR MATRIMONY OF OUR OFFSPRING: 438
ON SCIENTIFIC REALISM AGAIN 438

STRUCTURAL DIVIDER: THE REMAINDER OF THIS BOOK IS MEANT TO ESTABLISH MY CLAIM FOR A NEW AND FUNDAMENTAL RE-ORIENTATION OF OUR PERSPECTIVE ON REALITY ITSELF –AND TO ESTABLISH A NEW BASIS FOR SCIENTIFIC REALISM! 441

CHAPTER 14: OVERALL CONCLUSIONS & OPINIONS 443

CONCLUSIONS: 445
THE PERSPECTIVE OF BIOLOGY AND EINSTEINIAN REALISM 448
CASSIRER'S SYMBOLIC FORMS 450

A Claim for a Fundamental Re-orientation in our Conceptions of Reality: 452
An afterthought: 453
Back to my claim: 454
My Ultimate Conclusions: 455
Specifically Biological Conclusions: 459
So where do we go from here? 460
Devil's Advocate: 461
So Why Bother? 463
How do we live? 465
My *Own* "Act of Faith": 465

CHAPTER 15: EPILOGUE 467

APPENDIX A: THE DENNETT APPENDIX AND THE COLOR PHI, (FROM IGLOWITZ 1995) 473

Towards a Working Model of *Real* Minds: Dennett, Helmholtz and Cassirer 473
Cassirer on the Color Phi: 479
An Extension of the Schematic Model: A Brief Sketch 481
A Thought Experiment 482

APPENDIX B: LAKOFF, EDELMAN, AND "HIERARCHY" 487

George Lakoff: 488
The Classical Concept 488
Cassirer and Lakoff's Logic 493
Putnams' Requirements 500
Lakoff's ICM's 504
An Important Distinction: Biology as a Symbolic Form 505
Maturana: 505
Edelman: 507
God's and Edelman's Eye 511
On "Presentation" 511
Re-entrant Maps 512
What Edelman has not solved: the problem of the Cartesian Theatre! 515
On Epistemology: 518
Appendix B Conclusion 523

APPENDIX C: SOME FURTHER THOUGHTS ON CASSIRER AND SOME FURTHER THOUGHTS ABOUT THE MODEL 525

Another, Related Thought: 526
The Limitations of Cassirer's Perspective: 527
Some Random Further Thoughts about my Proposed Model Itself: 528

APPENDIX D: ON NIELS BOHR AND CASSIRER'S "SYMBOLIC FORMS" 531

A REMARKABLE PARALLELISM 535
BEGIN EXCERPTS FROM NIELS BOHR INTERVIEW 536
CONCLUSIONS: 541
BACK TO THE INTERVIEW: 542
BOHR AND KUHN ON PHILOSOPHERS OF SCIENCE: 543
CONCLUSIONS: A CONFIRMATION OF MY EPISTEMOLOGICAL
FOUNDATIONS 550

APPENDIX E: (CASSIRER SPEAKS TO SHAPIRO & MAC LANE) – A CONTINUATION OF CHAPTER 2 551

APPENDIX F: AN OUTLINE OF MY OVERALL ARGUMENT ... 565

BIBLIOGRAPHY 575

ENDNOTES 579

INDEX..597

A Note for Impatient or Skeptical Minds

Impatient or skeptical minds should probably begin at the New Précis, ("In a Nutshell") —at the very beginning of Chapter 1 and then skip to Chapter 12 which shows two relatively contemporaneous criticisms of my ideas and my answers to them.[1]

Those violently opposed to materialistic approaches beforehand should probably turn to the opening pages of Chapter 8 to get some foretaste of my ultimate position on this issue. My answer is not so bleak as you may presume.

If you want to go straight to the heart of my conception however, you should probably skip directly to Chapter 4 which is the actual beginning of my thesis. I must warn you, however, that in doing so you will probably *not* understand it as you will have no understanding of its origins and context! (The first three chapters are presented to give you a basis from which, hopefully, you may interpret it. See the "Kantian Caution" to follow shortly.) It is a radically new and very different conception of the fundamental problem of cognition itself.[2] It ends with a claim for a new grounding for "scientific realism" in biology itself rather than in physics!

---

1 (Note: This is a new and definitive rewriting of my original book: "Virtual Reality: Consciousness really Explained". The original book was completed in 1995, (revised 1998). Though this book lacks some of the detail of the former, it incorporates a later and richer perspective with much new material and elucidates my second thesis far better than the original. I do not think it changes, but rather enriches the substance and sense of the earlier book -and clarifies its rationale. It also includes the new and important D'Espagnat Chapter 13 on quantum physics and an appendix devoted to Niels Bohr which I think prove my essential case.

This book uses a mix of footnotes and endnotes. The footnotes, (in ordinary numerals), are necessary for immediate clarity, but I felt the material included in the endnotes, (in Roman numerals), interrupted the flow of thought. Hence they were relegated to their endnote status.)

2 As I quoted on the cover, Arnold Leiman, a respected neuroscientist, said he thought it was an entirely "new and original theory of cognition".

I should probably also state at the very outset that my solution incorporates a conception of "epistemological relativism" at its very foundations. But consider the meaning of those words! It is a very specific and particular *kind* of epistemological relativism that I will argue! It is a precise and *scientifically rigorous* position deriving from the actual mathematical invariants[3] of our most successful scientific theories, (i.e. from those that Roger Penrose would deem "SUPERB"[4]).

It would perhaps have been better to have renamed this characterization of my ideas as something else: to have called it "the epistemological relativism of scientific invariants" for instance, but I have kept the original term as it is the one used to characterize Ernst Cassirer's position and I wish to keep the association.[5] This is *not* "anything goes relativism", "cultural relativism" or "irenic relativism"[6] however, but a specific, rigorous and purely scientific conception.[7] As such, it has a natural affinity with Einstein's equations of special relativity or Galileo's laws of motion. Ultimately, my usage is derived as an extension of Cassirer's "Theory of Symbolic Forms"[8] –but taking the latter within a very special and delimited context and with a very specific and rigorous meaning. That specific meaning is *itself* explanatory to and operative *within* my conception.

---

[3] i.e. it incorporates any given theory's fundamental *relationality*, but considers it in a *context-free* perspective. This will become clearer in Chapters 7 & 9.

[4] His CAPS

[5] Kant did something very similar when he characterized his ideas as "critical idealism" – see Chapters 4 & 6. I think it was the greatest mistake he ever made, but I will make the same gamble he did –with the codicil above- in hopes that my reader is more sophisticated than his.

[6] i.e."make everybody feel good (peacemaking) relativism"

[7] (George Lakoff had the same problem in decontaminating his own brand of "epistemological relativism" –"ICM's"- to deal with, but I think I have gone far beyond his conception.)

[8] See Chapter 8: Contra Cassirer: (What are the Real Parameters?)

A Note: A Note for Impatient or Skeptical Minds

A Kantian Caution for your Reading:

At this beginning point please let me add a final and specific caution for *your own* reading of this work. Kant defines your personal and hardest problem *as a reader* quite explicitly:

> "If in a new science which is wholly isolated and unique in its kind, we started with the prejudice that we can judge of things by means of alleged knowledge previously acquired –*though this is precisely what has first to be called in question* –we should only fancy we saw everywhere what we had already known, because the expressions have a similar sound.
>
> But everything would appear utterly metamorphosed, senseless, and unintelligible, because we should have as a foundation our own thoughts, made by long habit a second nature, instead of the author's." (Immanuel Kant, Prolegomena, my emphasis)[9]

What I propose is just what he characterized: i.e. "a new science which is wholly isolated and unique in its kind". I propose a new science of mind!

The reflections above should resolve many difficulties before they start. This is a very difficult subject to present with any perspective other than the standard ones: i.e. the very ones that have already clearly failed! Give me some space and I'll try to make a revolutionary out of you! I think the answer is important as I think it is our one chance of restoring "humanity" to the human brain!

On the other hand, let me insert an apology at this point. I am currently 73 years old, and have had several strokes which have impaired my abilities. And yet I consider the new content

---

[9] Note: The d'Espagnat letter of Chapter 13, ("D'Espagnat Replies"), almost exactly mirrors this passage as you will see!

of this book important. What it is lacking is an overall stylistic form of sufficient refinement to do it justice as my concentration has been narrowed to specific problems which I have responded to and which I think make my perspective clearer.[10] Some of the citation references might need "tweaking", but that should be achievable with minimal effort given sufficient interest. There is also a certain amount of redundancy —partly from a lack of sustainable concentration, and partly because this MS may perhaps be read in parts wherein certain citations must be explicit and considered in context. This is the "hard problem" and you'd better begin by expecting it to be so.

Jerome Iglowitz, 2012

P.S. Some of my illustrations require a higher resolution than is possible within this book format. I am therefore supplying them in a hi-resolution form at: http://foothill.net/~jerryi/Illustrations. You are free to examine or print them, but are required to acknowledge their source in any subsequent usage. Also please note that I have not included a Glossary because of the easy usage of the internet —which will probably serve your special needs better!

P.P.S. If you need a C.V. to cause you to evaluate even the very plausibility of these ideas, then you had probably best go elsewhere as you will not do well here. I abandoned academia long ago as I felt it was not possible to fully explore this huge problem within its rigid confines.[11] I think my completed answer

---

[10] Purely from the standpoint of organization, my first book is clearly superior to this one. From the standpoint of understanding however, I feel this rendition is a marked improvement.

[11] Note: Upon the conclusion of your reading, I will make a challenge to your ingenuity. I will challenge you to compress the content of this book into an acceptable paper suitable for a journal, (10,000 words). I tried innumerable times,

validates this presupposition. Is it complete and final? Of course not! This is the very beginning of a dialogue and I have repeatedly asked for help, but it will take more courage than I have found in academia to go beyond trivial answers, risk association with a maverick mind, and face up to the real problem like a man, (woman)! No sexism intended.

---

always received backhanded compliments on the part it specifically addressed, but was always derided for not covering the other issues which that particular reviewer considered the ultimate problems! (See, as just one example for instance, the "JCS review" incorporated as part of chapter 12.) The concept itself is just too big for such a format!

Preface:

There is a wonderful though longish passage by the famous logician W.V.O. Quine[1] which I will quote in its entirety to serve as an introduction:

"The totality of our so-called knowledge or beliefs, from the most casual matters of geography and history to the profoundest laws of atomic physics or even of pure mathematics and logic, is a man-made fabric which impinges on experience only along the edges. Or, to change the figure, total science is like a field of force whose boundary conditions are experience. A conflict with experience at the periphery occasions readjustments in the interior of the field. Truth values have to be redistributed over some of our statements. Reevaluation of some statements entails reevaluation of others, because of their logical interconnections- the logical laws being in turn simply certain further statements of the system, certain further elements of the field. Having reevaluated one statement we must reevaluate some others, which may be statements logically connected with the first or may be the statements of logical connections themselves. But the total field is so underdetermined by its boundary conditions, experience, that there is much latitude of choice as to what statements to reevaluate in the light of any single contrary experience. No particular experiences are linked with any particular statements in the interior of the field, except indirectly through considerations of equilibrium affecting the field as a whole.......
Furthermore it becomes folly to see a boundary between synthetic statements… and analytic statements…Any statement can be held true come what may, if we make

---

[1] (recently deceased)

Preface

drastic enough adjustments elsewhere in the system...
Conversely... no statement is immune to revision... even
the logical law of the excluded middle... and what
difference is there in principle between such a shift and the
shift whereby Kepler superseded Ptolemy, or Einstein
Newton, or Darwin Aristotle?"[1]

And another much shorter quote from another of his writings which displays the full extent of his horizons:

"One could even end up, though we ourselves shall not, by finding that the smoothest and most adequate overall account of the world *does not after all accord existence to ordinary physical things.....Such* eventual departures from Johnsonian usage", (Samuel Johnson is said to have demonstrated the reality of a rock by kicking it!), "could partake of the spirit of science and even of the evolutionary spirit of ordinary language itself."[II]

This has always been my personal goal – i.e. of "finding .... *the smoothest and most adequate overall account of the world"* –but to include *my own mind* as well! But it will involve a conceptual framework as large as Quine's.

Piaget had a relevant comment which I think is applicable. The famous child psychologist was interested in the foundations of mathematics as a secondary interest. He evaluated mathematical Platonism, and concluded, (paraphrasing):

"If a mathematician (thinker), were to arrive at some conclusions that neither he nor his readers were able to fully understand, and if he were to write these conclusions down, (that is, to date stamp them)[2], and if, furthermore,

---

[2] Note: if you should doubt the temporal accuracy of my claims, let me refer you to the online site: "The Wayback Machine" which features images of websites and their contents for a given time frame for the referenced citations: http://www.archive.org/web/web.php

Preface

they were found to be correct at some future time –then the conclusive case for Platonism would be made."

I think the argument is applicable to ideas in general. If I am right in my conclusions, (and I do not dogmatically claim that I am), then the future of science will come to my perspective asymptotically. When and if that happens, hear me again! I will probably be gone, but my cause will not be.

Finally, let me cite Kepler regarding his profound revelations in astronomy:

> "Now, since the dawn eight months ago, and since a few days ago, when the full sun illuminated my wonderful speculations, nothing holds me back. I yield freely to the sacred frenzy; I dare frankly to confess that I have stolen the golden vessels of the Egyptians to build a tabernacle for my god far from the bounds of Egypt. If you pardon me, I shall rejoice; if you reproach me, I shall endure. The die is cast, and I am writing the book –to be read either now or by posterity, it matters not. It can wait a century for a reader, as god himself has waited six thousand years for a witness."[III]

Take care, and good luck,   Jerome Iglowitz 2012

Note: I will respond to any decently offered questions or comments at jiglowitz@rcsis.com . Please put some verbiage equivalent to "In Response to your Thesis" in the subject line as, otherwise, I will probably delete it, unopened, as "spam"!

Preamble: (On Realism!)

Let me state at the outset that I am as much a realist as any one of you —maybe more so. I enjoy, and fear as well, my naïve reality at least as much as anyone[1]. It is the *foundations* of realism I question. But so does realism itself. Science continually changes the rules of the game. The world is no longer truly made up of the simple atoms of Democritus, nor is it made up of the subatomic particles of Bohr and Heisenberg. It is made up of whatever it is that was most recently proposed —and seems to work, (quarks, bosons, superstrings,…) -as "substance" or "material". Supposedly hierarchy and emergence resolve the difficulty, but is this, in fact, true?[2] (See footnote and Chapter1 –

---

[1] I have lived more on the "rough side" of life probably more than most of my expected readers, though certainly less so than a multitude of others who have been forced to deal with unimaginable horrors.

[2] "Emergence" supposedly solves the problem of hierarchy in materialist explanations of the mind-brain problem. It purportedly explains how new phenomena "emerge" from more fundamental explanations. These new emergent phenomena are said to embed themselves hierarchically in ontic material -taken at the deepest level. The conception seems to derive from, or at least be analogous to the embedding of mathematical explanations –or of computer languages, (high vs low level languages). In point of fact, however, we are allowed to embed some higher level axiom system, (or computer language), in some more fundamental or different axiom system or language *if and only if* we can prove/derive each of the axioms, (or new computer language terms), of the higher system from the lower one. But that implicit level of proof is always there. No new "phenomena" are allowed to exist in the former that cannot be reduced to perhaps more complicated implications of the grounding system. (One need only replace any usage of the axioms, (terms), of the higher system with its proof system in the lower to derive the same result.) Nothing radically new comes from such an approach. The rationale for instituting the higher system derives from operational simplicity. Nothing emerges –hierarchy will not allow it. In the computer language example, all the computer itself ever sees is machine language!

it is a total misuse of legitimate concepts drawn from other disciplines! You might also look at Chapter 13: d'Espagnat, wherein he also rejects such a "multitudinous" viewpoint based on recent experimental results from modern physics.)[3]

It is the *phenomenology* of realism —those relations that work -and the "naïve realistic world" itself —that hard, cold, violent, passionate and very concrete reality we all must live in and survive in that must be preserved. But the ever changing substance of the "objects" *per se* of realism is at constant peril. I wish to severely question realism's ultimate "objects" themselves to resolve the deepest dilemma of mankind: i.e. the mind-body relationship.

But I must do so in a way that preserves the realism of science, the realism of the naïve world, and the reality of the mind which perceives them both. This is the core and the center of my conception. I think that all of us, deep down, accept these perspectives as our most fundamental realist presuppositions. It is in the attempt at their mutual resolution that this pervasive paradox endures.

It has been said of my work[4] that I am simply repeating Kant. This is fair in one perspective —I am very much like Kant insofar as the "What" of reality is concerned, though we differ

---

Materialist explanations of consciousness of the usual sort all have this flaw. As I will state the problem later: "how can a (biological) machine/mechanism whose parts are discrete in time and space ever know anything *whatsoever*? But I mean "knowing" in a different sense than simple mechanical, "zombie-like" performance, and I think you wish it to be taken so too. "Consciousness" could never arise in any normal sense of the word! It would constitute too great a divide from the current, and specifically (*meta*)physical models of brain function.

[3] "In fact, what I can only say is that, while almost all biologists and researchers on the nature of consciousness hold firmly to ontological naturalism (to the idea that science is on the way of revealing to us reality as it is *per se*) you quite definitely rejected this idea, and that I very much approve of your doing so. Clearly this is one basic point on which we agree." Bernard D'Espagnat, personal correspondence.

[4] By an anonymous JCS reviewer who questioned my claim of the novelty and the "outrageousness" of my proposal.

about the categories and ethics, and fundamentally about epistemology. My particular thesis consists in supplying the actual "How" and the "Why" –and the "Where"- of Kant's profound insight however, and which he never even attempted to explain. I think I have accomplished that goal. If you would argue with me, argue with me here.

Chapter 13: Discovering Bernard d'Espagnat and Manjit Kumar addresses the problem of Scientific Realism directly and explicitly from the standpoint of contemporary physics. Neither Albert Einstein nor Niels Bohr, (arguably amongst the most brilliant men in the history of the world), could find a way to support ordinary "materialism"[5] or "scientific realism" as viable perspectives on ordinary or ultimate, (ontic), reality in light of recent scientific advances within physics itself –and the subsequent actual *experimental verifications* of that fact!

We, (you and I), therefore have our work cut out for us! I do not claim to be as "smart" as these giants, but I have the advantage of an entirely new perspective on the very problems themselves. I do not say that you must agree with my answer, but you must do the work, (admittedly difficult), to understand it before you can legitimately reject it. As I had originally stated on the back cover[6]: If you *really* want an answer to this problem, you must do the work!

---

[5] "multitudinism" in d'Espagnat's terminology –i.e. the ordinary existence of our normal "naïve" *or even of our discrete scientific objects*!
[6] Since removed!

Chapter 1. Introduction and New Précis: "In a Nutshell"

In my conclusion I will argue that you will have to come to the same conclusions about the mind and the brain, (but not necessarily my own), no matter which perspective you start from initially —whether from materialism, from dualism, from idealism... provided that you do it rigorously enough.

Provisionally accepting that conclusion then,[1] let me start from the *easiest* perspective therefore. Let me approach the problem as a strict materialist would see it.

First though, a codicil: all materialist explanations of science and in this instance of the mind-brain relationship must necessarily start with mechanics.

To quote Maturana:

> "The key to understanding all this is indeed simple: as scientists, we can deal only with unities that are structurally determined. That is, we can deal only with systems in which all their changes are determined by their structure, whatever it may be, and in which those structural changes *are a result of their own dynamics* or *triggered* by their interactions."[2], (my emphasis).

In this case we must start with the structure of the brain per se, and ultimately reduce it to mechanics —in this instance to the biological and physical mechanics of brain process at some fundamental level.[3]

Computer people do essentially the same thing in their quest for artificial intelligence. (I took a half dozen computer

---

[1] which we will broaden presently
[2] Maturana & Varela: tree of knowledge, [96]
[3] See Marchal, 2004 for a somewhat similar approach.

Chapter 1: Precis

classes long ago to try to see if the "brain-is-a-computer" people had anything important to say at this fundamental level. When I came to the "systems" course, I concluded that they didn't. It all came down to microcoding of the CPU which entailed essentially nothing other than "nots'" and "ands" chasing each other around the CPU at unimaginable speeds, but adding nothing new to content and no new insight to the essential problem.)

Emergence

Let me start by promoting the footnote made early in the Preamble of this book which has something to say on this subject:

"Emergence" supposedly solves the problem of hierarchy in materialist explanations of the mind-brain problem, (e.g. P.S. Churchland's[4]). It purportedly explains how new phenomena, (e.g. the "mind"), "emerge" from more fundamental explanations, (e.g. the mechanics of the brain). These new emergent phenomena are said to embed themselves hierarchically in ontic material -taken at the deepest level. The conception seems to derive from, or at least be analogous to the embedding of mathematical explanations —or of computer languages, (high vs low level languages), into deeper systems.

In point of fact we are allowed to embed some higher level mathematical axiom system, (or computer language), in some more fundamental or different axiom system -or computer language but *if and only if* we can prove/derive each of the

---

[4] Comment: It is important to note that neurophilosophers like, e.g. the Churchlands, cite philosophers of general science —like Nagel —who themselves embed neurophilosophy as substantiation for their own conclusions about emergence. E.g. "But he", [an omniscient archangel], "could not possibly know that these changes would be accompanied by the appearance of a smell in general or of the peculiar smell of ammonia in particular, unless someone told him so or he had smelled it himself."[Nagel, 1961]

This is a blatant circularity, more serious for the neurophilosophers who should know better.

axioms, (or new computer language terms), of the higher system from the lower one. But that implicit level of proof is always there.

No new "phenomena" are allowed to exist in the former that cannot be reduced to perhaps more complicated implications of the grounding system. (One need only replace any usage of the axioms, (or terms), of the higher system with its proof system in the lower to derive the same result.) Nothing radically new comes from such an approach. The rationale for instituting the higher system in fact derives from *operational* simplicity! Nothing "emerges" –*hierarchy will not allow it.* In the computer language example, all the computer itself *ever* sees is machine language[5]: i.e. ones and zeros!

Materialist explanations of consciousness of the usual sort all have this flaw. As I will state the problem later: "how can a (biological) machine/mechanism whose parts are discrete in time and space ever know anything *whatsoever*?" But I mean "knowing" in a different sense than simple mechanical, "zombie-like" performance, and I think you wish it to be taken so too. "Consciousness" could never arise in any normal sense of the word! It would constitute too great a divide from the current, and specifically (*meta*)physical[6] models of brain function.

A Very Basic Argument –An Argument from Fundamentals
(In mathematics, the *strongest* argument!)

In light of my opening comments, (i.e. my assertion of the ultimate irrelevancy of the particular choice of beginning perspective), let us therefore begin our dialogue at the materialist

---

[5] *the lowest level language!*
[6] i.e. "functionalist"

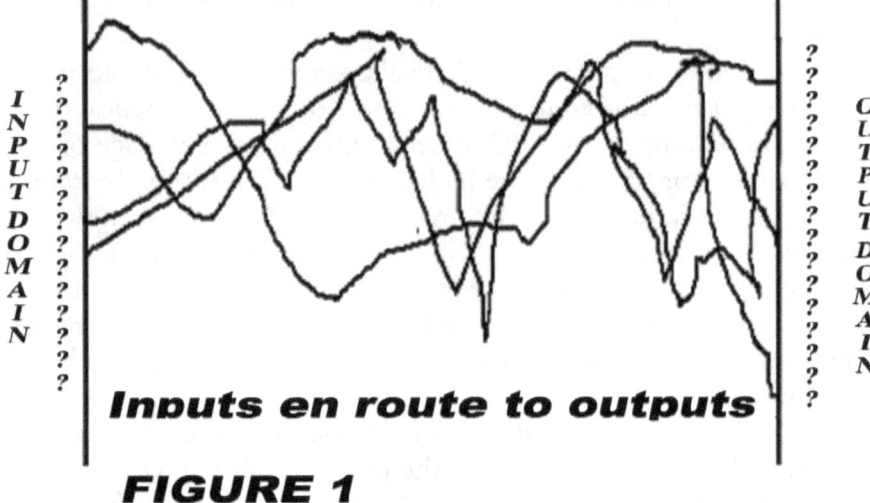

Inputs en route to outputs

FIGURE 1

level of mechanism. Let us begin at the level of the machine we call the brain.[7]

Consider your opinions and your objections well —as I will expect you to follow them to the limits of reason.

1. First of all I assert that no machine can ever "know where it is"! Now this may seem silly, but a machine only processes inputs on route to outputs. This is Nagel's "brain in a vat" argument. If we could simulate any input with a high enough level of sophistication, the machine could not tell the difference, (reversing the sense of the "Turing test").

The machine therefore lives in a space of what I will call "ontic indeterminacy". It cannot know where or what it is! (See fig.1) It is a complicated linear sequence from start to back consisting of pure mechanics —"gears and levers", chips ... It does not cognate the space which supplies its input nor does it cognate

---

[7] Important Note: I will continue within this particular perspective until we reach Chapter 8 as it greatly simplifies the problem of the communication of my ideas!

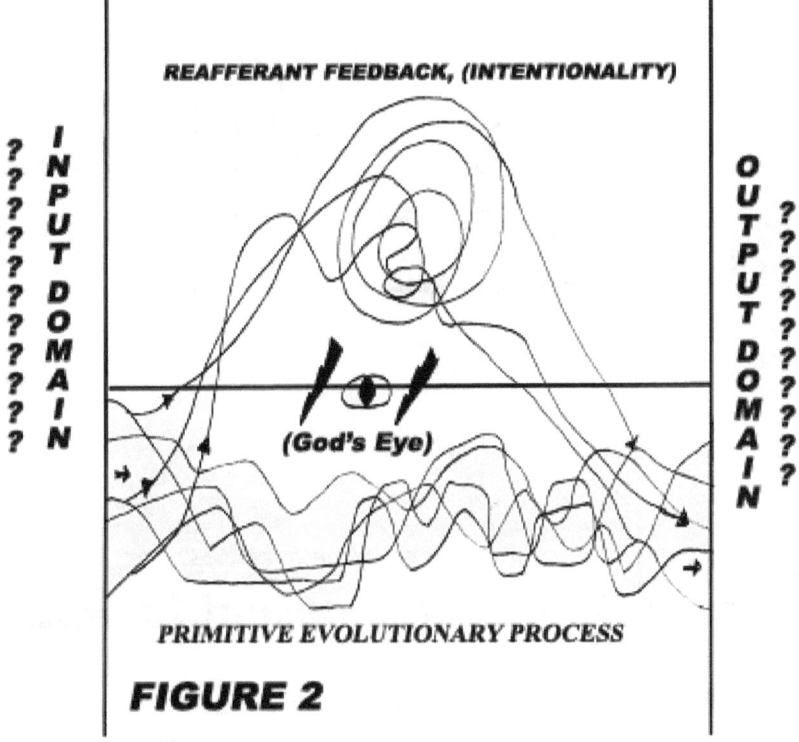

FIGURE 2

the space wherein its output is received. And it doesn't "care"! There is nobody home! (Important Technical Footnote[8].)

2. But for higher order, better functioning machines, we would want some form of feedback to allow them to "learn". That "learning", however, must be understood solely in the sense of a progressive optimization of the initial process, (see figure 2) But again there is nobody home!

---

[8] To repeat a prior reference, some of my illustrations require a higher resolution than is possible within this book format. I am therefore supplying them in a hi-resolution form at: http://foothill.net/~jerryi/Illustrations . You are free to examine or print them, but are required to acknowledge their source.

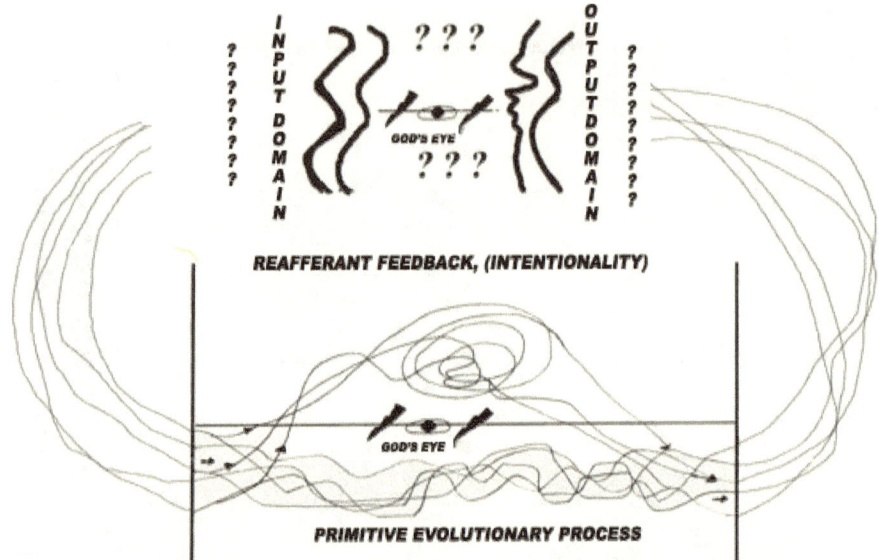

Fig. 3: Connecting output to input

3. A significant point occurs at this stage however. The "learning" in the step just above leads us to bend the linear diagram into a circle. (See Fig. 3) What good would feedback do if it were not imprinted right back onto the very output which itself then again re-affects its input? It implies *some connection* between its input and its output domains. This is the one good thing I found in Merleau-Ponty.

To quote W.J. Freeman:

> "In particular, Maurice Merleau-Ponty in "The Phenomenology of Perception" [2] *conceived of perception*" [itself] "as the outcome of the "intentional arc", by which experience derives from the intentional actions of individuals that control sensory input and perception. Action into the world with reaction that changes the self is indivisible in reality, and must be analyzed in terms of "circular causality" as distinct from the linear causality of events as commonly perceived and

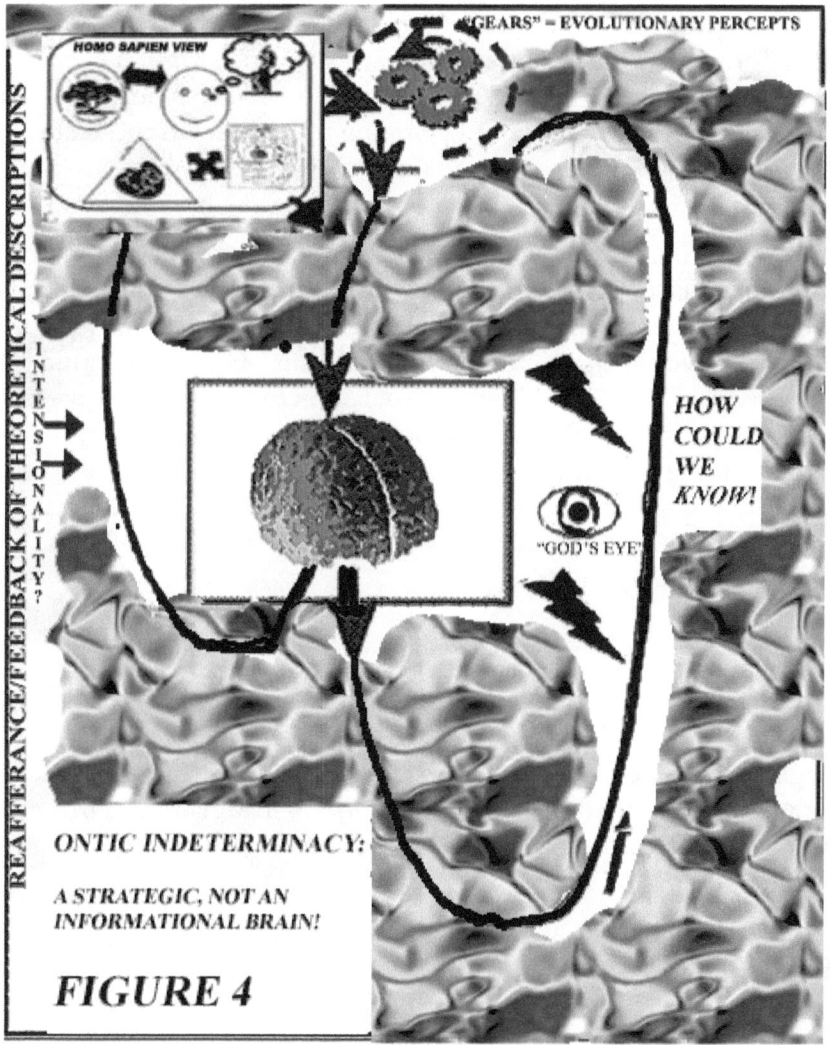

FIGURE 4

analyzed in the physical world." W.J. Freeman, 1997, my emphasis.

But this is essentially the same conclusion I derived in the first version of my paper "Mind-Brain: the Argument from Evolutionary Biology". (See Fig. 4.)

4. But the "where" and the "what"–the "what and which" of the input/output domains remains just as indeterminate

at this step, (Figs. 3 and 4), as it was in steps one and two. There is *still* nobody home! And it is *still* just an automaton!

5. This, however, is precisely *the particular model* I propose as the initial stage in beginning to understand the brain mechanism! If you are a materialist, I think you must necessarily accept it.

On the face of it, this result seems profoundly damning to even the very possibility of "mind" in all the normal senses of the word. But I assert that this model is fully rigorous and fully legitimate within the confines of materialism. How then could there even exist a "mind" within such a picture? Where is there even the *possibility* of such a thing? Mechanisms just *do*; by definition they cannot "know" in the sense we all mean the word and in the sense of the materialist picture sketched above. So it seems I have just disproved the possibility of "mind" in all our intuitive conceptions of it.

The "Hard Core":

This is the "hard point" around which my conception begins and centers -and becomes meaningful however! I should emphasize here that this is a problem for *all* materialists. Their best answers to date are vague and ambiguous at best and duplicitous at worst.

Ernst Cassirer:

[Note: at almost at the very beginning of my mature intellectual life −at 18 years old, I was lucky enough to be exposed to Ernst Cassirer's brilliant and persuasive reconceptualization of the very word "concept" itself.[9] He defined it as a complex *rule*,

---

[9] As a foretaste −against the classical concept based on abstraction/set intersection: "Similarly, we could not maintain the general concept of "animal", if we abandoned in it all thought of the aspects of procreation, of movement and of respiration, because there is no form of procreation, of

as "a new form of consciousness", and it has reoriented my entire world from that time to the present. The whole of this discussion must be conceived in that context, i.e. of "the concept" *as a rule*! See the very opening lines of Chapter 3 and Chapter 5 where I explore this conceptual revolution in depth.]

David Hilbert:

6. Early on when studying mathematics, I had a revelation pertinent to this issue of the "hard core".[10] There was precisely one sense I concluded, (and I challenge you to suggest some other), wherein an *actual* possibility consistent with science —and with the materialist picture above –arose. There was one case, I found, wherein a *purely operative* system, "a machine" *could* in fact know something!. It could know its *own* "objects"! I discovered the possibility in David Hilbert's profound, (though purely mathematical), concept of "implicit definition" for mathematical axiom systems. Was it a vague correlation, did it need deepening and reorientation to this specific problem? Of course it did. (See Chapters 2, 3 and 5 for a full discussion of the idea and an explanation of my interaction with it.)

Though solely mathematical entities of course, Hilbert's "axiom systems" actually *define* the specifically mathematical "things" they contain, (their "objects") —and they actually *know* them in the profoundest sense of "meaning" itself! In the context of his novel axiomatization of Euclidean Geometry, what in fact is a "line"? What is a "point"? These concepts arise from the *whole* of an axiom system, (see citation below), and it is only *as a whole* that they can know them —and they actually do!

Here is a quote from Hilbert answering an objection to his conception made by Gottlob Frege:

---

breathing, etc., which can be pointed out as common to all animals." (My emphasis) See Chapters 3 and 5.
[10] See Chapter 3

> "It is impossible to give a definition of 'point', for example, since only the whole structure of axioms yields a complete definition. A concept can be fixed logically only by its relations to other concepts. These relations [are] formulated in certain statements [which] I call axioms, thus arriving at the view that axioms are the definitions of the concepts." (Hilbert via Shapiro[I])

And another:

> "I do not want to assume anything as known in advance. I regard my explanation ...as the *definition* of the concepts point, line, plane ... If one is looking for other definitions of a 'point', e.g. through paraphrase in terms of extensionless, etc., then I must indeed oppose such attempts in the most decisive way; one is looking for something one can never find because there is nothing there; and everything gets lost and becomes vague and tangled and degenerates into a game of hide and seek." (ibid, my emphasis)

Here was Moritz Schlick's[11] early characterization of Hilbert's brilliant original conception:

> "[Hilbert's] revolution lay in the stipulation that the basic or primitive concepts are to be defined just by the fact that they satisfy the axioms.... [They] acquire meaning only by virtue of the axiom system, and possess only the content that it bestows upon them. They stand for entities[12] *whose whole being* is to be bearers of the relations laid down by the system."[II]

Otherwise stated: its "objects", i.e. its "entitites" –are a function of the system/structure *itself*; the system is *not* a function of its objects! These latter are, in fact, clearly and specifically

---

[11] The founder of the famous "Vienna Circle" of philosophy
[12] Please note his use of the word "entities"

virtual objects![13] They "acquire *both* meaning *and* content "only by virtue of the axiom system"! The discovery of this conceptual possibility opened the keyway to the solution of my particular "hard problem", (defined above), that I had sought!

Maturana and Freeman:

If the mechanics of the brain were biologically analogous to such an "axiom system", (think of nerve nets –W.J. Freeman's "equivalence classes" perhaps, as "axioms" as I will later suggest, or more simply, as the most primitive operational "rules" of the brain[14]), and if the "we", ("my mind"), were taken to be the *whole* of that system of the brain, (remember Hilbert's reference to the "whole of the axiom system" above), then it would indeed be possible for "us", (the "me"), to *actually know* something, (*sans* any necessity of a homunculus), in something like our usual meaning of the word. We, (I), could know our "objects" in the *profoundest* sense of "knowing"! "Meaning", "knowing", I think, are Hilbert's special gifts to us.

The bad part of this, however, is that the only thing we, (I), would be *capable* of knowing would be the implicitly defined objects of the biological "axiom system" itself –i.e. its virtual objects/artifacts –themselves relevant *only* to the mechanism itself.[15]

7. This latter was the huge problem I addressed in my first hypothesis, (Chapter 4), wherein I argued that the brain is *organizationally* rather than referentially defined. I argued that our very "objects of perception" themselves, (our "gears and

---

[13] See Resnick's discussion of mathematical structuralism in Chapter 2 which essentially reaffirms this interpretation.

[14] i.e. in terms of the rules of Maturana's "structural coupling" which is the subject of Chapter 6

[15] See Chapter 13: "D'Espagnat Provides a Possible Solution to Some of my Own Difficulties" especially on the issue of singularities and the "static problem".

levers" = our naïve objects), are organizational and virtual only – that they are the evolutionarily derived, implicitly defined metaphorical and virtual reflections *solely of process itself*[16]–and not representations! Taking "axioms" in a biological/mechanical sense however, seeing them as the most primitive operative units of brain biology[17], we are allowed therein for the very first time to *legitimately* conceive, (i.e. as materialists), of an *actual* physical mind!

But how can "axioms" be *physically* conceived?

8. It does not seem difficult to conceive of "axioms" *as rules per se* –even within pure mathematics, so the conception of the most primitive operative units, (rules), in the brain in the same light –via Maturana's "structural coupling" *as rules* [18]- does not seem problematical. Then, expanding our scope to include Hilbert's "implicit definition" of an overall *system*[19] of purely mathematical axioms, (rules), to what I propose is an analogous case for the self-organization of the rules of the brain seems to me to be straightforward. But *this* system of rules *self-defines* its own "objects"!

9. (Some of you, I am sure, have some limited knowledge of Hilbert and his concept of implicit definition. I had a reviewer totally mischaracterize it in his response as solely a formalistic theory of mathematical proof, but it was profoundly larger and different from that, (see chapters 1- 3). True, I believe that Hilbert later went astray, but I think the young Hilbert saw

---

[16] See Chapter 4 heading: "Turning our Perspective Around –a Model of Process!"

[17] In the sense of Maturana's "structural coupling" –See Chapter 6

[18] and in reference to Cassirer's "Mathematical Concept of Funceiton" mentioned earlier. See Chapters 3, 5 and 6.

[19] See his references in Chapters 1 & 2 to *the whole of the axiom system* as necessary to the implicit definition of its terms.

something that he later forgot. I think he was, in the language of Chapter 2, clearly the first "mathematical structuralist"!

10. One last point here and it is highly relevant to our base problem: I believe in "other minds", (and I think you do too) – which, I think defines much of the rest of our problem.

These minds, I believe, see through the exactly same evolutionarily derived "gears and levers" that I do. That our conclusions about reality should therefore agree neither surprises nor impeaches me, (contrary to Durant's similar negative commentary on Kant. See Chapter 12 re: Durant). I believe we *all* see with the same indeterminacy that Figure 5, (below), shows, but through the *same* parameters, i.e. through the *same* "gears and levers"[20]! [III]

---

[20] i.e. through the same perceptual artifacts

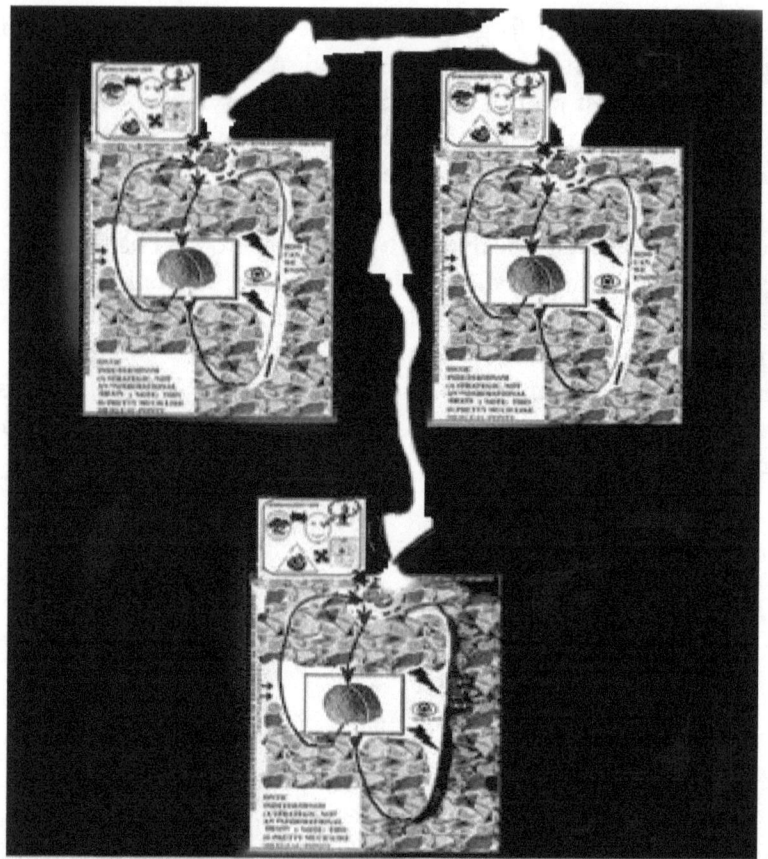

FIGURE 5

This is the model I propose for human reality, but it is lived through the "gears and levers" of our evolutionary artifacts, the latter being understood from the perspective of Biology, itself being just one of Cassirer's multitudinous "Symbolic Forms". This, I believe, is the homo sapien cognitive reality!

(Please note how closely Figure 5 resembles the picture of philosophical idealism. But the "black space" is *not* non-existence; it is ontic *unknowability*!) (See technical footnote to Figure 1.)

This, I assert, is the reality of our human linguistic and cognitive world: we all speak the same language, but we are all equally ontologically blind! Therefore the totality of our dialogue must be interpreted "heterophenomenologically", (using Dennett's word).

Kant

I guess I could quote Kant ad nauseum at this point, but I will not. I consider my ideas an extension and a completion of much of his conception. I feel that Kant was, and still remains the deepest thinker on the mind-brain problem.

End note. :

A more explicit Nutshell summary extracted from an early webpage rendition follows below:

Old Precis Follows:

The Brain: A Materialist Perspective:

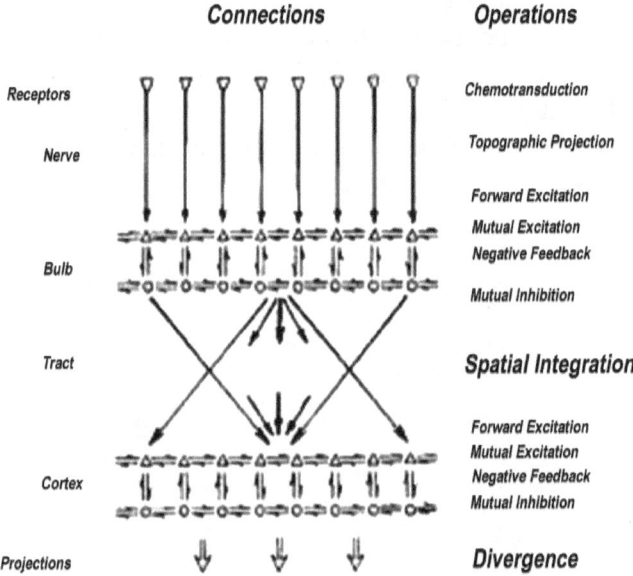

FIGURE 6, (Freeman's Figure 2)

1. From the materialist perspective,[21] what I propose is that "mind" is specifically a function of the organization of behavior itself, and not a function of knowledge. Loosely stated, I propose that the brain/mind is the evolutionary result (by a multicellular organism) of an optimization of process itself. It is the result of the *self-organized* evolutionary optimization of primitive "atomic" and reactive process —but an optimization of *blind* behavior per se and not one of knowledge!

---

[21] To be broadened presently —see Chapter 8

In that process, I maintain that our naïve perceptual "objects" are non-representative, purely behavioral, (i.e. organizational and virtual), *artifacts*, but stable ones. (This, though biologically plausible, is a *very* radical hypothesis, but I believe it is the only viable scientific pathway to the solution of the *other* leg of the problem –i.e. my second hypothesis.)[22]

I propose that these artifacts/"objects" are *re-used* in the "intentional arc", (re: Merleau-Ponty), to test our (behavioral) hypotheses –i.e. both scientific and non-scientific. They are the ground for the whole of cognition.

But these artifacts, (our naïve objects), need not correlate *hierarchically* to absolute reality, (see W.J. Freeman for instance, - Freeman's fig. 2—my Figure 6 above wherein he reveals a specifically *non-hierarchical* mapping of sensory input into, (not onto), the cortex). It is necessary only that these naïve "objects" be locked into the re-entrant loop between action and perception which passes we-know-not-where!

(Note how closely this perspective of "circular causality" fits with modern quantum theory –i.e. in the Schrödinger equation vis a vis "measurement"![23])

> "But there is something very odd about the relation between the time-evolved quantum state', (the Schroedinger equation), "and the actual behaviour of the physical world that is observed to take place. From time to time –whenever we consider that a 'measurement' has occurred – we must discard the quantum state that we have been laboriously evolving, and use it only to compute the various probabilities that the state will 'jump' to one or another of a set of new possible states." (Penrose, 1989, pps. 226-227)

But each new instance of a measurement causes yet another "loop"! The mind, I assert, is a similar looping and circular probability machine –in this case utilizing the

---

[22] See Chapter 5
[23] You might also consider it in the light of the Raichle discussion of Chapter 3.

feedback/intentional aspects of the brain. It must countenance each "measurement" against our biologically innate, (and specifically *stable*), evolutionary objects/artifacts and then recompute its overall picture and strategies. But that "picture" is one of process itself, not of "representation". This is what cognition is.

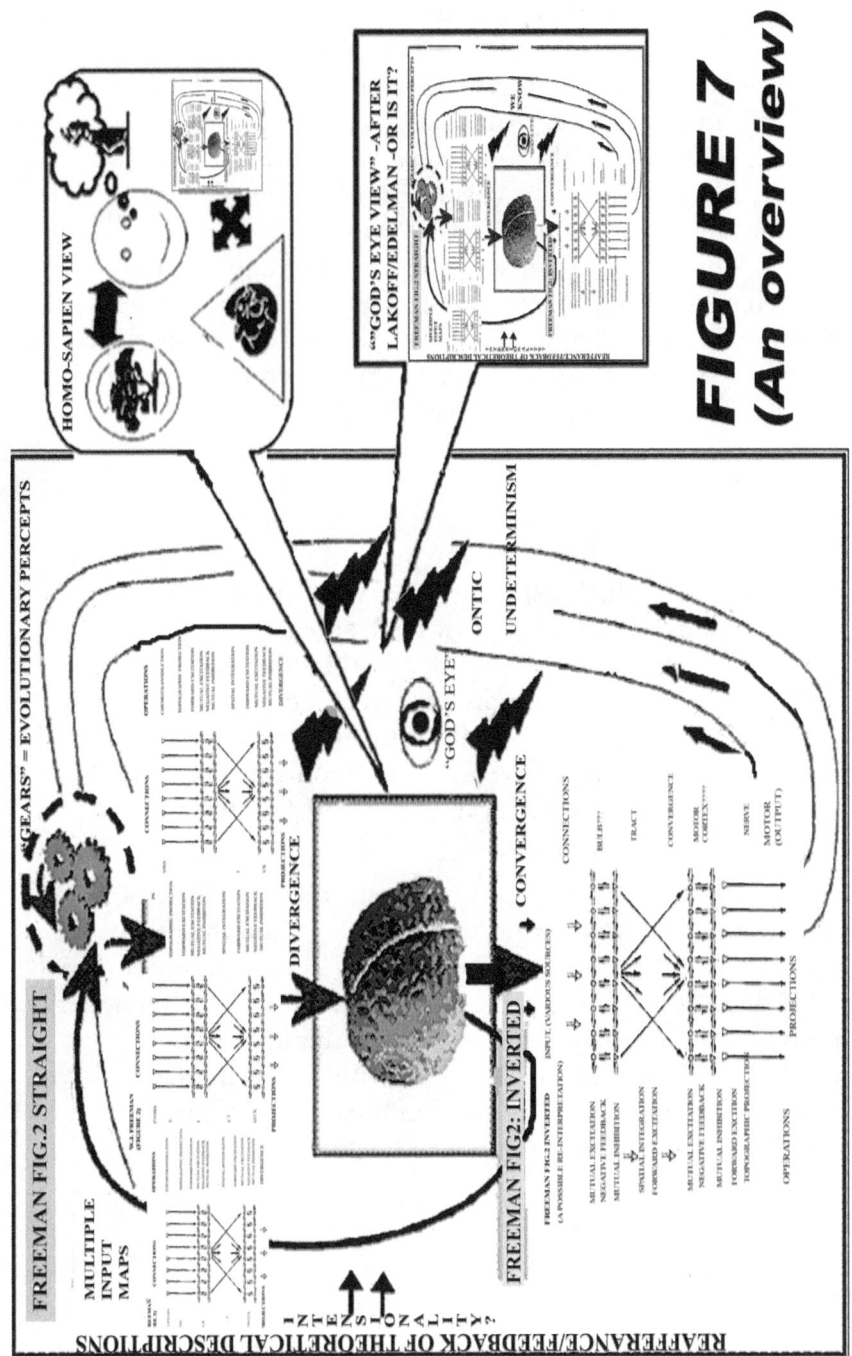

I maintain that our mental, (i.e. naïve), "objects" are the evolutionary yardstick we carry. They function to crystallize and organize our input, and to crystallize and organize our output. But they must be *rigidly* maintained as the "working gears", (alternatively the "A/D converters", or, better still, as the hierarchical/non-hierarchical converters), of perception. I argue that they are organizational artifacts only!

This is the answer to the question of how a non-hierarchical mapping, (e.g. Walter Freeman's chaotic dispersive mapping, or Edelman's non-topological "global mapping"), could specifically function in cognition. I think it also gives a very pointed clue to Penrose's problem.[24]

> "In particular, Maurice Merleau-Ponty in "The Phenomenology of Perception" [2] conceived of perception" [itself] "as the outcome of the "intentional arc", by which experience derives from the intentional actions of individuals that control sensory input and perception. Action into the world with reaction that changes the self is indivisible in reality, and must be analyzed in terms of "circular causality" as distinct from the linear causality of events as commonly perceived and analyzed in the physical world." W.J. Freeman, 1997 {22} (as cited previously)

This particular thesis, (my first of three hypotheses), supplies the necessary perspective of biology and the brain. It is our very own "cave of shadows", (Plato), -but it need not even be projective, (as a "shadow")! I propose that it is the evolutionary result of a self-organized and virtual optimization of pure response. It is instead as a G.U.I., (graphic user interface), rather than as a "shadow" or a "projection" that it functions. And G.U.I.'s actually have the potential for this.

---

[24] i.e. the *direction* of time!

This potential *per se* was a specific target of my argument in my paper: "Why: Mind- the Argument from Evolutionary Biology, (Virtual Reality -A Working Model)". It culminated in my discovery and interpretation of the experimental neurological researches of the noted neurophysiologist Walter J. Freeman which validate *exactly* that possibility. I argue that our "objects" are deep metaphors of process, and virtual; they are not objects, (even indirectly), of representation!

The "Mental" Perspective:

(2) Mind as the functional organization of behavior, (as proposed above), gives us the first viable answers to the other profound questions of mind. It gives answers to the "homunculus" problem, to the "Cartesian theatre" problem, to the problem of "meaning", and to Leibniz's pentultimately profound question: how can the one know the many, (that is, how is it possible that a unity, (e.g. a unified mind), could somehow actually know the (localized) "objects" it contained – i.e. "the many")?

These answers are found in the specifically *operative* interpretation and application of David Hilbert's mathematical thesis of "implicit definition" as applied to the working "axioms"[25] of the brain. Implicit definition allows an operative knowledge specifically of functioning itself, (sans a homunculus); it does *not* allow "representative knowledge".

But this is "knowing" in all the crucial aspects we require as mechanisms! This "mental" perspective is that of "mind" itself, (rather than of "brain"), and constitutes my second and central hypothesis. We can know our "objects" if (and *only* if), they are specifically (and purely) operative objects! This is the whole sense of Hilbert's sally as interpreted within the context of materialism. Mathematics has already solved this problem in principle!

---

[25] i.e. the most fundamental and primitive "rules" of the brain

The Concordance:

(3) At this point I argued what I called "the concordance" between my first and second hypotheses above. I argued that hypotheses one and two are fundamentally *isomorphic*. It is proved by reconstruing *and embedding* logic, (*-per se!*), as a purely and solely biological and evolutionary faculty –which I argue is the *necessary and unavoidable* interpretation for any strict materialist. This reconception forces the identification of my first two theses. It is just such correspondences that are the most valuable clues to science.

(4) From there, my thesis gets harder, but justifiably so I think. It is very "sophisticated", (in the mathematical usage of the term), and relativistic, most of it lying outside the bounds of a short précis such as this, so I will merely sketch its outline. (See Chapters five through ten for a full explanation). It will take a very sophisticated mind to comprehend it adequately, but I think it actually does complete the project I initially set myself. I think it actually does answer the question: "What and where is the mind?"

"Symbolic Forms"

Employing a refinement and delimitation of Ernst Cassirer's "Theory of Symbolic Forms"[26], I argue a case of ontic indeterminism, (likened and clearly similar to the Input/Output Domain question discussed initially in this précis). I argue that it is a legitimate extension of Kant's beginnings. I propose that such a modification of Cassirer's thesis is the only plausible answer for what it is that we must finally consider ourselves, (that is, scientifically –from the materialist perspective again) *-as purely*

---

[26] Narrowing its scope to the physical sciences

biological organisms. Organisms, (aka Mechanisms), do not *know*, organisms *do* -organisms are "triggered", (after Maturana). Or rather, the only "knowing" of which we are capable is an *operative* knowing –following Hilbert's lead- of the artifacts of our very own process! Ontology is, and must always be, an indeterminate.

5. But Cassirer's "Theory of Symbolic Forms" provides much broader and deeper insights as well. It provides the means for the mutual reconciliation of the many perspectives on the mind-body problem promised in the opening sentences of this chapter. Cassirer argued that each of the perspectives of thought asks its own legitimate questions in the very beginning, "each from its own standpoint". Each employs an implicit *logical* context specific to itself! Without, *or in the act of relativizing* that specifically logical context, the "object-in-itself", (i.e. the ontological "object"), becomes "a mere X". (Cassirer)[27]

How close his conclusions are to our beginning materialist perspective –to the brain/machine's total inability to know its input/output domain and to the purely intentional, (specifically with feedback), functioning of that mechanism! The further implications I have drawn from Cassirer's "Symbolic Forms" reconcile these multitudinous perspectives and the broader perspectives of epistemology as well and makes them whole.

"The Interface"

6. In Chapters nine and ten, I suggest "interface" itself, defined abstractly and by necessity heterophenomenologically[28] as the invariant commonality, the "mathematical ideal",[29] of all materialistic interpretations of the sensory boundary. This, I argue is a necessary and legitimate realist ontological existence postulate in itself! Of those realist ontological existence postulates, I assert

---

[27] Please note that the "mere X" is specifically conceptual, not substantial! –see Chapter 2
[28] Using Dennett's word again
[29] See Chapter 9

there are exactly three –largely parallel to Putnam's postulates of realist belief, but substituting my "interface" postulate for his number 4, (of "real knowledge"), which I obviously cannot accept.[30]

I then propose "interface" as being essentially equivalent to the concept of the GUI[31] presented earlier. (This is my third and final hypothesis.) Each is "implicitly defined", and I argue that they are isomorphic! (See Chapter 10). Granting the actual *ontological* existence of this "interface", then, it in itself supplies the ontological reality of an *actual* mind! All the "hard problems" have been solved en route to this point.

An Argument for Validity

Chapters 13 & 14, along with Appendix D, (Bohr), constitute a strong argument for the validity of my thesis, and point to far stronger implications. I will let you discover these for yourself!

Conclusion:

Mine is admittedly a very long and a very complicated solution, but it is the nature of the problem and not my inclination which has made it so. I think you probably expected a 10,000 word answer to a 60,000 word problem. The normal size of scientific papers is about that word length, and I guess that most ordinary ideas could be covered in such a scope –at least in summary. But I think any even reasonably comprehensive, *mere statement* of an answer to this particular problem will require at least 60,000 words -and with a conceptual depth to match.

Kant made a highly relevant comment on this point, (of necessary explanatory "size"):

---

[30] See Appendix B: "Putnam's Requirements"
[31] See Chapter 4

> [The problem of the mind] "is a sphere so separate and self-contained that we cannot touch a part without affecting all the rest. We can do nothing without first determining the position of each part and its relation to the rest ... It may, then, be said of such [an argument] that it is never trustworthy *except it be perfectly complete*, down to the minute elements [of pure reason]. In the sphere of this faculty you can determine and define either everything or nothing." ("Prolegomena", P. 11, my emphasis)[32]

Now finally, hear Cassirer:

> "A glance at the history of physics shows that precisely its most weighty and fundamental achievements stand in closest connection *with considerations of a general epistemological nature.* Galileo's 'Dialogues on the Two Systems of the World' are filled with such considerations and his Aristotelian opponents could urge against Gallilei that he had devoted more years to the study of philosophy than months to the study of physics. Kepler lays the foundation for his work on the motion of Mars and for his chief work on the harmony of the world in his 'Apology for Tycho', in which he gives a complete methodological account of hypotheses and their various fundamental forms; an account by which he really created the modern concept of physical *theory* and gave it a definite concrete content. Newton also, in the midst of his considerations on the structure of the world, comes back to the most general norms of physical knowledge, to the *regulae philosophandi*
>
> ... But all these great historical examples of the real inner connection between epistemological problems and physical problems are almost outdone by the way in which

---

[32] It is this fact which forces me to the level of detail embodied in this book!

this connection has been verified in the foundations of the theory of relativity.... Einstein...appeals *primarily* to an epistemological motive, to which he grants...a decisive significance." [33] (Cassirer: "Einstein's Theory of Relativity", Pps.353-354, my emphasis.)

How could you think that our *particular* problem –the self-referentiality of the brain- would not require such epistemological considerations *more than any other*? Our conclusions must turn upon themselves to validate our very beginnings. They are progenitors and antecedents of theories. But these would have to be an integral *part* of the new science, not mere reflections upon it –as, in fact, were the epistemological presuppositions of the entire history of our greatest thinkers on our hardest problems. Philosophy, i.e. *constructive* philosophy in the service of science and integrated *within* the science must be our focus.

It is a current buzzword amongst neurophilosophers that the solution to this problem will be "multidisciplinary", but most of this is merely talk, supporting and applying mainly to the assumed "obvious truths" of naïve realism. My argument is that this is *truly* a multidisciplinary problem, involving radical departures across the *whole spectrum* of human thought. My thesis actually fulfills this core requirement within a plausible perspective. In some ways, surprisingly, my conclusions are very similar to our current deepest scientific worldview except that they substitute the idea of a non-hierarchical GUI for the notions of hierarchical embedding and emergence.[34] Our world is not a "shadow"; it is an algorithm! My thesis will require an intellectual sophistication that we are not normally required to maintain however. But whatever made you think that a solution to this

---

[33] Modern Quantum Physics is a further and an extreme example of this. See Chapter 13

[34] wave/particle duality in Quantum Mechanics– i.e. "complementarity" is an obvious exception to the contemporary schema

millennia-old problem would be simple? If you read it, I will answer.

    I believe the very act of the presentation of any adequate solution to this problem is probably the hardest (technical) writing problem that has ever existed. There are so many preconceptions and prejudices, so many "prior certainties", so much confusion over even the basic beginnings, that it is almost impossible -and the resulting initial reactions often strongly hostile or dismissive.

    There is also, I feel on the other hand, a built-in biological prejudice against a real answer. (Absolute dedication to the innate algorithm is clearly biologically essential.) I need, (and anyone with a similar case needs), active participation from my, (his), reader -and the realization of the necessity for a bravery to believe differently. The problem demands it.

    My original book stated my basic case, but there were crucial later advancements over the years in my online papers, "A Very Different Kind of Model: Mind, The Argument from Evolutionary Biology", and "A Shortcut to the Problem: Consciousness per se!" This new book gives the best overall rendition of my conception as it attempts to outline the origins of my own very different beginning perspective on the basic problem. That perspective is very unlike any you have ever seen before. The new book also contains an entirely new chapter on the work of the renowned quantum physicist Bernard d'Espagnat, and a new appendix on the philosophical and epistemological origins of Niels Bohr -the "father" of Quantum Mechanics. I believe these two sections actually confirm my thesis, and come from the most modern of modern science itself.

    I will ask that you examine my whole case before rendering a judgment. I start out with an extremely abstract approach, but reach very concrete and specific answers. I think this is the shortest and easiest path between this profound problem and its solution.

Chapter 2: Exotic Mathematics: What is it, and how is it relevant to the Mind-Brain Problem?

(A deeper look at Hilbert)[1]

I have been very careful in my choice of words in the chapter heading because I do not wish to be misunderstood. By "exotic mathematics" I do not refer to arithmetic, ordinary algebra or the slight but profound extension of the latter which is called "the calculus".

Nor do I refer to the manifestations of formal logic, ordinary set theory, statistics or ordinary topology. Instead I wish to refer to new developments in the very core of mathematics itself. I refer specifically to Mathematical Structuralism and to Category Theory which have been proposed as new and radical foundations for the whole of mathematics.

It is an old saying that reality is written in the language of mathematics. The question I propose to address is whether that language is only descriptive and pragmatic or is it fundamental to the problem —and to reality itself. And why would the *philosophy* of the foundations be relevant? It is relevant just in case the latter is true —i.e. it is relevant if the foundations of our reality are just the foundations of mathematics itself!

I believe there remains just one truly significant question in the mind-brain problem. It is a simple question, but it is as crucial as it is blatant, i.e. how can a (biological) machine/mechanism whose parts are discrete in time and space ever know anything *whatsoever*? That it can mechanically function is no longer in question, but mechanical functionality is not the same as "knowing". (Dennett addressed this problem

---

[1] Note: The essentials of this chapter are presented here but, since it required still greater depth and because, furthermore, I felt that such an extension would have interrupted the flow of my primary argument, I continued it as Appendix E which is therefore to be considered as part of this chapter.

quite well, but, he concludes, we are necessarily mechanical "zombies"!!)

This was the question that overpowered me over fifty years ago, and the one I still must reevaluate each day and every day -even today. I think it is the relentless, recurring and irrepressible question underlying all objections to any proposed solution of the mind-brain problem.

Long ago when I was very young I was extremely lucky in that I stumbled across what I saw as the beginnings of an answer.[2] In the summer of my nineteenth birthday I read Saunders Mac Lane's "A Survey of Modern Algebra".[II] This book, very little concerned with ordinary "algebra" per se, was significant not because it dealt specifically with Mathematical Structuralism or Category Theory as such, but because I believe the book's very own structure and content was framed within that context.

The very layout of its content inherently defined its origins —and I perceived them intuitively. (Mac Lane, of course, was the actual co-discoverer/inventor of mathematical Category Theory which he conceived long before writing this particular book.) I think the book proselytized its origins -*sotto voce*!

In my autodidactic digestion of this book in that summer over fifty years ago, it became apparent to me that its subject matter had more to do with structures themselves and very little to do with the content of the "objects" of those structures. I understood it furthermore as an exploration and extension of the mathematical possibilities of all abstract "orderings" rather than about ontology.

But I understood that "ordering" itself largely from the perspective of Ernst Cassirer's "Substance and Function"[3] which I had read earlier and whose relevance to this specific mathematical and logical problem we will examine later.

---

[2] The question was highly pertinent for me in that my mother had been recently diagnosed as paranoid schizophrenic.

[3] And specifically from the standpoint of his redefinition of the word "concept" itself as "The Functional Concept of Mathematics" —see Chapter 5

The strong implication I derived from Mac Lane's book, (under the perspective of Cassirer's ideas), was that the "objects" of mathematics were in fact only virtually defined "positions" – defined *implicitly* within the structures of their axiom systems.[4]

Resnik, one of the leading contemporary proponents of mathematical structuralism and recently discovered by me, says it this way:

> "The view's [Mathematical Structuralism's] leading ideas are that mathematics studies structures or patterns and that reference to mathematical objects figures in this study only as a means for depicting structures.
>
> Mathematical languages do not refer to determinate elements of fixed domains, as, to use Hellman's term, "objects-Platonists" hold, but rather, in so far as they refer at all, they refer to positions whose identities are fixed only through their relationships to other positions in the structure under discussion." [III]

Here is a relevant quote from my original early manuscript, (Iglowitz 1998), on the subtleties of the mind-brain problem which proposed an equivalent view:[5]

"The mere existence of a multiplicity of alternately viable calculuses, (sic), and the allowable incommensurability of their objects suggests an interpretation of the 'objects' of those models contrary to representation or denotation however. It suggests the converse possibility that the function and the motivation of the 'objects' of those models, specifically *as entities* per se, (in what I will call these "schematic models"), is *instead* to illustrate, to enable, –to crystallize and simplify *the very calculus of relation proposed between them*!

---

[4] –which, (via the axioms themselves), was solely and precisely the way I was actually obliged to work with them!

[5] In this current book, this argument is made in Chapter 4 for my "schematic model".

They are a byproduct of deep ordering![6] The "objects" of these schematic models, I propose, are manifestations of the structure; the structure is *not* a resolution of the objects. It is the structure which is ~~predictive~~, [or better, "operative"]; its objects merely ~~enable it~~, [enable the simplification of its "calculus"]![7] I continued: "The rationale for this move comes from Hilbert's profound mathematical notion of 'implicit definition'…" [Note September, 2008: Hilbert is the focus of much of the current debate over mathematical structuralism and Category theory –see Shapiro section to follow.] [8]

---

[6] See Chapter 3 for a detailed analysis of Cassirer's conception of the innate ordering of Concepts themselves.

[7] Consider this statement from Quine's opening perspective as cited in the Preface.

[8] I continued: "(Under this conception, the 'objects' of a mathematical system are not given beforehand, but are, in fact, a resolution of the operative rules of the system itself. [See prior Schlick quote.]   They exist solely as focuses, nexuses (sic), [i.e. *virtual expressions*] of the interaction of those rules. Now let us consider applying this notion of 'objects' to the 'objects of the mind'. If this shocks you –and I hope its implications will at least interest you- consider this: it presents the very first, truly scientific pathway around the problems of the homunculus, of knowing, and enables the first possibility of an actual scientific existence of an *actual* mind! If 'the (biological) system' were to know, and 'we' *were* the system itself, then it would indeed be possible for us to know our objects. But *only*, however, if those objects were like the 'objects' of implicit definition." This is the revelation I had 40 years ago.) There are many profound difficulties in it, I know, e.g. in regard to what I have referred to as "the static problem" which I addressed in the "Dennett Appendix" in my book, [Iglowitz, 1995, 2011], in the "Freeman Appendix" to [Iglowitz, 2005, and in the d'Espagnat Chapter 13 in this book. The later comments deal with the biological viability of such evolutionarily determined objects in a changing environment. See the reference in the latter to "A/D converters" [ibid]. (My conception raises deep logical problems as well. I will address these presently.) Contrarily, it seems to be the *only* possible pathway to "knowing" *per se* within the context of modern science.

Quoting Stefanik re: Resnik's structuralism:

> "These objects [for Resnik] serve as only positions within these structures, with their identity determined only by their relationships with other positions within that structure. His basic underlying metaphor[9] is that of geometric points, and he claims that we do not have knowledge of mathematical objects given in isolation but rather as 'pieces of structures.'"[IV]

Here is a much older viewpoint on Hilbert's related conception of "implicit definition" that I was aware of almost from the beginning and which helped form my initial conception. (From Moritz Schlick, physicist/philosopher and founder of the famous "Vienna Circle", and, according to Cassirer, the actual inventor of the phrase "implicit definition"). He grasped the deep implications of Hilbert's innovation early on:

> "[Hilbert's] revolution lay in the stipulation that the basic or primitive concepts are to be defined *just by the fact that they satisfy the axioms*.... [They] *acquire meaning* only by virtue of the axiom system, *and possess only the content* that it bestows upon them. They stand for *entities whose whole being* is to be bearers of the relations laid down by the system."[V]

I do not claim to be a mathematician, or even a philosopher of mathematics, but the perspective Mac Lane's book engendered was the one I held from the very beginning of my quest for an answer to the mind-brain problem. I saw the blinding possibility of the first truly viable answer to the core question cited above.

My purpose here is to solicit the help of real mathematicians and philosophers of mathematics in the

---

[9] Please note that the usage of geometry is a "metaphor" only. The objects of mathematics may be more complexly ordered than is possible for geometry.

completion of a formal and rigorous resolution of this problem capable of empiric verification.[VI] I think this is the ground where neuroscience must go. My own thrust for most of my intellectual life has been to pursue the profound further consequences of this idea —and they are huge and difficult.

In my mistaken youthful naïvety and humility, (which I renounce in this my old age), I had assumed that all minds bright enough to approach the deepest problems of our species, (the brightest minds of my elders and of my contemporaries), had long since understood and internalized these lucent, sophisticated ideas.[10] These ideas seemed to be an absolutely essential part of the working tools of the intellect.

It seems I was wrong. It seems clear from their writings that most neurophilosophers have very limited conceptions of the dimensions and scope of modern mathematics, (–and of modern physics[11] as well)!

Mac Lane's book suggested to me what I still believe is the only genuine possibility for an answer to the question posed above, viz: "How can a (biological) machine/mechanism, whose parts are discrete in time and space, ever know anything *whatsoever*?" It also suggested an answer to Leibniz' earlier and more purely philosophical question as well, i.e. "How is it possible for 'the one' to know 'the many'?" That is, how is it possible that a unity, (e.g. a unified mind), could somehow actually know the (localized) "objects" it contained –i.e. "the many"?

Equivalently, how is it possible that a "Cartesian theatre" could exist without a homunculus?

---

[10] This was the second of my naïve youthful intellectual assumptions. The first was that such minds —which I had idealized- had realized and adopted the profundity of Cassirer's redefinition of the formal, technical "concept" of logic itself—which he had reformulated in terms of "ordering" rather than of "extensionality"- and which had struck me with a profound force at about the same time.

[11] See Chapter 13: D'Espagnat

What Mac Lane's book spoke to me under my prior perspective of Cassirer's "Concept" in which I saw it,[12] was that a system of mathematical "axioms"[VII] *when taken as a whole,* (see the relevant Hilbert quotes following shortly), could actually *create* its "objects", (albeit virtually). It could actually "know them" in the profoundest meaning of that phrase moreover, rather than the other way around.

Tentatively taking a system of "axioms" in a specifically *operative* sense, [Note: 2010: (think of nerve nets –W.J. Freeman's "equivalence classes" perhaps, as "axioms")], –in a *biological* sense then –as the fundamental and "atomic" units of brain process[13]- it suggested that as a system, (i.e. *as a whole*), it could create and actually *know* its "objects". Meaning, (sans reference), was Hilbert's gift to conceptual possibility.

Here[14] is some very recent material I have found and which buttresses my early interpretation: Quoting Hilbert's response to Frege, (David Hilbert, of course, was one of the most famous mathematicians in history and the actual father of "implicit definition". Please note his emphasis on the "*whole* structure"):

> "It is impossible to give a definition of *point*, for example, since *only the whole structure of axioms yields a complete definition.* A concept can be fixed logically only by its relations to other concepts. These relations [are] formulated in certain statements [which] I call axioms, thus arriving at the view *that axioms are the definitions of the concepts.*"[VIII]

---

[12] –see later Cassirer sections –especially Chapters 3,4,5 and Chapter 8

[13] Please do not be put off in considerations of size. Mathematics has no problem with axiom systems of even infinite size. I do not propose that they are that large, (sic), but a few hundred billion, (probably too large), probably wouldn't raise many mathematical eyebrows.

[14] The following is a re-iteration of material covered in Chapter 1 –the précis!

Here is Hilbert's expansion of his perspective:

> "I do not want to assume anything as known in advance. I regard my explanation ...as the definition of the concepts point, line, plane ... If one is looking for other definitions of a 'point', e.g. through paraphrase in terms of extensionless, *etc.*, then I must indeed oppose such attempts in the most decisive way; one is looking for something one can never find because there is nothing there; and everything gets lost and becomes vague and tangled and degenerates into a game of hide and seek."[IX]

From my early intuitive interpretation of Mac Lane's book, I proceeded at that early stage in my life to tentatively assume "axioms"[15] as the fundamental operative units of the brain to see where the assumption led.[16] (The repercussions are enormous, I realize, but please bear with me.) This was the perspective I started with fifty years ago and have pursued it ever since. I believe it has been fruitful.

I concluded early on that the (virtual) "objects" of this particular system, (i.e. of the brain conceived in this way), could actually function as the perceptual and conceptual objects of the "mind" and that the problems which I later came to know by name as "the homunculus" and "the Cartesian Theatre" would be solved.[17] This was the *one* case in which a mechanism could *actually know* its "objects" –but *only* in precisely the case where those "objects" were, in fact, purely a manifestation of the ordering, (structure), of the mechanism itself.[18]

---

[15] i.e. "atomic processes"

[16] This is highly compatible with Maturana's conception which I later discovered –see Chapter 6.

[17] These problems are implicit in the core problem stated at the outset of course. I always saw the problems, but their naming came much later. See Chapter 3 :"Cassirer" for a rationale.

[18] It also clearly seemed to resolve the deepest problems of "meaning" as well.

The problem I then faced, and it was not a simple one, was that those "objects" would have to be mathematical objects *per se*! And how could this be?[x]

I spent many decades investigating and developing the biological and philosophical implications of my early insight with some success I think, though communication of them has been almost impossible for the reasons stated previously, I think.

After long consideration, (too long!), I have recently concluded that the difficulty has always been a direct result of my original naïve assumption. These tools are *not* in the workchests of neuroscientists or neurophilosophers!

As a whole I think these ideas make sense. Their ramifications are huge but admittedly raise substantial and profound doubts which are extremely difficult.

I will not minimize this fact. But I feel the latter objections are themselves essential to the problem however as they must constantly be clashed against the core problem raised at the outset. This was the hammer and the anvil on which my ideas were forged. Only recently did I discover the actual formal mathematics which elucidate my beginning insight and seem to validate it.[xi]

Mathematical Structuralism and Category Theory:

I claim no expertise in Category Theory whatsoever save through these references, but I have a greater confidence in my understanding of Mathematical Structuralism as I came to the same conclusions myself over fifty years ago. (See my youthful arguments in the next chapter regarding Georg Cantor's Diagonal Argument which led me to these conclusions.)

Let me begin by quoting pieces of the modern dialogue. I will not be able to truly elaborate this discussion, but my purpose here is to give you a flavor and to establish the legitimacy and the "legality" of my perspective.[xii]

First let me repeat the short quote of Resnik defining Mathematical Structuralism as a place for us to start:

"The view's leading ideas are that mathematics studies structures or patterns and that reference to mathematical objects figures in this study only as a means for depicting structures.

Mathematical languages do not refer to determinate elements of fixed domains, as, to use Hellman's term, "objects-platonists" hold,; but rather, in so far as they refer at all, they refer to positions whose identities are fixed only through their relationships to other positions in the structure under discussion."[XIII]

There is so much to present here, so many misunderstandings, so many presuppositions that it is almost impossible to even begin to communicate fluently on this subject. One aspect lies in the foundations of logic itself. Almost everybody, it seems, assumes set theory as the logical primitive of thought, (which assumption lies at the basis of some of the deepest problems here):

Benacerraf counters however:

"Very often philosophical logicians are really logicists who are promoting the program of reducing mathematical objects to logic and set theory. This is clearly distinct from the activities of mathematicians who are not interested in a reductionist program..."

....We will see that category theory has been proposed as an alternative to set theory as a foundation of mathematics. "The search for urelements, fundamental objects of the mathematical universe, is a mistaken enterprise that underlies an absolute theory of identity and the platonic philosophy of mathematics. ..."[XIV], [XV]

Here is another perspective:

"Mac Lane[XVI] correctly states that many interesting questions cannot be settled on the basis of the Zermelo-Fraenkel axioms of set theory."

'Various additional axioms have been proposed, including axioms which ensure the existence of some very large cardinal numbers and an axiom of determinacy (for certain games) which in its full form contradicts the axiom of choice.

This variety and the undecideability results indicate that set theory is indeterminate in principle: There is no unique and definitive list of axioms for sets; the intuitive idea of a set as a collection can lead to wildly different and mutually inconsistent formulations.

On the elementary level, there are options such as ZFC, ZC, ZBQC or intuitionistic set theory; on the higher level, the method of forcing provides many alternative models with divergent properties. *The platonic notion that there is somewhere the ideal realm of sets, not yet fully described, is a glorious illusion.*' " [XVII]

Stefanik continues:

"Mac Lane believes that this situation is similar to that of geometry after the proof of consistency for non-Euclidean geometry demonstrated that there are many geometries, and not just one. In a similar manner, *the intuitive idea of a collection leads to different versions of set theory.* For Mac Lane, this is sufficient reason to consider alternatives to set theory as a foundation for mathematics. The alternative that he proposes is category theory."[XVIII]

And a few more:

"Category theory is essentially anti-platonistic, for it undermines the received idea that the meaning of any mathematical concept is fixed by referring it to the context of a unique absolute universe of sets."[XIX]

"...it becomes natural, indeed mandatory, to seek for the set concept a formulation that takes account of its *underdetermined* character, that is, one that does not bind it so tightly *to the absolute universe of sets with its rigid hierarchical structure."* [19] [20]

I studied the Zermelo-Frankel set-theoretic foundation for numbers -a well-accepted and pretty much standard interpretation briefly, (long ago). Surprisingly, I found that there is precisely *one* thing that actually, (i.e. ontologically), exists for mathematics under this set-theoretical interpretation: "the empty set"! Everything else *–all else–* is grounded in sets and sets of sets and sets of sets of sets of....that selfsame "empty set" –i.e. the set which has no members!

The actual (ontological) existence of that empty set –that empty basket in the real world- is proved moreover solely on the basis of a logical contradiction.[21] Assume it does *not* exist and a logical contradiction arises![22] Therefore it, *our most crucial ontological logical building block* is "proved" thereby to *actually* exist! I think this is not a viable beginning for anything; much less as the very foundation for the logic we apply to the very core of our world.[xx] It is a self-serving and circular argument.[23] The solution I have evolved is harder, but I think it works.

---

[19] Bell, 238,my emphasis –please note his specific reference to the violation of hierarchy!

[20] Note: W.J. Freeman and I both specifically argued against hierarchy in the compositing of the brain. He has supplied a physical model. See Chapter 4 and Iglowitz, 2005

[21] The proof is grounded in "material implication"..

[22] Please think back to the initial Quine quote in the Preface! If "even the law of the excluded middle" is ultimately questionable, then why is "material implication" not so? See shortly –d'Espagnat and modern physics on the subject of "existence" in Chapter 13!.

[23] I say it is circular because the logicians stand on the foundations of mathematical logic, and the mathematical logicians stand on the foundations of philosophical logic. I had a well known philosophy professor who used to sit in on my beginning mathematical logic classes and take ferocious and copious notes even at that level. So an appeal to philosophical logic to support mathematical logic seems very strange to me.

Chapter 2 — Hilbert & the New Mathematics

D'Espagnat's[24] comments on Scientific Realism are pertinent to this discussion here, (paraphrasing): -that the very actual, (ontic), *existence* of "multitudinous" localized objects, (and "sets" of them?), in our universe has actually been *disproved* on the basis of modern physics. Bohr always thought so, and Einstein, in his later years, finally agreed with him in this –"he was prepared to give up localization to keep realism".[25] But what then constitutes the model and the philosophical basis for Platonic mathematics?

Back to the sources of mathematical structuralism and category theory:

Resnick:

> "... As positions in structures, they have no identity outside of a structure. Furthermore, the various results of mathematics which seem to show that mathematical objects such as the numbers do have internal structures, e.g. their identification with sets, are in fact interstructural relationships."[xxi]

Repeating Stefanik:

> "These objects [for Resnik] serve as only positions within these structures, with their identity determined only by their relationships with other positions within that structure."

But how could "a structure" serve a biological mechanism without the further inclusion of actual "objects"[26] within that

---

[24] See Chapter 13

[25] See the d'Espagnat chapter, and especially Kumar's treatment of Einstein's final correspondence.

[26] How about a structure of neural connections, for instance?

structure?[27] On the other hand, how could "a structure" serve a biological mechanism *at all?*

Either it is conceived to exist externally in the unknown input/output domain in which case it is merely *manipulated*, (and still unknown), or it is conceived internally to that mechanism itself, in which case it is a part of the intentional feedback loop of the mechanism described early on, but here, as argued in the very beginning, there is *still* nobody home!

The better answer is that *virtual* "objects", (and I think it is pretty clear that the objects of structuralist mathematics are truly virtual), could serve the organism as a highly effective and optimizing *organization* of response[28], alternatively as a cohesive operative metaphor[XXII] of its primitive structural units, serving as a "higher level language" for intentionality perhaps,[XXIII] and this is just what I propose for the human brain.[XXIV]

I will argue in Chapter 4 that our "perceptual objects" are a byproduct of the deep evolutionary self-organization (ordering) of our primitives! Our "objects", I will propose, are manifestations of the structure; the structure is *not* a resolution of the objects. It is the structure which is predictive and operative; its "objects" merely enable or optimize it![XXV] There is considerably more to the solution of this problem than I have discussed so far,[29] but, surprisingly many of these further deep ramifications are actually mirrored in the current (mathematical Categorial[XXVI]) dialogue as well!

---

[27] Considering it as a connectionist structure –as we will do– will make more sense.

[28] See Iglowitz, 2005, for a precise biological elaboration of this possibility, and the later citation of my letter to Rosen in Chapter 3 as an expansion of my ideas on this subject.

[29] Briefly, I have argued for evolutionary "objects" as primitives, but I also strongly suggest the inclusion of intentional axioms as well –as an answer to the "static problem" –see Freeman Appendix of Iglowitz 2005, the "Dennett Appendix" in Iglowitz, 1995, and Appendix C, and the d'Espagnat Chapter 13 of this current book, Iglowitz, 2012.

This is our deepest and ultimate problem and we must expect the implications to be vast on all fronts.

One of these implications, and it is very real, relates to my investment long ago in Ernst Cassirer's "Symbolic Forms" which incorporates an absolute *relativism* of epistemology.[30] (This is clearly consistent with our present, albeit conditional, materialistic perspective!)

The absolute ontological object for Cassirer becomes "a mere X".[31]

> "Even in 'nature', the physical object will not coincide absolutely with the chemical object, nor the chemical with the biological —because physical, chemical, biological knowledge *frame their questions* each from its own particular standpoint and, in accordance with this standpoint, subject the phenomena to a special interpretation and formation[32]....
>
> The end of this development ["critical idealism"[XXVII]] seems to negate its beginning —the *unity* of [ontological] being, for which it strove....The One Being,", [i.e. the absolute ontological object], "to which thought holds fast and which it seems unable to relinquish without destroying its own for, eludes *cognition*. The more its metaphysical unity as a 'thing in itself' is asserted, the more it evades all possibility of knowledge, until at last it is relegated entirely to the sphere of the unknowable and becomes a mere 'X'".[XXVIII]

---

[30] Which fits very nicely with my base ideas, of course —i.e. with the necessities of the brain as "machine". But please remember that I delimit Cassirer's thesis as outlined in Chapter 1 and expanded in Chapter 8: "Contra Cassirer"!

[31] Ontology is clearly relevant to the mind-body problem which occurs as a self-referential question within it.

[32] i.e. each discipline incorporates its own specialized logical structure —without that particular structure the "object" becomes faceless

Simply put, Cassirer argued that each of the forms of science incorporates its own specialized logical perspective, ("each *frame(s) their questions* each from its own particular standpoint"), and, if that unique, purely logical, and specialized framework were removed -*or relativized,* all that would remain of the ontological "object" would be "a mere X"! [xxix]

But note how closely Cassirer's conception of reality matches the materialist perspective I sketched in the Precis! There is no way that a machine, (and this is the precise characterization of the brain implicit in materialism), could have knowledge of the space which constitutes its input or its output domains!

Cassirer's conception seems to be precisely mirrored mathematically in Bell's "Category Theory and the Foundations of Mathematics",[33] [J.L. Bell]. (Citation shortly)

But first, in close parallel to my comments above, Stefanik cites Benacerraf:

> "[Hilbert] argues that what constitutes an object *varies* from theory to theory, category to category, and that Frege failed to realize this fact. It is a thesis that is supported by the activity of mathematicians, and is essential to the philosophical perspective underlying category theory, as we shall discuss later.
>
> The search for urelements, fundamental objects of the mathematical universe, is a mistaken enterprise that underlies an absolute theory of identity and the platonic philosophy of mathematics.', "[and of neuroscience as well I propose],"... 'It [logic][xxx] remains the tool applicable to all disciplines and theories, the difference being only *that it is left to the discipline or theory to determine what shall*

---

[33] If these, (structuralism and category theory), are, in fact, the foundations of our mathematics, then they are also the foundations of our logic. This should be confirmed with even a casual glance at the present mathematical state of logic. As such they strongly imply that Cassirer's conclusions are relevant to logical thought at the very bottom level. This seems to be confirmed from Bell's perspective.

count as an 'object' or 'individual.' [Benacerraf, 288, my emphasis]

Quoting Stefanik, (solely for definitional purposes here):

"Any topos[34] may be regarded as a mathematical domain of discourse or 'world' in which mathematical concepts can be interpreted and mathematical constructions performed. Bell develops an analogy between mathematical frameworks and local coordinate systems of relativity theory. Each serves as the appropriate reference frame for fixing the meaning of mathematical or physical concepts respectively." [Stefanik, 1994]

Here are the promised relevant citations from Bell:

"The topos-theoretical viewpoint suggests that the absolute universe of sets be replaced by a plurality of 'toposes of discourse', each of which may be regarded as a possible 'world' in which mathematical activity may (figuratively) take place. The mathematical activity that takes place within such 'worlds' is codified within local set theories; it seems appropriate, therefore, to call this codification local mathematics, to contrast it with the absolute (i.e., classical) mathematics associated with the absolute universe of sets.

Constructive provability of a mathematical assertion *now means that it is invariant*, i.e., valid in every local mathematics."[XXXI]

"There is an evident *analogy* between mathematical frameworks and the local coordinate systems of relativity: each serve as the appropriate *reference frames* for fixing the meaning of mathematical or physical concepts

---

[34] "Topos": In mathematics, a **topos** (plural "topoi" or "toposes") is a type of category that behaves like the category of sheaves of sets on a topological space. (Wiki)

respectively. Pursuing this analogy suggests certain further parallels.

For example, consider the concept of *invariance*. In relativistic physics, *invariant physical laws* are statements of mathematical physics (e.g. Maxwell's equations[35]) that, suitably formulated, hold universally, i.e. in every local coordinate system. Analogously, *invariant mathematical laws* are mathematical assertions that again hold universally, i.e. in every mathematical framework." [Bell, 241]

The trick is to understand that Cassirer's epistemological relativism is based, (at least insofar as it refers to the physical sciences[36]), like Bell's[37] and Einstein's, in the *absolute* preservation of connective –i.e. *translatory* invariants.[38] I came to these conclusions by an independent route -by a continual reexamination of the original core question, (matched against Cassirer's ideas and those structuralist implications I inferred from my early exposure to Mac Lane's book), which I feel lead inexorably to Cassirer's ultimate perspective.

How is it possible for a pure mechanism to actually know, (in the sense of ontology), anything whatsoever?" As I argued in the Precis, the ultimate answer is that, other than the "objects" of its own operationality, it *cannot*! But it is possible for a mechanism to have "beliefs" –i.e. operative strategies/organizations each of which addresses the invariant core

---

[35] but especially the Lorentzian transformations -or even Galileo's for that matter

[36] repeating a prior footnote: Discussing Hilbert, Cassirer says: "The procedure of mathematics here", (implicit definition), "points to the analogous procedure of theoretical natural science, for which it contains the key and justification." Cassirer, 1953 p.94

[37] "suitably formulated"

[38] For Cassirer, these invariants are preserved in "the phenomena".

of raw experience instead, (transformed through our evolutionary artifacts)! [39] [40]

I believe that invariant core, (of experience), its primitive objects are, in fact, evolutionary artifacts implicitly defined by the structure of brain process![XXXII] These artifacts, (which are our naïve objects), I think, are our primitive "percepts".[XXXIII]

They function, as I said in one of my papers, as fixed "A/D" converters,[XXXIV] (or, better: hierarchical/chaotic converters), so to speak, mediating between input[41] and central brain process[42]. (These are the "schematic artifacts" of the "GUI", (graphic user interface), that I argue in Chapter 4.) I believe this was how evolution organized and optimized the behavior of the seventy trillion celled megacollosus called man!

Here is a bit from my original manuscript on the subject of "objects: (note: where I have used the term "implicit definition" you may substitute the terms "mathematical structuralism" or "category theory" if you like. I don't automatically agree, (that is, I do not automatically agree in a "knee-jerk" sense. I don't know enough.) It should work however.

"I propose that the boundaries -the demarcations and delimitations of these schematic objects, (their "contiguity" if you will) -are formed specifically to meet the needs of the operations themselves. I propose that they exist to serve the structure, (the rules of the "calculus") - not the converse.[XXXV]

I propose that the 'objects' of these schematic models – specifically as objects *qua objects* - serve to organize process, (i.e. analysis or response). They are not representations of *actual*

---

[39] See Chapter 9 for a rationale!

[40] The "invariants" for Cassirer's "Symbolic Forms" are our raw percepts themselves sans an interpretation.

[41] "Triggering" to use Maturana's more profound perspective

[42] which, I propose in agreement with W.J. Freeman and consistent with the categorial perspective as well, is not organized hierarchically

objects or *actual* entities in reality.^{xxxvi} This, I propose, is why they are taken [utilized] as "things" in the first place.

They functionally bridge reality in a way that physical, (i.e. ontological), objects do not and I suggest that they are, in fact, materialized metaphors of analysis or response. The rationale for using them, (as any good "seminarian"[43] would tell you), is clarity, organization and efficiency. ^{xxxvii}

"Real Space"

As a side issue, remember that axiom systems may embody "space" itself. Hilbert's original axioms in the "Grundlagen", (itself only a small part of the beginning of his massive revelation), also deal with "between", "dimension", "point", "motion", etc – i.e. his conception includes spatiality itself. So I think do the axioms of the brain! Is spatiality then external? Or is it just an extremely useful part of the reactive and pragmatic organization of brain process?[44] ^{xxxviii}

I believe we will never know as I concur with Cassirer that ontology *in its entirety* "is a mere X"[45]. Maturana's perspective is absolutely relevant here. Briefly Maturana proposes that we do not pass or receive *information* from externality, we are simply "triggered" by that externality to preserve our "Autopoietic entity", –just as externality in its turn is merely "triggered" by us.[46]

But isn't this just the conclusion that we might have expected from our opening discussion regarding a mechanism?[47]

---

[43] The intended humor will become apparent in the discussion at the early stages of Chapter 4 –it applies to "the training seminar".

[44] The self-referential aspect of this viewpoint is addressed in my adoption of a modification of Cassirer's epistemological relativism –again based in invariants- as my third hypothesis.

[45] See Damascius's conception of *"pante aporeton"* in Chapter 13

[46]. See Chapter 6. Maturana 1987.

[47] I addressed this issue as "the (intentional) axiom of externality" in my MS. (Iglowitz, 1995 and in the current MS in Chapter 6.)

And another:[48]

>"Even idealism and dualism do not resolve the underlying logical problem however -the *how* of Leibniz's "expression of the many in the one", for even then how could *this* part of even a *mental* "substance" know *that* part?[xxxix] These are [precisely and profoundly] logical problems [*per se*] -the problem of the "homunculus" and the problem of the "Cartesian theatre". Where does there exist even the possibility of a solution?
>
>Implicit definition, virtual existence -and logic as biology[49]- this triad is the *only* example within our intellectual horizons that seems to hold even *any* promise for sentiency in this our ordinary sense of it. It suggests the only scientifically plausible solution to "the mind's eye" and the "Cartesian theatre" and the only non-eliminativist, (for "mind"), [biological] answer to the homunculus problem.
>
>These are answers which *must* exist if mind in our ordinary sense is, in fact, to be real. Implicit definition, taken operatively, permits knowing as a whole what are, in some real sense, our distinct and separate parts. This is precisely because those parts, ("objects"), are in fact non-localized and virtual (logical) expressions specifically *of* the whole. It opens the first genuine possibility, therefore, for a resolution of this essential requirement of "naïve" consciousness.
>
>But that pathway, (implicit definition), *does not make sense from the standpoint of representation*! For implicit definition solves the problem *logically* -from the standpoint of constitutive logic -and speaks to nothing other than its own internal structure. Repeating myself: "objects", (under implicit definition), are known *to* a system, (i.e. universally/globally), *only* because they are specifically expressions *of* the system.

---

[48] Since this is just me quoting me, I don't think I have to apologize for the length of my citations.

[49] See Iglowitz, 1995 and the heading to follow shortly in Chapter 3: "Logic as Biology".

It becomes a viable and natural solution to the problem of awareness, therefore, only when the objects of consciousness themselves are conceived operationally and schematically, (and specifically, *logically*[50]), rather than representatively.[51]

When our objects are taken specifically as schematic representations of process itself however, (as per my first thesis [Chapter 4 of current MS]), the solution becomes both natural and plausible -the specifically logical problem of sentiency is resolved.[52] I assert that no other actual solution, (other than a denial of the problem itself), has ever been suggested. This is the argument from the second to the first hypothesis -and different from the argument from the first to the second presented earlier [in the original MS, Chapter 4 here].

But this conclusion is greatly strengthened by the arguments I will propound in Chapter 4 …and by the conclusions of several eminent contemporary biologists.[XL]

My biological thesis, *considered biologically*, (i.e. aside from its admittedly profound, but purely epistemological difficulties -which I will make good in Chapters 5 through 10 in an extrapolation of Cassirer's "Symbolic Forms"), is exceedingly strong. How could evolution organize -*as it had to organize*- the reactive function of this colossus of seventy trillion cells?

Even this formulation of the question disregards the yet more profound complexity of the reactivity of the individual cells -also organisms- themselves! It was the overwhelmingly crucial issue in the evolution of complex metacellulars. My thesis of schematism is both viable and plausible in this context. But what

---

[50] and "bio-logically"

[51] That the objects of this constitutive logic would further "*represent*", however, would be a genuine assumption of the miraculous -possible but difficult. See P.S. Churchland: "and then a miracle happened…." [Churchland, 1988]
Representative objects are not the right *sort* of mathematical objects to be applicable here. Representative objects are based in reference, denotation and not in connectivity. At the very beginning they resurrect the homunculus.

[52] Though not the *substance* problem. That is a separate epistemological and metaphysical issue addressed by my third thesis.

does this evolutionary development and organization of the reactive process of complex metacellulars have to do with [actual] *'information'* ?[53] There is still, again, "nobody home"!

As an aside: I dealt with this "information" problem from the perspective of Humberto Maturana in Chapter 3 of my original MS, (Chapter 6 of the current writing). I believe this brilliant, if slightly flawed work[XLI] is the modern equivalent of Kant's "Prolegomena" and is clearly relevant to the problem at hand.

Back to the mathematical problem:

Presently I will introduce Shapiro's critique of the structuralist claims. I think it is important because I think it seriously challenges my position and causes me to deepen and clarify it. It will lead me to a discussion of the other main thrust of my conception, starting with Ernst Cassirer's reinterpretation of the deepest problem of all —of the fundamental technical logical "concept" itself. This in turn will lead to a reconsideration of even the fundamental concept of the "class" which grounds modern set theory.

Remember, I asserted previously that our problem here is a profound problem of logical *possibility per se!* This is the ground in which Shapiro's discussion, (and most of mathematics' preconceptions), must necessarily be evaluated.

---

[53] "Information" is a subject that must be discussed, obviously. Both the materialists and myself see the function of the brain in the light of optimized efficiency. From their standpoint, this is accomplished by the incorporation of a realistic model of externality within it. From my standpoint this is an impossibility —it goes against the whole grain of the evolutionary perspective. Evolution works by the selection of processes. But the subsequent extension into "information processes" invokes a miracle. How did, and how could it start? Maturana attempted it in his "structural parallelism", but I find that this aspect of his arguments is faulty. [See Chapter 6]

But first hear Goldblatt:

> "Now, since category theory, through the notion of topos, has succeeded in axiomatising set-theory, the outcome is an entirely new *categorial foundation of mathematics!* The category theorists attitude that "function"[54] rather than "set membership: can be seen as the fundamental mathematical concept has been entirely vindicated."[XLII]

(Do you hear a distant echo of Cassirer's "mathematical concept of function" here? I think you will when we enter Chapter 3!)

Back to the Sources of Category Theory and Structuralism:

Shapiro Part One

I will cite just one more perspective from mathematical structuralism[XLIII], and then go on to present my own solution to the Mind-Brain problem. Shapiro's perspective on mathematical structuralism exposes what I think is a critical defect in our thinking about mathematics and "reality" generally -and about the

---

[54] "One of the primary perspectives offered by category theory is that the concept of *arrow*, abstracted from that of *function* or *mapping*, may be used instead of the set membership relation as the basic building block for developing mathematical constructions, and expressing properties of mathematical entities. Instead of defining properties of a collection by reference to its members, i.e. *internal* structure, one can proceed by reference to its *external* relationships, with the other collections. The links between collections are provided by functions, and the axioms for a category derive from the properties of functions under composition." Goldblatt, Robert, Dover 1984, p.1 Yes, I do see the problem, but what, then, are the "other collections" between which these "external relationships" are defined? It seems that mathematics is approaching Cassirer's re-interpretation of the concept itself, but, apparently, unknowingly!

mind-brain problem specifically. I think it derives from the presumed foundations of the classical logical "concept" itself.

Shapiro dealt fairly deeply with Hilbert's original conception of "implicit definition". But he argued that the issue of consistency/coherence is the more critical question. He distinguished strongly between the "young Hilbert" and the "older Hilbert". (And, most definitely, *so do I!*)

The former was the father of "implicit definition" who proved consistency only algebraically and relativistically. The latter sought an answer in Formalism with its "finitary arithmetic", "tokens" and "assertatory statements" and Shapiro seems to have approved.

My own perspective based on fifty years of contemplation in a much broader and very different context is that the young Hilbert was closer to the truth than the older Hilbert. This was the young man who was called "the king of invariants" and I think his breeding showed in his apotheosis as embodied in his concept of "implicit definition".[55]

> "In this note, I hope to shed a little light on the question, or questions, by relating the present debate to a clash that took place over a hundred years ago, between two intellectual giants, Gottlob Frege and David Hilbert. I propose to focus on the role and function of meta-mathematics[56], which, I suggest, does not fit smoothly into Hilbert's algebraic perspective at the time.

---

[55] I think he was later seduced by Cantor's easier, but highly alluring perspective. "No one will drive us from the paradise which Cantor created for us" [Hilbert]

[56] Please note and remember that it is metamathematics *per se* which is Shapiro's focus throughout this paper. But, as he later states: "For one thing, *the meta-theory is not axiomatized in the Grundlagen*, and so there is no implicit" [or explicit] "definition of the meta-theoretic notions." I think they originated in Hilbert's native but superb, mathematical and logical perspective, (as the "king of invariants"), and not from his later (Cantorian) perspective. I think his conversion was one of the greatest mistakes in intellectual history.

> The problem was directly remedied in the subsequent development of the Hilbert program some decades later, where it is explicit that the proper meta-mathematics is finitary arithmetic. But, the story goes, this resolution was undermined with the incompleteness theorems, thanks to Gödel. So there is some unfinished business in the original debate, at least from Hilbert's side of it."[57]

> "A crucial aspect of the axiomatization is that the system is what I call 'free-standing'. Anything at all can play the role of the undefined primitives of points, lines, planes, *etc.*, so long as the axioms are satisfied. Hilbert was not out to capture the essence of a specific chunk of reality, be it space, the forms of intuition, or anything else.

> Otto Blumenthal reports that in a discussion in a Berlin train station in 1891, Hilbert said that in a proper axiomatization of geometry, 'one must always be able to say, instead of "points, straight lines, and planes", "tables, chairs, and beer mugs".[58]' [ibid 64]

To further quote Shapiro:

> "the early pages of Hilbert [1899] contain phrases like "the axioms of this group define the idea expressed by the word 'between...'" and "the axioms of this group define the notion of congruence or motion.....we think of ...points, straight lines and planes as having certain mutual relations, which we indicate by means of such words as 'are situated', 'between', 'parallel', congruent', 'continuous', etc. The complete and exact description of these relations follows *as a consequence of the axioms of geometry".* [ibid]

---

[57] Shapiro, 2005
[58] This is pretty nearly equivalent to Wilder's "permissive". See Wilder 1952

But "*Anything at all* can play the role of the undefined primitives"! He quotes Hilbert:

> "... it is surely obvious that every theory is only a scaffolding or schema of concepts together with their necessary relations to one another, and that the basic elements can be thought of in any way one likes. If in speaking of my points, I think of some system of things, *e.g.*, the system love, law, chimney-sweep ...and then assume all my axioms as relations between these things, then my propositions, *e.g.*, Pythagoras' theorem, are also valid for these things ... [Any] theory can always be applied to infinitely many systems of basic elements.
>
> One only needs to apply a reversible one-one transformation and lay it down that the axioms shall be correspondingly the same for the transformed things. This circumstance is in fact frequently made use of, *e.g.*, in the principle of duality ... [This] ...can never be a defect in a theory, and it is in any case unavoidable." [Hilbert via Shapiro]

But what precisely *could* the Pythagorean Theorem mean for "beer mugs", "love", "chimney sweeps" for example? These terms are proposed as *co-equal* to "line", "between" ... as *primitive* terms, not as derivational from *other* primitive terms! If we were to translate the question into one of the *positions* of beer mugs for instance, we would only have come back to the very "points", "lines", etc. that we started out with and begged the question, as we would not have done an actual substitution of the basic terms. It would have been a circular argument and a completely trivial conclusion. Hilbert was certainly brighter than that!

In my earlier papers, I had adopted Wilder's interpretation –i.e. in "Consciousness: a Simpler Approach to the Mind-Brain Problem."[XLIV] This was a completely workable answer I believe for the purposes of my central thesis, but in Chapter 3 I will explore a different and deeper interpretation of Hilbert's remarks which will broaden our context considerably.

In those earlier papers I had interpreted Hilbert's comments in the most minimal sense –that these "objects" were, using Wilder's terminology, "permissive and presumptive only", that is, semantically neutral.

> These objects, (of its domain –and "existence" terms generally), are assumed *only*, (as Wilder points out) "presumptive(ly)" and "permissive(ly)" however. We are told *nothing* about them in an objective sense."[XLV]

I began with an interpretation of Hilbert where the "objects" of a system are taken in a purely *impartial* sense.

> "... it is surely obvious that every theory is only a scaffolding or schema of concepts together with their necessary relations to one another, and that the basic elements can be thought of in any way one likes. If in speaking of my points, I think of some system of things, *e.g.*, the system love, law, chimney-sweep ...and then assume all my axioms as relations[59] between these things".

Shapiro continues:

> "It is hard to be definitive on what his view was, or should have been, but I suggest that the meta-theory—the mathematical theory in which the consistency of an axiomatization is established" [ontologically established – *but where?*] "—is not to be understood algebraically, not as another theory of whatever satisfies its axioms.
>
> Instead, the statement that a given theory, such as Euclidean geometry, is consistent[60] *is itself assertory*. [My

---

[59] But *what kind* of relations? What is the conception of relation itself that Hilbert had in mind?

[60] I have no idea whether Euclidean Geometry is indeed consistent. All I know of it is that it works exceedingly well –and that is the most we can demand of an

emphasis] The notion of consistency is a contentful[XLVI] property of theories[XLVII], and is not to be understood as defined implicitly by the axioms of the meta-theory.

For one thing, *the meta-theory is not axiomatized in the Grundlagen, and so there is no implicit definition of the meta-theoretic notions.*[61] [my emphasis] This, of course, is not decisive. It would be a routine exercise for a graduate student in mathematical logic to axiomatize the meta-theory of the Grundlagen.

Given the structural analogy between natural numbers and strings, [See footnote[62]], the meta-theory would resemble elementary arithmetic. However, if a Hilbertian algebraist did think of the axiomatized meta-theory as algebraic, then she would have to worry about its consistency. How would we establish that? The ensuing regress is vicious to the epistemological goals of the Grundlagen. "[XLVIII]

---

evolutionary artifact. See my illustration: "Bounds and Limits" in Iglowitz, 1995. Relative consistency is all we can demand -but this is the actual meaning of "invariance"!

[61] But why would there have to be? I don't think that Hilbert, at this stage, intended one. I think, in Quine's words, he meant to "kick away the [Fregean] ladder".

[62] There may be a "structural analogy between natural numbers and strings", but there is no such analogy between natural numbers and meanings!

His claim assumes the adequacy of current formal (set-theoretic-based) logic to Hilbert's (then) perspective. I think it is suspect. It is not string representations, but *meanings* –which may differ- and which are significant here. Within a rigidly abstractive and hierarchical worldview, these are essentially the same. But within a non-hierarchical conception of the mind and brain, they are most definitely not. See the W.J. Freeman quote to follow (~p.53). It is "Alice down the hole", i.e. the non-parallel distributive mapping and a non-hierarchical meaning for each individual recipient brain, but these can be quite different. This is a wholly new perspective on this mathematical problem.

Or, to quote Edelman: "*certain symbols do not match categories in the world* . ... Individuals understand events and categories in more than one way and sometimes the ways are inconsistent."

It is quite clear that Shapiro and I have *quite different* ideas of the meaning and the goals of epistemology! As a brief excerpt from the footnote immediately above: "Or, to quote Edelman: "*certain symbols do not match categories in the world*. ... Individuals understand events and categories in more than one way and sometimes the ways are inconsistent."

> "In the later Hilbert program (e.g., [1925]) relative consistency gives way to absolute consistency. There, the meta-theory is finitary proof theory, focused directly on formal languages themselves. It is explicit that finitary proof theory is not just the study of another structure, on a par with geometry and real analysis.
>
> Finitary proof theory has its own unique subject matter, related to natural numbers and formal syntax, and it is ultimately founded on something in the neighborhood of Kantian intuition. [The older] Hilbert said that finitary proof theory is contentful. In present terms, the theorems of finitary proof theory are *assertory*,[XLIX] not algebraic."[L]

I think that Shapiro falls into the trap of his own philosophical prejudices in failing to understand and accept the *thoroughgoing* relativism of Hilbert's original idea. In Quine's words, "we must kick away the ladder" in order to appreciate the full brilliance of Hilbert's insight. "Relative proofs only"? How could they be anything but?[63] I think the young Hilbert understood that, but lost his insight in his new passion.

"Hilbert's claim that a concept can be fixed only by its relations to other concepts is a standard motivation for structuralism." [ibid]

Shapiro now exhibits a viewpoint, (fundamental for him), which I will oppose through the rest of this book!

---

[63] To tie in with the early pages of the present book, how could a mechanism, (brain), ever know —i.e. assert absolute truth to features of its environment. This is the "assertatory" that Shapiro demands!

"Nowadays we have a rough and ready distinction which we can apply here. The algebraist says that a group is anything that satisfies the axioms of group theory; a ring is anything that satisfies the ring axioms, *etc.*" [but] *"there is no such thing as 'the group' or 'the ring'"*.[LI] "Hilbert says", [Shapiro says –but *which Hilbert? Young or Old?*], "the same thing *about* geometry, and, by extension, arithmetic, real analysis, and so forth."

If "the concept" –and "the class" below and within it- is truly all about extensionality, then I think Shapiro stands on solid ground. If it is about something more, (as Cassirer's ideas, and my own thesis of the "schematic object" will suggest), then I think his ground becomes far less secure.

Is this discussion really finished? Of course not! I have and will argue that the fundamental problem of cognition entails a re-examination of *all* of our disciplines across the board –to include mathematics itself.

Mac Lane and Shapiro maintain that "the *general* group", "the *general* ring", "the *general*....does not exist. I think they do exist –but that they do so as *concepts*, (in Cassirer's sense rather than as classical concepts). And, if they do not exist, then *what* is it that they do think *does* exist?

There are two lines of argument that I could take here. Let me take what I consider the weakest –but simplest- first:

(1) since my argument centers in and disputes their claims, and since it results, (as argued in chapters 13, 14 and in Appendix D), in a viable picture of the *whole* of reality, it is validated, I think therefore, -by demonstration.

I really don't especially like this line of argument, as it smacks too much of an argument by contradiction![64] It isn't

---

[64] i.e. it is an argument based on the principle of the excluded middle as Quine discussed in the very beginning paragraph of the preface. See Chapters 13, 14, and Appendices D and E for a broader and deeper perspective. But if we, temporarily taking ourselves as machines, cannot know the reality outside of us, how can we make assertions about it. I think I wrote my first essay against this

exactly that, of course, but it has the same "flavor" which is not to my taste.

(2) There is another route however, but it involves much work. It will involve examining the very foundations of mathematics itself, and of Cantor's, ZF's, ... set theory, of category theory, of structuralism, et al. But this is already happening at this very moment, isn't it?

I had thought to include a great deal of additional material here, but decided to put it into a new appendix, Appendix E instead so as not to break the flow of my overall argument and which will go much more deeply into this question with some very strong consequences. It will take up the argument exactly where we have left it here![65] I have titled it "Cassirer speaks to Shapiro and Mac Lane"[66]

---

idea as a freshman in high school. Ms. Strandberg, wherever you are, smile down upon me please!

[65] If you choose to skip directly to that appendix, you may, of course, do so, but I would recommend against it on your first reading as it will break up my overall line of argument which is already complex –though necessarily so for this ancient and profound paradox. After that initial reading, I would recommend that you read both sections together as a connected whole.

[66] See Appendix E!

Chapter 3: Adventures of the Mind: A Critical Turning Point and the Origins of my Conception

At this point I feel I must interpose another necessary but somewhat lengthy tangential discussion of fundamentals so that you may have some understanding of my very different and unique beginnings from which I approached these problems.

Hopefully it will help you to better understand my conclusions. Then I will return to and try to answer Shapiro's objections, (and even Mac Lane's similar ones to which I will come presently).[1] Since I have structured this paper on my own intellectual history, let me continue to do so here as well. I think it is the most efficient way to pursue this new logical perspective on the problem.

I said very early in this paper that even before my exposure to Hilbert's "implicit definition", I had encountered Ernst Cassirer's radical reformulation of the very *definition* of the "concept" of logic itself.[1] It was through this filter that I interpreted the mathematics and modern algebra, (which I saw as an extension and fulfillment of the profound possibilities of Cassirer's notion of ordering *within* concepts[2] and which we will come to very soon) -and through which I interpreted Hilbert's conception of "implicit definition" that I was exposed to shortly thereafter as well.

---

[1] This reformulation of the logical "concept" by Cassirer's is different and distinct from his "Theory of Symbolic Forms", another powerful insight which I will go into presently.

[2] I conceived mathematics then as the study of Concepts, not of Classes, (of objects), nor of Sets, (of objects) –I think this is a relevant and defensible perspective in light of the discussion to follow. It enables, for instance, the concept of "the (general) class" -or the concept of a particular class. Also, obviously, it enables the concept of "the (general) 'set'" and the concept of a particular 'set' as a subspecies! I think, shortly, that this will become clearly relevant to the Shapiro discussion begun above.

Cassirer and Logic:

> "... Every attempt to transform logic must concentrate above all upon this one point: all criticism of formal logic is comprised in criticism of the general doctrine of the construction of concepts."[II] (Ernst Cassirer)[III]

When I went as a beginning freshman to the University of Chicago, I rode the train from the north side of the city of Chicago –about an hour's commute in all. In the very beginning, I visited the campus bookstore, and acquired a copy of Cassirer's doubly bound volume: "Substance and Function" and "Einstein's Theory of Relativity".

I won't say I read it all at that point, but I started at the beginning while on my daily commute and discovered Cassirer's radical re-assessment of the very *meaning* of the word "concept" as used in logic, (and in everyday thought as well), in the opening chapter. It shaped my understanding of everything that followed.

He reinterpreted the formal logical "concept" quite differently from the classical Aristotelian understanding of the idea. He did not interpret it hierarchically –by the abstraction and inclusion of properties of objects, but reformulated it instead as "the mathematical concept of function".[IV]

Please forgive the longish quotes, but please try to follow his argument. It was my actual starting point and understanding it is crucial to your understanding my ultimate perspective. I think, in conjunction with just a few other steps, it lays the groundwork required for a scientific understanding of Consciousness itself!

He characterized the traditional, Aristotelian concept this way:

> "A series of presentations", ["things" with characteristics=properties], "with characteristics: (a,b,c,d), (a,c,d), (a,c,e), for instance, is held to bring forth the classical concept: {a,c}.[3] From *mere abstraction*, (via attention), the whole of the doctrine of the classical

---

[3] The specific order within a class is not relevant, of course.

Concept follows from these simplistic origins."[4] It is a concept based on and fully resolved in the extensionality of its properties. It is a concept based on the properties of real things, i.e. of "objects".

Under that classical Concept it follows that "every series of comparable objects has an ultimate generic concept, which comprehends within itself all the determinations in which these objects agree, while on the other hand, within this supreme genus, the sub-species at various levels are defined by properties belonging only to a part of the elements."[v]

The end result of this process is horrific to normal thought however. Hear him carefully!

> "But the successive broadening of a concept necessarily correlates to a progressive lessening of its *content*, so that finally, the most general concepts we can reach no longer possess any definite content.", [*at all!*]. The ultimate genus -"something"- is totally (and logically) devoid of specific content![5]

The Concept in this classical form, however, is clearly not adequate or consistent with scientific, or even with *ordinary* usage however:

> "When we form the concept of *metal* by connecting gold, silver, copper and lead, we cannot indeed ascribe to the abstract object that comes into being the particular color of gold, or the particular luster of silver, or the weight of copper, or the density of lead; however, *it would be no*

---

[4] But are not the quantifiers of more modern logic an exception? "Set membership" →f(a,b), (a rule) , but membership here is not *primitively* defined by *abstraction* as it is in the Aristotelian concept; it is defined by a rule instead. What is {x: x memb s} where "s" defines a rule? Where does the rule come from? It certainly does not come from abstraction. See later Lakoff and Cassirer references on "cue validity".

[5] It is interesting that d'Espagnat cited the same passages in his "Physics and Philosophy".

less inadmissible if we simply attempted to *deny* all these particular determinations of it."

It would not be sufficient to characterize "metal", for instance, "that it is neither red nor yellow, neither of this or that specific weight, neither of this or that hardness or resisting power"; but we have to add that "it is colored in *some* way in every case, that it is of *some* degree of hardness, density and luster." Similarly, we could not maintain the general concept of "animal", "if we abandoned in it all thought of the aspects of procreation, of movement and of respiration, *because there is no form of procreation, of breathing, etc.,* which can be pointed out *as common to all animals.*" (My emphasis)

These few paragraphs sum up what I considered, and still do consider to be *Mirabile dictu,* (i.e. I don't think it could be said better or more convincingly). I think it exposes the critical flaw at the ultimate foundations of logic. I believe his reformulation of that concept is more appropriate to both ordinary and to formal logical thought as well than is the classical concept.

Cassirer's new "Concept"[VI] was reformulated instead as a *functional* rule, a complex rule of series but that rule *per se,* he concluded, was necessarily generated *internally* to the mind,(/brain), and not from the percepts themselves ! That is to say: it is *not* primitively defined (specifically as a rule) *in the presentation* itself.[6] It comes from elsewhere! It is *not,* he argued moreover, abstractive!

---

[6] There is an uncanny parallelism of argument throughout between Lakoff's and Cassirer's treatment of logic. Consider, as an example, the following: "Category cue validity defined for such psychological (or interactional) attributes might *correlate*", (his emphasis), "with basic-level categorization, but it would not *pick out* basic-level categories; *they would already have to have been picked out* in order to apply the definition of category cue validity so that there was such a correlation." (Lakoff: P.54, my emphasis) This is almost an exact parallel to one aspect of Cassirer's argument against the classical concept, and the "theory of attention", (see Chapters 2 and 5), –and for a "new form of consciousness". Discussing Erdman, Cassirer writes: "...instead of the community of 'marks,' the unification of elements in a concept is decided by their 'connection by

He characterized his reformulated Concept as "a new form of consciousness" as I will discuss presently. He proposed instead an alternative and considerably more plausible basis for a *different* technical logical Concept -borrowed from mathematics. He called it "the *Functional* Concept of Mathematics":

> "Lambert pointed out that it was the exclusive merit of mathematical 'general concepts' not to cancel the determinations of the special cases, but in all strictness fully to retain them. When a mathematician makes his formula more general, this means not only that he is *to retain* all the more special cases, but also be able *to deduce* them from the universal formula."

But this possibility of deduction does not exist in the case of the scholastic, (Aristotelian), Concepts, "since these, according to the traditional formula, are formed by *neglecting* the particular, and hence the reproduction of the particular moments of the concept seems excluded."

> "The ideal of a *scientific* concept here appears in opposition to the schematic[VII] general presentation which is expressed by a mere *word.* The genuine concept does not disregard the peculiarities and particularities, which it holds under it, but seeks to show the *necessity* of the occurrence and connection of just these particularities. What it gives is a universal *rule* for the connection of the particulars themselves.... Fixed properties are replaced by universal *rules that permit us to survey a total series of possible determinations at a single glance.*"

---

implication.' And this criterion, here only introduced by way of supplement and as a secondary aspect, proves on closer analysis to be the real logical *prius",* (his emphasis), "for we have already seen that 'abstraction' remains aimless and unmeaning if it does not consider the elements from which it takes the concept to be *from the first* arranged and connected by a certain relation." Cassirer, "Substance and Function", p.24

Of critical importance is the fact that Cassirer's new formal Concept *is no longer logically derivable from its extension*, (its membership), however:

> "The meaning of the *law* that connects the individual members is not to be exhausted by the enumeration of any number of instances of the law; for such enumeration lacks the generating *principle* that enables us to connect the individual members into a functional whole."

> "If we know the relation according to which a b c . . . are ordered, we can deduce them by reflection and isolate them as objects of thought. "It is *impossible*, on the other hand, to discover the special character of the connecting relation from the mere juxtaposition of a,b,c *in presentation*."

And again:

> "That which binds the elements of the series a,b,c, ... together is not itself a new element that was factually blended with them, *but it is the rule of progression*, which remains the same, no matter in which member it is represented. The function F(a,b), F(b,c), ..., which determines the sort of dependence between the successive members, *is obviously not to be pointed out as itself a member of the series*, which exists and develops according to it." [7] (My emphasis) See footnote.

This is the *definitive* argument against "abstraction" as the *general* case and against "presentation", ["things"], as an ultimate foundation for logic.[VIII]

---

[7] cf. Stewart, 1995, "Fibonacci Forgeries". Stewart's article illustrates the case. The "insufficiency of small numbers" leads to an indeterminability of any finite series.

He continued:

> "We do not go therefore from a series: a-alpha1-beta1, a-alpha2-beta2, a-alpha3-beta3... directly to their common element a, (Cassirer argues), but replace the alphas *by a variable* x, and the betas *by a variable* y. Therein we unify the totality in the expression 'a-x-y' ", (actually w-x-y, where "w" is the constant function w(p) = a, for all "p" of the "generic concept"). This expression can be changed into the "concrete totality" of the members of the series by a continuous transformation, and therefore "perfectly represents the structure and logical divisions of the concept"![IX]

Cassirer's "series" may be ordered by *radically variant* principles however: "according to equality", (which is the special case of the "generic concept"), "or inequality, number and magnitude, spatial and temporal relations, or causal dependence"[X] –*so long as the principle is definite and consistent.*

But where does this principle, *this rule*, come from? Any finite series of presentations, *no matter how long*, is not definitive to establish a general case.[XI] I could, for instance expand the series "E": 1,3,5, ... to the googleplexth element, (GP,– i.e. 10 to the 100th power), and then insert any arbitrary series behind it. 1,3,5,7, ...EGP-1, EGP, 99, 47, 20075, ... The rule itself is never inherent in the *presentation* of the series. To a mathematician moreover, any finite number is pretty much as relatively small as any other –that is what it means to say that it is *finite!*

Googleplex is not significantly different in its fundamental nature from "34", for instance –i.e. they are both finite, each could be raised to the GP^{th} power, and for each and every such resultant, it could itself be treated likewise ad infinitum! If rules *per se* are not inherent in *presentation,* then, where do they come from? I will propose that they come from the billions of years of biological self-organization which is itself based in pragmatism – i.e. in an optimization of the functioning of its neural primitives, and, of course, in the organism's subsequent survival!

Cassirer continues: "The distinction between the concept and its extension, therefore, *is categorical* and belongs to the 'form of consciousness'". It is "a new expression of the characteristic contrast *between the member of the series and the form of the series*".[XII] But the rule itself is now internal to the mind(/brain) itself! It is "a new form of consciousness"![XIII]

Thus he fundamentally reconceived the formal Concept, *this our ultimate logical building block*, as "the "Functional Concept of Mathematics". It is the *functional* rule, F(x,y,z, ...), which organizes and embodies the totality of its extension.

For Cassirer the association of the members of a series by the possession of a common "property" is only a *special case* of logically possible connections in general, but it is the *sole* possibility allowed by abstraction alone. I believe it encompasses what I will term "Diophantine logic" which I believe constitutes the essence and the focus of contemporary logic.[XIV] But the connection of the members "*is in every case* produced by some general law of arrangement through which a thorough-going rule of succession is established." This is the general and comprehensive case. He argued that it is "a new form of consciousness".

He posited it –*his* "Concept"- at the very bottom of our mental world. I saw, I understood, and I agreed. This was my starting point.[8]

Cassirer's "concept" was so natural and so real that it affected my thinking forever after. I believe his Concept is the "concept" we actually use in our thinking. It still sits at the bottom of my understanding and it colored my understanding of the mathematics which followed it.

My next mark was my beginning calculus course. I was very fortunate in that I was exposed, (even at that level), to three very famous mathematicians: Saunders Mac Lane, (cited earlier), Paul Halmos, and lastly, but most importantly for me, to Isaac

---

[8] (Note: this is a very truncated version of my earlier analysis of Cassirer's ideas. There is a much fuller analysis of Cassirer's conception which I have presented in Chapter 5. I think it is important for a better understanding of his ideas, and of my subsequent expansion of them, but I also thought it would have interrupted the flow of my argument here.)

Wirczup who was my primary instructor. I was concurrently and exhaustively reading three texts on the calculus. I spent a minimuml of six to eight hours a night on calculus alone –but *trying to understand it within Cassirer's conceptual framework* – my other classes were essentially neglected. It was so beautiful that I changed my major to mathematics. I was in love!

Wirczup was a kind and marvelous teacher who taught me rigor. But most of all, (from my current perspective), he was important to me because I think he was a "closet intuitionist". He defined the word "infinity" for me strictly in terms of the delta / epsilon relation, (the precisely defined relations of limits), and not in terms of "size" or "magnitude". It made great sense to me, I think, because of my newly acquired Cassirerian perspective.

For Wirczup, (at least as I understood him), "infinity" meant simply and solely that we could not bound the epsilon, no matter how small the delta –and that was the totality of the *meaning* of the word "Infinity"! Any sentence using the word "infinity" became a statement purely and strictly about the relationships between limits, (defined logically, relationally and conceptually); it was not a statement about objects or sizes!

I was able to interpret the whole of the Calculus I encountered at that level from my newfound Cassirerian/Wirzcupian perspective and it made total sense to me in those conceptual terms[9] –whether that was ultimately to be the correct perspective or not. During the following summer, I finished Mac Lane's "Modern Algebra" by myself.[XV] I saw the latter with the help of my prior acquaintance with Wilder's characterization of the objects of axiom systems as "presumptive and permissive" only[XVI]. That is, I saw it in purely conceptual terms.

I saw Mac Lane's book specifically and solely as a comprehensive exploration of all the possible forms of abstract

---

[9] Thinking it over, I guess I always saw mathematics in terms of concepts rather than in terms of sets –see later.

mathematical ordering, (in Cassirer's sense of the ordering of the concept discussed earlier), and that was certainly consistent with the perspective of modern structuralism which I discovered (*by name*) only recently.

For me, Mac Lane's book had nothing to do with actual objects at all[10] –it had to do with the possibilities of "ordering" specifically. It had to do with the possibilities of ordering, (taken in Cassirer's sense), *within a concept itself.*

Benacerraf's characterization of structuralist mathematical objects as "positions in a structure", (cited earlier), is certainly corroborative to such an interpretation.

Mac Lane's book also exposed me for the first time to Cantor's famous proof showing the indenumerability of the reals: i.e. that the real numbers cannot be set in one-to one correspondence with the natural numbers -or even to the rational numbers for that matter. Cantor's proof was obviously profound, (and ingenious), but the specific conclusion that he drew from it was not acceptable from my (Wirzupian?) perspective –and led to another critical revelation which is absolutely pertinent to my ultimate perspective on the problem of the mind/brain whose examination we, together, have undertaken to understand here..

I conceived a fundamental objection at that point in time to Cantor's "diagonal proof" which had argued against the commensurability of the rational numbers and the reals wherein he purported to show that the reals are of a *larger order* of infinity, (size), than the rationals.

My interpretation was based specifically in my prior conception of Modern Algebra as being fundamentally about "ordering" within concepts –from what would now be called a structuralist perspective. Even though it was the objection of a very young man, I think it has held up well over time.

---

[10] This latter perspective seems to be the only way that the subject could be approached via the classical, generic concept. Upon reflection, this seems to be the cusp of our difference and seems to refer directly to Shapiro's and Mac Lane's demand for the non-existence of "the group", etc. mentioned earlier.

On Cantor's Diagonal Argument –written 50 years ago!

(Copied pretty much verbatim from my notes 50 years ago –this is just me quoting my own very old writing)

"The uncountability of the real numbers would not seem to be derived from 'size' or 'magnitude' discrepancies between the rationals / integers –and the reals, but rather, would seem to boil down to a fundamental question of *order* – i.e. of the impossibility, (even in theory), of setting up a procedure, a continuing intellectual (ordering) procedure which would present each and every real number. They cannot all be presented serially, (even in theory), in spite of the fact that we can (in theory) present in a serial list *any* given real, and any *list* of reals.

Cantor's proof displays a (specific but variable) real number (via his diagonal/slash procedure), appropriate to any serial list, (and continuing intellectual procedure) which is not, and cannot be contained in that list. Its construction is derived from the specific serial nature of the particular list itself. It is also, of course, related to the serial (decimal) presentation of any given real number itself.

Consider this *alternative* visualization: (I will claim exactly the same kind of freedoms that Cantor was allowed in his diagonal proof here.) It seems I could set up a serial presentation of all reals in this manner: set up a hypothetical line segment of unit length one, (which, of course, is fully sufficient as this segment can be shown to be in one to one correspondence with the whole real line).[11]

Select a hypothetical dart with a zero magnitude point, (equivalent in principal to Cantor's hypothetical ability to write out the whole of a real as an infinite decimal –i.e. exactly, precisely). Throw the dart at the line, (axiom of choice?), and the

---

[11] Simply consider the semi-circle based at the origin whose length equals 1, and then radiate the lines from the origin through the semi-circle to some horizontal line. This yields the necessary correspondence.

point hit is then r1. Continue throwing the dart for r2, r3, etc. If the dart hits a previously speared point, throw again for that member of the list. No point is privileged or exempt *a priori!* If I were infinitely lucky —which is *theoretically* possible, (but infinitely improbable), it seems that I might derive such a list in an infinite time. (Cantor gave himself an equivalent time in the writing of his list!)

Certainly, though, this shows that the difficulty is not as usually thought and that such a sequence would be possible except, I believe, for one *fundamental* reason, not to do with 'size' or 'magnitude'.

Rather, I believe it derives from the inherent *impossibility* of setting up such a procedure in the first place. Here, -of setting up an intellectual procedure which will assure that *every* point on the line would *definitely* be accounted for —even assuming infinite luck and time, (because the *number of the throw* is always an integer value and its "value" submits itself to the diagonal/slash procedure.)[12] Thus, the difficulty would seem to derive from fundamental differences *in structure* of the reals and the integers/rationals —i.e. of the real and the rational fields.

You could not predetermine the placing or even if there actually exists a definite placing —given a particular real in the line —and this seems to be inherently so. This is contrary to the situation of the rationals, (wherein a denumerable correspondence is easily demonstrable using the "zig-zag technique"[13]), and, I think, is the essence of the difference.

---

[12] i.e. You can't *sequence* the correspondence between the two manifolds!

[13] Lay out the integers in two dimensions a and b, then consider the intersections as the ratio of a / b. Come back and start at the origin 1/1, (which is the first element of the sequence), traverse in one of the two dimensions to the second element, then traverse the diagonal, then sideways or down, repeat the diagonal, etc. This gives a unique ordering to the rationals and a one to one correspondence with the integers.

Chapter 3: Adventures of the Mind   Cantor & Structuralism

A reiteration of my later reflections on Cantor's argument:

(A much more recent return to the subject):

"Infinite sets are unique in that it is possible that they *can* be put into 1 to 1 correspondence with (some) subsets of themselves. (This is not to say that some given subset may *not* qualify —e.g. the classic case of the rationals inside of the reals with which we are concerned here —or to the trivial subset {1,2,5,11,3}.) We may even leave out huge – *even infinite subsets*. (e.g.: 1>2, 2->4, 3->6, ... -all the odd integers are left out of the second set of integers), but each of these sets is still infinite! "Largeness" is not the issue.

Consider Cantor's definition: Two infinite sets "are of equivalent size" precisely if and only if they *can* be put into 1 to 1 correspondence with each other.

(Within the reals themselves this would correspond, for instance, to the possibility of the 1 to 1 correspondence between the unit interval into the whole of the real line. O.K. so far- but suppose they *cannot* be, (*rejecting* the "precisely" in the definition). Suppose we are not talking about "size".

Now consider Cantor's 'diagonal slash' argument. Suppose this reveals the fact that the rationals and the reals cannot be put into 1 to 1 correspondence *not* because they are of different *sizes,* but because the reals cannot be *ordered* like the rationals.

Suppose this is an argument about *possible ordering* rather than about *size, - i.e. that the reals are incapable of a natural ordering!* ("Ordering" had become a big word to me by that time as it became the focus of my orientation of modern algebra which I saw as the progressive development of all the possible orderings of ideal and abstract mathematical objects.)

It is certainly amazing that the whole of the rationals —and not just the integers- can be ordered countably as is clearly known and easily demonstrable, but it is an amazing fact nonetheless! But consider: between any two rationals there exists another rational. Between any two rationals there exists a real. But *between any two reals —no matter how close- there exists a rational as well!*

Consider the open interval from x to x + Δ, (delta), for any non-rational real x, and consider the limit in that interval as

Δ, (delta), approaches 0 of the truth value of the statement: "There exists a rational number in the interval." The limit is the *constant*: "true"! It is not viable *at* the limit however, i.e. it is discontinuous, approaching from either side of x.

Are we talking then about "size" or about something else? If this is, in fact, not a discussion of "size", then there need exist only *one* "infinity" – one unbounded "quantity"-reflecting a statement about the delta/epsilon relationship; it is not a statement about magnitude! But then ordering and structure become the crucial issues!

Nowhere does Cantor's diagonal argument have anything at all to do with "size" *per se* save in his conclusion. *Everywhere* it does have to do with order and ordering –even in his specification of the problem itself. So why take the dubious, more complicated conclusion over the leaner and clearly justifiable one? Why not invoke Occam's razor right here? Why not recharacterize Cantor's argument specifically as an argument against imposing a natural ordering on the reals and stop right there?

That two *finite* sets are equal "in size" just in case they may be ("may be" = "can be") put in one to one correspondence with each other is clearly justifiable. But to make the *same* assertion for infinite sets does not seem to be anywhere near as plausible.

It is trivial, (and definitional) that any infinite set may be set in one to one correspondence with some, (but not *any arbitrary*), proper subset of itself, (by definition). Are they then of "the same size"? Under Cantor's definition, of course, they are because of the correspondence. And yet the original set contains elements, (perhaps even an infinite "number" of elements), not in its proper subset.

It seems to be an equivocal assertion, then, to assert the converse -that just because two (infinite) sets *cannot* be set in one to one correspondence that they are therefore of "*different* sizes". The simpler, (leaner –invoking Occam's Razor), though more abstract conclusion would seem to be the better one: simply to assert the raw result itself: i.e. that *they cannot be set in one to one correspondence*!

This directly converts my claim about the possibility of imposing an order. It seems to me that Cantor's proof is a profound revelation about "ordering" and about "correspondences", not about size. It elucidates the impossibility of a natural ordering of the reals. (But whence then his transfinite sets? Where have the "alephs" gone? DNE? [14])

A Powerful Argument for the Structuralist Perspective

Consider this: Cantor could not fulfill even the very first natural and anticipated step after his initial conclusion –i.e. the proof, (or disproof), of the Continuum Hypothesis[15]! Indeed Gödel proved it is impossible within the standard axiomatizations of set theory. And yet it is accomplished *merely in the addition of a single axiom* to the axioms of the rationals: i.e. the "Least Upper Bound" axiom!

This is a very powerful argument for "ordering" vs. "size" in our conception of numbers[16] and for the actual root of that ordering as being in the logical connectivity, (the structure), of the very axioms themselves, (think Hilbert's "whole of the axiom system", or, in a biological setting -"nerve connectivity"), rather than deriving from some *a priori*, pre-supplied "objects" of the system. You might also look at Appendix E: "Cassirer speaks to Shapiro & Mac Lane", wherein Cassirer makes a very similar case based in relationality itself –based in Leibniz's conceptions of geometry. Cassirer argues that the latter is the actual basis of number rather than the converse and so constitutes a strong argument against Cantor's perspective.

Adopting Hilbert's conception, it also supplies the meaning and the existence of its conceptual objects themselves. I think it is a powerful argument specifically for the structuralist

---

[14] Math-speak for "Does not exist"?
[15] i.e. that the Reals are the "next larger size" beyond the Rationals!
[16] Or of *any other* abstract mathematical object

perspective itself and one of the reasons I became enmeshed in my own version of it 50 years ago! I had figured it out for myself.

You see, I think the young Hilbert[17] had it right in the first place. The "properties", the *ordering* of his "things" comes *from the whole of the axiom system* (itself) , not from just a part. Without the Least Upper Bound Axiom or its equivalent, we cannot attain the Real Continuum, though we may attain the Rationals. To reiterate Schlick's comments:

> These elements "acquire meaning only by virtue of the axiom system, and possess only the content that it bestows upon them. They stand for entities *whose whole being* is to be bearers of the relations laid down by the system."

This is not about size, this is about ordering, about structure, about the connectivity of *meanings* of the axioms themselves!

My Conclusion:

"Ordering", I concluded, is a function of all the *axioms* of an abstract axiom system – of the complex, (semantic), rather than the simplistic, (Aristotelian), logical connectivity of the axioms themselves; it is about the connectivity of *meaning!* There may be "a natural analogy between natural numbers and strings", (requoting Shapiro), but there is no natural analogy between natural numbers and meanings! (Edelman's quote is worth repeating here: "certain symbols do not match categories in the world … Individuals understand events and categories in more than one way and sometimes the ways are inconsistent! "

Ordering is not a function of the properties of its "objects" which are specifically *virtual* reflections of its underlying structure. "Ordering" is not a function of these "positions in a structure, it is

---

[17] Contrary to his "older brother" who was seduced by Cantor and lost his virginity thereby!

not a function of these solely "permissive" and "presumptive" virtual "objects" from which it supposedly "arises"!

The "rule" of the continuum, (in the sense of Cassirer's usage of the "rule" of a concept), here lies in the logical interplay of the meaings, the mechanics and the connectivity of the axioms themselves. It is a conceptual, (in Cassirer's sense), rather than a reductive set-theoretical one. It refers instead to its own axiomatic structure in Benacerraf's sense which generates its objects as (virtual) "positions in that structure".

(On the level of biology, its components may be understood analogously to the intentional functions in the sense of figure 3 early in the first chapter,. These are strategic rather than informational functions. But then again, what *else* could we demand from a "machine"?) [18]

This is a new conception of "order" itself! (This perspective will find validation in both Quine's remarks and in the structuralist perspective of the very concept of "class" itself!)

To quote a prior paper of mine: "the 'objects' of these schematic models, I propose, are manifestations of the structure; the structure is *not* a resolution of the objects."[19]

To give a bit more credence —and to dissuade you from the tempting conclusion that the foregoing was just an instance of the arrogance and ignorance of youth, let me insert a couple of citations from some famous contemporaries of Cantor, (I discovered these citations much later). They argue the same case I made!

Poincaré[20], a famous contemporary of Hilbert and certainly Hilbert's equivalent as one of history's most significant mathematicians said:

---

[18] Consider W.J.Freeman: "The only knowledge that the rabbit could have of the world outside itself was what it had made in its own brain!" (W.J Freeman, 1995)

[19] See Chapter 4 where that argument is presented.

[20] Poincaré actually conceived of the theory of relativity before Einstein did. His mistake was that he only conceived of it as a possibility! (Penrose, 1989)

> "Actual infinity does not exist. What we call infinite is only the endless possibility of creating new objects no matter how many exist already"

Poincaré again: "set theory is a disease from which I hope future generations will recover."

Hermann Weyl, another famous name:

> "...classical logic" [itself!] "was abstracted from the mathematics of finite sets and their subsets...Forgetful of this limited origin, one afterwards mistook that logic for something above and prior to all mathematics, and finally applied it, without justification, to the mathematics of infinite sets. This is the Fall and original sin of [Cantor's] set theory ... "

And more current quotes: (William P. Thurston):

> "Set theory is based on polite lies, things we agree on even though we know they're not true. In some ways, the foundations of mathematics has an air of unreality."

Morris Kline:

> "[The pure mathematicians] have followed a gleam that has led them out of this world...the work of the idealist who ignores reality will not survive."

Cantor's set theory, it is true, has come to lie at the core of contemporary mathematics –largely, I think, because of its easier conceptualization and fruitfulness, but it probably will not remain so.

To repeat Thurston, it is based on things we agree on even though we know they're not true. Consider just the Banach/Tarski theorem embedded in this perspective, for instance. If one were *really smart*, it would be possible to dissect the moon to fit it into one's pocket! Think about that! Does this correspond in *any* sense to our reality?

For science generally, progress goes in "fads", (not to be taken in a mean sense). Calculus was the predominant (and most successful), mathematical fad from Newton till about 1900 when it was determined to be not rigorous enough.

It was regrounded in set theory which was able to supply that logical rigor. The problem, I feel, is that it supplied too much, both for mathematics and for logic –leading to the stalemate and the paradoxes that mathematics finds itself in currently. There is a new "fad", mathematical structuralism and category theory, which is taking mathematics back to function over set membership.

> "One of the primary perspectives offered by category theory is that the concept of *arrow*, abstracted from that of *function* or *mapping*, may be used instead of the set membership relation as the basic building block for developing mathematical constructions, and expressing properties of mathematical entities. [XVII]

(Now can you hear the distant echo of Cassirer I foreshadowed in Chapter 2?)

The Concept of Implicit Definition [21]

When I was exposed to Hilbert's "implicit definition"[22], either at this point or in Cassirer's "Einstein's Theory of Relativity"[23] [XVIII], I had a final and completing insight –I hope you will find it relevant.

If mathematics was fundamentally all about ordering as I had concluded that it was, (and if the root of that ordering resided in the infrastructure of the axioms themselves as Hilbert argued

---

[21] See Chapters 2 and 5 for an earlier presentation of this idea.

[22] Let me repeat the quote from Schlick, (cited in Cassirer): "[Hilbert's] revolution lay in the stipulation that the basic or primitive concepts are to be defined *just by the fact that they satisfy the axioms....* [They] *acquire meaning* only by virtue of the axiom system, *and possess only the content* that it bestows upon them. They stand for entities *whose whole being* is to be bearers of the relations laid down by the system."

[23] Actually have come to believe recently that I first encountered it in Cassirer's "Substance & Function", (Cassirer, 1923) –see Chapter 3 of that work!

rather than as a consequence of supposedly prior (but only permissively assumed) "objects"/ "concepts" in Frege's sense), then the profound plethora and the richness –and the depth- of such orderings already extant in mathematical axiomatic systems conversely suggested a radical extension of Cassirer's rule-based "Functional Concept of Mathematics". It suggested an expansion to a new and radical -and larger notion and rule of "concept" itself. It suggested the expansion to what I have called "the concept of implicit definition", (C.I.D.).

This latter is based at the deepest level in the axioms themselves and represents what I believe to be the broadest possibility of the formal logical Concept. Cassirer's "functional concept" is based in a set of denumerable and essentially dimensional functional rules[24] –in manifold rules of individual series. But axiom systems, as systems per se, have rules too –more complex, more profound constitutive rules of ordering deriving from their profound structural connectivity as just discussed in my conclusions regarding Cantor's argument.

They have comprehensive overall *unary* rules, (of the whole of the system of axioms itself –see Hilbert's comments earlier –"*since only the whole structure of axioms yields a complete definition.* "- and my just finished discussion above),

The logical infrastructure of such axiom systems is not, in fact, itself dimensional, (in Cassirer's sense -about properties of percepts or objects –f(x,y,z)) -but profoundly and interconnectedly *logical* and *meaningful* instead to the structure of the axioms in the system itself.[25] This is the import of my Cantor argument expressed above, and how I originally conceived the notion.[26]

---

[24] i.e. f(a,x,y,...)

[25] which is the way I interpreted the Cantor diagonal proof

[26] Consider: sets, at the fundamental level, are always conceived "raw"! Think about the elements in Cantor's original diagonal slash listing, for instance, from the standpoint of Cassirer's "Mathematical Concept of Function". There may be *only one entry which possesses* an "x" dimension, and another which may have as its sole entry a "y" dimension –etc. [Note: this could be made more

The "rule" of the continuum, (in the sense of Cassirer's usage of the "rule" of a concept), here lies in the logical interplay, the mechanics of the axioms themselves. It is a new conception of "order" itself!

I propose to interpret logic in the same manner that I think the "young Hilbert" intuitively did, and to which I have referred before. I propose to interpret it within the context of "the concept of implicit defintion"!

This "Concept of Implicit Definition" suggested an extension of Cassirer's "functional concept of mathematics" into a conceptual rule, (an ordering) grounded in the unary rule[27] of an overall axiom system.[28] I ultimately related this, under Hilbert's "implicit definition", (and Cassirer's "new form of consciousness" perspective –to which I assert it is a legitimate heir), to a constitutive and specifically operative ordering totally internal to the brain.

This new form of consciousness could specifically reflect the structure and the operationality of that brain –its own rules and connectedness[XIX], its "triggering" to use Maturana's more pregnant conceptualization,[XX] as well as its (now virtual) objects.

But within such a system the elements, (the perceptual "objects" *themselves*, (i.e. our naïve objects) –as well as the conceptual "objects"), could be "implicitly defined" after Hilbert's conception. This, then, was how I was able to conceive even our "percepts" themselves as internal to such a model. I conceived them solely as "positions in a structure", as specifically *virtual* objects!

---

rigorous by enlargement of the base number.] This may be a list *of specific and totally unrelated entries*, grouped in no other way. The "functional concept of mathematics" may be the only route to a, (any), rule of series and the resultant "concept" derived therefrom. Abstraction just won't work!

[27] Recall Hilbert's remarks that it is the whole of the axiom system which defines its objects!

[28] I ultimately identified "mind" with the operative, unary rule of the brain –i.e. with its overall rule of "structural coupling". See Maturana in Chapter 6.

I conceived them as purely conceptual objects, implicitly defined by the "axioms" which embody the physical operationality of the brain.[29] I conceived "percepts" themselves as *metaphors* of the brain's own process! I conceived them as specifically virtual and operational objects! (See Chapter 4)

But How can we conceive of purely operational objects as correlating with the real world?

But how can we possibly conceive the objects of our ordinary but very concrete naïve world as solely operational objects? Certainly, if you call yourself a materialist, you must admit that "percepts" do not actually, (physically), exist as they seem; science *already* sees them quite differently. Do we perceive mathematical magnitudes, (wavelengths), of light waves or "colors"? Do we perceive molecular density or "hardness"? Do we perceive mean molecular energy or "heat"?

Consider moreover the best of our current physical theories. Consider the parallel between Penrose's comments on the Schroedinger equation and the functioning I propose between the re-afferent brain and action into the world.

Repeating an earlier section of this paper, quite pertinent here,[30] consider the parallel between the *most* SUPERB[1], (according to Roger Penrose- his CAPS), of modern physical theories, and my own conclusions:

> "There is a very precise equation, the *Schroedinger equation*, which provides a completely deterministic time-evolution for this [quantum] state. But there is something very odd about the relation between the time-evolved quantum state and the actual behavior of the physical world that is observed to take place.

---

[29] W.J. Freeman's "equivalence classes" might be a reasonable beginning here.
[30] I will come back to this passage again later. I think it is highly pertinent and a strong argument for my conceptions.

From time to time –whenever we consider that a 'measurement' has occurred –we must discard the quantum state that we have been laboriously evolving, and use it only to compute various probabilities that the state will 'jump' to one or another of a set of *new* possible states." (ibid, P.226, his emphases)

In this "more optimistic" view, it is only "*in relation to the results of 'measurements'*" that concrete reality emerges –i.e. that a specific rendition of space-time is enabled.

Now compare this to the re-afferent model I have already sketched and which I will formally present in the Freeman Appendix of Chapter 4, (alternatively my Figure 3 of Chapter 1 is a reasonable referent). wherein it finds a striking parallel. Each evolves a "state" equation and then performs a "measurement", (action into the world), which then causes a *new* state equation, (Schroedinger/Merleau-Ponty), to be formed until the next "measurement" is performed. How close these conceptions are!

I think my perspective is in fact legitimate and answers the basic *biological* question. The biggest remaining problem that I have is the one from organism to externality and I think that Maturana and Varela, (see Chapter 6), have framed the essential problem very, very well.

There remains one fundamental objection to my thesis which I have long considered, do not consider trivial, and which is exposed throughout this dialogue however: why then, does our model work *so well*? I have thought this over deeply, and perhaps the best answer that I can make is the analogy to a "hive of bees" completing their hive, (cited in Chapter 12). That is, I think good science is a self-fulfilling prophecy. The trick, however, is to understand it entirely relativistically –ie. to understand it in its *entirety* heterophenomenologically[31]!

---

[31] Using Dennett's word yet again!

The Anthropic Principle

Or, to put it in a more respectable setting, I think it may be the ultimate fulfillment of the concept that I first saw in Penrose's book: i.e. of the "anthropic principle".

But the usage I imply here is a deeper sense and meaning of the words. It is not that "if the world were not as it is, then we would not be here to see it", (Penrose, paraphrase), but rather in a sense where "our seeing it that way" allows an algorithmic interaction with a nameless reality. Put more simply, we can only see what —and in the precise manner that we are "designed" = "configured" to see.

We are, however, allowed to extend and expand that vision. But our current perspective must be understood as a specifically biological perspective under Cassirer's "Symbolic Forms" ,(see Chapters 8 & 9). I will broaden this perspective in Chapter 8 to attain the full vision.

We are clearly *already* dealing with a model even within rigorous science itself, and my hypothesis seems to fit very well with what we know so far. I propose that the mind/brain is even more of a model than we suspect however -to include *our "objects"* themselves in the sense of Quine's earlier comment!

Back to Mac Lane Again:

Here was a brief (though negative) comment that Saunders Mac Lane was gracious enough to make about my conception: "the idea that axiomatics amounts to an 'implicit' definition is no longer generally accepted. It fits well with class axiomatics (e.g. for geometry) where there is just one intended model. It doesn't fit for axioms for groups or space, where there (sic) are many models."

Just what *does* exist for Mac Lane? "Categories"? But what are they if not founded on the Platonic existence of sets, (of real things?), themselves? And are not these furthermore conceived in analogy to the multitudinous naïve reality we are accustomed to? D'espagnat, Einstein, Bell and I thought differently!

That may be precisely the point. If there were, in Mac Lane's words, "just one intended model" –then the intended model would be the physical brain itself.

I believe that mathematics itself must fall under my contention that this problem is *so basic* that radicalism in *all* of our disciplines is demanded –to include mathematics! Just what *is* real for Mac Lane? In the words of Chapter 13, mathematics, even Mac Lane's mathematics, is still founded in "multitudinism" as d'Espagnat defines –and refutes it from the standpoint of modern experimental physics!

The problem arises, however as to just what sort of an axiom system might enable the kind of complexity found in the human brain.

In reading Gerald Edelman a possibility suggested itself drawn from his theory of immune response, (for which he won the Nobel Prize). He treats the whole subject of immune response as "an information system". Antibodies are originally and autonomously made, (i.e. before the fact), for all possible antigens.

Combining this broadness of spectrum with his discussion of the phenomenon of "neural pruning", (the massive destruction of the early connectivity of the fetal brain which he pursues in "Bright Air ..."[XXI]), it opens a useful line of thought regarding the "a/d converters", (or better "hierarchical/non-hierarchical converters"), mentioned earlier and in Chapter 4, which I believe constitute our actual perceptual "objects".

Despite the obvious differences in conceptualization, the "objects" of the mind are treated somewhat similarly by me – perhaps as the massively enabled and massively pruned a/d converters of the pure process, the connectedness of the brain! Perhaps they follow a similar developmental path.

D'Espagnat suggests another perspective on the problem which I will explore in Chapter 13 –I think it is a very pregnant beginning.

This problem is huge, and I don't claim to have fully solved it, but let me remind you of the raw neuronal "size" of the brain. If we are dealing with axioms, or Freeman's "equivalence classes", then we may have billions of them, and that opens new possibilities.

A final comment by Cassirer relevant to this Current Problem

Consider Cassirer's commentary on the fundamental nature of the percept:

> "For example, if we conceive the different perceptual images, which we receive from one and the same 'object' according to our distance from it and according to changing illumination, as comprehended in a series of perceptual images, then from the standpoint of immediate psychological experience, no property can be indicated at first by which any of these varying images should have preeminence over any other.
>
> Only the *totality* of these data of perception constitutes what we call empirical knowledge of the object; ... No one of the successive perspective aspects can claim to be the only valid, absolute expression of the 'object itself;' rather all the cognitive value of any particular perception belongs to it only in connection with other contents, with which it combines into an empirical whole."
>
> "...In this sense, the presentation of the stereometric form plays *'the role of a concept'*", (my emphasis), "'compounded from a great series of sense perceptions... This ordering *by a concept* means, however, that the various elements do not lie alongside of each other like the parts of an aggregate, but that we estimate each of them according to its *systematic* significance...."[XXII]

Please note Cassirer's focus specifically on "the concept" – his *reformulated* "functional concept of mathematics"–at the very center of his percept. My extension of Cassirer's functional concept of mathematics into the concept of implicit definition, (C.I.D.), will be the final step into an understanding of my ultimate perspective.

Cassirer's "functional concept of mathematics" and my newfound deeper anti-Cantorian conception of ordering, (seeing the latter as residing in the intrastructure of the axioms themselves rather than in the properties of their "permissive" objects), *when combined with Hilbert's "implicit definition"* enabled a profound "logical leap" to "the concept of implicit definition"[XXIII] which is a new thing. It enabled for the first time an *explicit* conception of a "constitutive concept" in the sense of Kant.

[CLARIFYING NOTE: December, 2011

If Cassirer's "concept" is a rule, and exists as a "new form of consciousness", then those things "implicitly defined", (in Hilbert's sense), *by that same rule* are part of that rule and exist in the very same sense *within* that "new form of consciousness"! The question, therefore, is that of "existance" itself, and whether or not we "hypostacize" his new concept-which I address in Chapter 5, ("Contra Cassirer"), and conclude that we must do so because of the implications of Maturana's profound conception of "structural coupling".]

The Concept of Implicit Definition, (C.I.D.), supplied a totally new rule of "ordering" beyond Cassirer's beginning extension of the meaning of "Concept". It goes from Cassirer's internalized ordering of the series, ("the new form of consciousness"), to a more complex[XXIV] ordering: the Concept of Implicit Definition, itself *also* a "new form of consciousness" and consistent with the former - imposed by the specifically semantic structure[XXV] of an axiomatic system under implicit definition, (and probably to the foundations of structuralism). In fact, I will go further –I believe it is the only *possible* form of consciousness!

My conclusions from my "Cantor diagonal" paper, along with my ("ordering") conclusions from my study of modern algebra, (all seen through the filter of Cassirer's reformulated concept), supplied the genesis of this notion.

The question remains only whether such a leap is justifiable or necessary. The thrust of my overall thesis argues that it is. It suggests the first actual non-eliminative resolution of the mind-body problem!

It suggested the first possibility of a solution to the problem I stated at the opening of this paper: "How can a biological mechanism ever know anything at all? Answer: it could if its "objects" were *purely operative*[XXVI] *–and virtual-* objects like the objects of implicit definition, (or of structuralist mathematics).

These "objects" could be defined internally and *known* to the organism/mechanism itself, (which would be its model) – likened to and extending "the new form of consciousness" claimed by Cassirer for his "mathematical concept of function" but obviating his necessary external referent and substituting instead Maturana's much more precise "structural coupling".[32] (My third and final thesis of "ontic indeterminism", coupled with Maturana's "structural coupling" explains and answers the obvious materialist epistemological objections.)

Mac Lane and category theory

Perhaps I misunderstood Mac Lane's book, but I believe it did preach the doctrine of structuralism implicitly. Structuralism was "implicitly defined" by the import of the whole of the book. Mac lane was still working within the confines of "objects" and referents however, and this is where I think he went wrong.

Listen to Quine once more:

> "One could even end up, though we ourselves shall not, by finding that the smoothest and most adequate overall account of the world *does not after all accord existence to ordinary physical things......Such* eventual departures from Johnsonian usage", (Samuel Johnson, again, is said to have demonstrated the reality of a rock by kicking it!), "could partake of the spirit of science and even of the evolutionary spirit of ordinary language itself."[XXVII]

Quine, generally acknowledged as one of the leading logicists of the 20th century, was able to conceive of an account of the world that *"does not after all accord existence to ordinary physical things"*. But how *could* we account for the world *without* "accord(ing) existence to ordinary physical things", - without "accord(ing) existence to some- *"thing"* –i.e. without objects?

---

[32] See Chapter 6

I think the mind is about concepts –intentional concepts; it is not about referents. It is not about classes or sets, (necessarily of "things" abstracted from dogmatically accepted "sense impressions"), except within a conceptual framework. (See Benacerraf comment shortly)

Cassirer's reformulated "Functional Concept of Mathematics" is wholly based in rules, contrary to the case in Aristotelian –or in Cantor's logic which is derived from it.

If the possibility of the non-existence of "objects" themselves that Quine asserted is, in fact, a *real* possibility, if it is truly plausible, then what could classes and sets refer to? And why would we, in fact, need them at all? I think we do need them, but as *specialized* concepts. I think these specialized concepts, (i.e. classes, sets), are generated to fulfill specialized perspectives, (see my first hypothesis of "schematic artifacts" to follow in Chapter 4.). You might also revisit chapter two of this paper to note Benacerraf's and Bell's comments. But consider each within the context of Cassirer's "Symbolic Forms" cited earlier. Cassirer:

> "…because physical, chemical, biological knowledge *frame their questions* each from its own particular standpoint and, in accordance with this standpoint, subject the phenomena to a special interpretation and formation"

–i.e. "each discipline asks its questions from its own perspective.":

> "[Hilbert] argues that what constitutes an object *varies from theory to theory*, category to category, and that Frege failed to realize this fact… It [logic] remains the tool applicable to all disciplines and theories, the difference being only *that it is left to the discipline or theory to determine what shall count as an 'object' or 'individual.'* [Benacerraf, 288, my emphasis.]

And Bell, (my emphasis):

> "The topos-theoretical viewpoint suggests that the absolute universe of sets be replaced by a plurality of 'toposes of discourse', *each of which may be regarded as a possible 'world' in which mathematical"* [and logical] *"activity may (figuratively) take place."*

But the concepts of the mind are strategic concepts, I believe, not referential ones, (of "objects").

Walter J. Freeman contributes a relevant perspective here:

> "This book had its origin ... in an experimental finding....I was tracing the path taken by neural activity that accompanied and followed a sensory stimulus in brains of rabbits. I traced it from the sensory receptors into the cerebral cortex and there found that the activity vanished, just like the rabbit down the rabbit hole in 'Alice in Wonderland'. What appeared in place of the stimulus-evoked activity was a *new* pattern of cortical activity that was *created* by the rabbit brain... My students and I first noticed this anomaly in the olfactory system... and in looking elsewhere we found it in the visual, auditory, and somatic cortices too... the only knowledge that the rabbit could have of the world outside itself was what it had made in its own brain."[XXVIII]

What makes sense of this perspective, (i.e. its seeming self-contradiction) is Maturana's stark and beautiful conception of "structural coupling", itself combined with Cassirer's other brilliancy: "Symbolic Forms"[XXIX], but the former must be taken in its broadest sense. Equivalently, I have called it "ontic indeterminism". It allows us to *act*, (pragmatically), without *knowing*.[33]

Reconsider Schlick's characterization and interpret it through the *young* Hilbert's eyes:

> "[Hilbert's] revolution lay in the stipulation that the basic or primitive concepts are to be defined *just by the fact that they satisfy the axioms*.... [They] *acquire meaning* only by virtue of the axiom system, *and possess only the content* that it bestows upon them. They stand for entities *whose whole being* is to be bearers of the relations laid down by the system." This is what I propose the "A/D converters", the naïve "objects" of the cortex do.

---

[33] It also allows an entirely new reassessment of the problem of "consciousness".

Back to Shapiro:

"Frege insisted that arithmetic and geometry each have a *specific* subject matter, space in the one case and the realm of natural numbers in the other. And the axioms express (presumably self-evident) truths about this subject matter.

Following a suggestion of Hellman's, let us say that for Frege, the axioms of arithmetic and geometry are *assertory*, and for Hilbert, they are *algebraic*. Sentences that are assertory are meant to express propositions with fixed truth values. Algebraic sentences are schematic, applying to any system of objects defined by them –that meets certain given conditions"[34] [ibid, my emphasis]

"[Young] Hilbert's *Grundlagen* provided consistency and independence proofs by finding interpretations that satisfy various sets of axioms. Typically, he would interpret the axioms of a theory in terms of constructions on real numbers. This approach, now as common as anything in mathematics, runs roughshod over Euclid's definition of a 'point' as 'that which has no parts'. When we interpret a 'point' as an ordered pair of real numbers, we see that points can indeed have parts.

This free reinterpretation of axioms is a main strength of contemporary mathematical logic and a mainstay of mathematics generally. It drives the structuralist, algebraic, perspective on mathematics. And it runs counter to the Fregean perspective."

---

[34] My disagreement with this characterization onto "systems of objects" should no longer need any elaboration. Think once again about Wilder's characterization of the "objects" of axiom systems "as presumptive and permissive only".

> [*But*]"...It seems clear that for Hilbert and just about anyone else, consistency is itself a mathematical matter. His methodology indicates that in order for us to be assured that certain mathematical objects exist; we have to establish the consistency of an axiomatization."

I differ with both parts of this sentence —both proof of existence and of consistency. The ultimate question is "how would it even be possible"! I think Hilbert was speaking a different language —of invariants and of relativity.

> "In the *Grundlagen*, Hilbert discharged this burden, *at least in part*, by providing relative consistency *proofs*." [ibid, my emphasis]

But not necessarily "in part" only. The system need only *be* consistent. Shapiro's comment confuses human logical certainty with reality. Hilbert's relative consistency proofs are of a different order entirely. They elaborate the notion of invariants themselves and are consistent with such. (I think relative consistency proofs are the only ones possible for the machine we call the brain!)

Those invariants must go across the board however –i.e. the whole of one system must be mirrored in the other –as in the principle of duality. As far as our assurance that "certain mathematical objects [must] exist" goes, however, this is a limitation in Shapiro's own epistemology. From Cassirer's perspective, this is something we will never know.

The Rosen Letter: (a Reflection on Shapiro's Position)

But what of the "beer mugs" conception? Must the organization of one system be mirrored *simply* in the other? Or may the translation be complex? Here is an extract from my (fairly recent) letter to Robert Rosen's daughter Judith Rosen[xxx]. I had just learned that he had died, (sadly before I even "discovered" him), and I wanted to express my sympathy to her as well as my excitement in newly discovering his views.

As part of my letter I discussed a theme her father had addressed to approach an understanding of "invariance", (which I think was Hilbert's focus). It so happened that it was a significant theme in my own work as you should be able to recognize by now, (and of Thomas Kuhn's as well). I talked of the mathematical equivalence *of physical, mechanical models* of the Ptolemean and the Copernican universes, easily seen by the arbitrary choice of our anchor point for the respective models. (The following is just me quoting me!)

"…The motions of the planets and the wildly gyrating stars of the one translate into the picture of the stable universe we are ordinarily used to! Mathematically, I hope you can appreciate the beauty and the inherent *mechanical necessity* of the absolute mathematical translation between these models. The version I had been thinking about used our own, (modern), view of the universe vis a vis the Ptolemaic system. (Kuhn used the Tychonean Model). Conceive again of a rigid mechanical model of our solar system revolving about our linearly moving sun, embedded in the field of stars –just as we normally conceive of it and sitting on your (large) desktop.

But let us reach down from some other dimension, (just to stay out of the way), and grasp the now moving and spinning earth firmly pinching it tight so that it becomes motionless, lifting the model off its prior base, and establishing a new "center" in the now unmoving earth, (with its now wildly gyrating extraterrestrial adjuncts).

The point is that the two perspectives must necessarily be *absolutely* mathematically and observationally equivalent – established by the *purely mechanical,* ["gear driven"] nature of the model itself! All of their *relative motion* is absolutely invariant! From the standpoint of an observer anywhere in that universe, all observations and measurements would necessarily be the same[xxxi], [though the language expressing them would be radically different!]!

These, then, are *purely mathematical translations,* (albeit complex ones), confirmed by the *purely mechanical* nature of the model. From this standpoint no observable data whatsoever is gained from adopting one viewpoint over the other.[XXXII]

I think this translation of perspective, (this invariance), illustrates a deeper interpretation of Hilbert's "beer mugs" assertion.[35]

"The problem, however, lies in the 'laws of nature'. All laws, (gravity, inertia, the speed of light, et al), would have to be rewritten to be *place specific* under the (Ptolemaic or, as I later saw from Kuhn —who used a very similar construction- using the Tychonean transformation instead[XXXIII]).

Laws of motion that hold on the earth would not necessarily hold in such simple form on the moon, (in fact, as seen through an all powerful, earth-based telescope, they *would not* under these non-Copernican [and pre-Galilean] perspectives!)"[XXXIV]

This situation is relevant to my suggestion made in reference to the lack of preconceived necessity for a preservation of hierarchy in Hilbert's "Pythagorean theorem" assertion. Think about the purely mathematical and necessarily *definitive*[36] nature of the translations involved between our models.

Modern Ptolemean Physics

Suppose, purely hypothetically of course, that some brilliant but esoteric mathematician of the Ptolemaic school had discovered the dualistic translatory laws for these (new, i.e.

---

[35] It gives a hint to the "how" of Hilbert's statement quoted earlier: "If in speaking of my points, I think of some system of things, *e.g.*, the system love, law, chimney-sweep ...and then assume all my axioms as relations between these things,[35] then my propositions, *e.g.*, Pythagoras' theorem, are also valid for these things ... [Any] theory can always be applied to infinitely many systems of basic elements. One only needs to apply a reversible one-one transformation and lay it down that the axioms shall be *correspondingly the same* for the transformed things." The current discussion is precisely about the translation of invariants, but more complex ones than normally considered.

[36] because they are "gear-driven"

"Copernican") laws of nature – but who conceived those translatory laws *as mathematics only*, (like Heisenberg's matrices perhaps –and, in fact, as the Pope supposedly advised Copernicus himself to do). Mathematically this discovery would have involved the implicit (though not necessarily a conscious and explicit) reorientation of the universe back to its "original" (Copernican) state, (a la Schrödinger?)

He would have implicitly reformulated and discovered new [more easily accessible] laws and implications in that context, (which would have been his mathematical "scratch pad"), and subsequently retranslated them, (and the new laws directly evolving from them –perhaps in a single combined, but possibly "blind" compositional act), back to the original, fixed earth formulation, skewing but precisely reflecting even the new laws.

But, (following our story just a bit further), this could very well have been a "blind", *purely mathematical and compositional* discovery –involving only purely mathematical translations *and without a necessary cosmology or insight.*"[xxxv] [Heisenberg's concept of "Matrices" supplies a reasonable parallel.]

I continued: "The point is that these are solely and precisely mathematical translations! All laws would be absolutely preserved and correct, (all motions would be exactly the same, of course). I think this is a very pretty idea with profound consequences. The biggest problem, however, would be in the discovery of new laws –i.e. the *fecundity* of the model! But, again, these might well be implicit in the transformations.

What does this mean for our problem? It means that our central problem is not one of data, (that is a distinct problem), but of organization! The observational data *per se* holds constant, (*by mechanical necessity*), in this example. Rather, it becomes a problem of organization for current understanding and for the organization needed *for future scientific progress*. It is not a problem of data or the necessary consistency of data as my example demonstrates. This data is obviously absolutely consistent.

(But mine –this present discussion –is a *relative proof* in the very form for which Shapiro so roundly criticized Hilbert!)

Aesthetics

As such, it relates to some of the issues raised by Penrose in his criteria for theories,[xxxvi] but in a deeper context. It relates to what I will call "centrality" and "shape" [theoretical "beauty" if you like], which are surely intentional attributes. Rules and principles are normally more "central" to theories than the language of their data. Galilean Relativity, gravity, the speed of light, Kepler's laws... are preferentially stated in their simplest and most intelligible mathematical form, not in skewed transformations. This is Occam's razor, but more finely honed. We centralize principles, (and, I argue along with Cassirer, intentional principles specifically as well), for *organization!* We then organize the data to fit! This is why we are most of us Copernicans rather than Ptolomeans!

> Hear Roger Penrose on this idea:
> "The above ... brings me to another issue concerning inspiration and insight, namely that *aesthetic* criteria are enormously valuable in forming our judgements. In the arts, one might say that it is aesthetic criteria that are paramount. Aesthetics in the arts is a sophisticated subject, and philosophers have devoted lifetimes to its study. It could be argued that in mathematics and the sciences, such criteria are merely incidental, the criterion of *truth* being paramount. However, it seems to be impossible to separate one from the other when one considers the issues of inspiration and insight. My impression is that the strong conviction of the *validity* of a flash of inspiration (not 100 per cent reliable, I should add, but at least far more reliable than just chance) is very closely bound with its aesthetic qualities. A beautiful idea has a much greater chance of being a correct idea than an ugly one. At least that has been my own experience, and similar sentiments have been expressed by others (cf. Chandrasekhar 1987). For example, Hadamard (1945, p.31 writes:
>
>> '...it is clear that no significant discovery or invention can take place without the *will* of finding. But with Poincaré, we see something else, the intervention of the sense of

beauty playing its part as an indispensable *means* of finding. We have reached the double conclusion:

that invention is choice
that this choice is imperatively governed by the sense of scientific beauty.'
Moreover Dirac (1982), for example, is unabashed in his claim that it was his *keen sense of beauty* that enabled him to divine his equation for the electron (the 'Dirac equation) ...while others had searched in vain ..."
(Penrose, 1989, p. 421, his emphases)

And again:
"It is remarkable that *all* the SUPERB theories of nature have proved to be extraordinarily fertile as sources of mathematical ideas. There is a deep and beautiful mystery in this fact: that these superbly accurate theories are also extraordinarily fruitful simply as *mathematics*." [37]

Theories have "shape" in the same sense that great music has "shape" -not only in its individual themes, but as an overall composition. Occam's razor, (least assumptions), is only the tip of the iceberg.

Cassirer's Symbolic Forms

Early on, in my early 20's when I first found Cassirer's "Theory of Symbolic Forms", (see Chapter 7), I visualized his brilliant conception of the equipotence of varied *specifically scientific* but *different* perspectives[38] as a network of "rubber bands" -as representative of their innate relationality![39] Given such a network, each perspective, each beginning was like grabbing a given nexus in the network and pulling it towards me and making it the focus and starting point for theorization. But

---

[37] Op cit, P. 174
[38] ignoring for the present, the larger panorama of Cassirer's perspective
[39] but this "relationality" must be taken in a context-free, *invariant* sense!

any nexus is a candidate —as long as we understand Cassirer's perspective on the specifically scientific, and viable, (with experience), theories. Each theory *must preserve* the invariants of relationality! These viable theories I identify with Penrose's "SUPERB" theories! Cassirer's scientific epistemological relativity was for me the natural extension of Einstein's Special and General Relativity, and it took some time before I was able to fully conceive its brilliance.

In Chapter 13, I will approach modern physics from this perspective and make a strong suggestion for a fundamental reorientation, comparing its current picture of reality, (foreshadowing just a bit), with that of the fictional King of Petrolia who, as you will see in Chapter 7, was forced to make continual *ad hoc* repairs to his worldview as contrasted with the case of the nuclear technician he was arguing with. This is not a totally negative judgement, but it definitely relates to the problem of scientific realism as I will propose in that chapter.

It seems that quantum formalism must itself likewise keep supplying *ad hoc* (cruder) answers and "work-arounds" as the questions deepen —decoherence, the dividing line between the macro and the quantum world, the question of scientific realism, et al.

My own thesis answers these questions from the very beginning. (It is an explicit requirement that it must always preserve the invariant relationality of the *other* SUPERB theories however, but possibly in a "distributed" sense as illustrated above.) The question becomes, then, one of "theoretical beauty", (i.e. "musicality"/ simplicity of organization), as seen from Penrose's criteria for theories.

That each (SUPERB) theory must be data-viable, (so to speak), and that each must be able to account for the other, (or at least not lead to contrary conclusions), is pretty much a given, but it is a question of "shape", and simplicity of organization which arises here.

To repeat myself: Theories have "shape" in the same sense that great music has "shape" -not only in its individual themes, but as an overall composition. Occam's razor, (least assumptions), is only the tip of the iceberg.

What does this "interlude" mean specifically for the problem of the brain? Does the mind do this? Is its organization based on aesthetic and intentional grounds as well? I propose that the problem of the organization of the mind is fundamentally like the problem of the organization of theories. I propose that this was how the megacellular colossus organized its process. (But what then, are the naïve "objects" of ordinary consciousness?)

By this discussion I have tried to introduce the kind of complexity that I think we are dealing with, and the profundity of Hilbert's approach. This, I think, is the kind of thing that Hilbert was thinking about with his remark about "beer mugs" and "points". It is *all* about invariants.

Another Look at Hilbert

In my discussion of Chapter 2, I noted that I had incorporated Wilder's interpretation of Hilbert's "objects" in my earlier writings, and promised a further perspective on the issue in this chapter.

In those earlier papers I had interpreted Hilbert's comments in the most minimal sense -that these "objects" were, using Wilder's terminology, "permissive and presumptive only", that is, semantically neutral.

> "These objects, (of its domain -and "existence" terms generally), are assumed *only*, (as Wilder points out) "presumptive(ly)" and "permissive(ly)" however. We are told *nothing* about them in an objective sense."[xxxvii]

I began with an interpretation of Hilbert where the "objects" of a system are taken in a purely *impartial* sense.

> "... it is surely obvious that every theory is only a scaffolding or schema of concepts together with their necessary relations to one another, and that the basic elements can be thought of in any way one likes. If in speaking of my points, I think of some system of things,

e.g., the system love, law, chimney-sweep ...and then assume all my axioms as relations[40] between these things".

In thinking it over, I have come to the conclusion that Hilbert had something much deeper in his mind. Hilbert himself did not interpret "these things" as semantically neutral in this specific instance but gave them names and meanings! (This is not the blatant contradiction it would appear to be. It depends, as I have said in another of my writings "on which end of the telescope you look through.")

Here he first assumes some "system of objects" but then he assumes "all my axioms as relations" [are] "correspondingly the same for the transformed things" –i.e. "*between* these [prior] things*'*! Here he does not begin with the axioms as the logical prius but rather begins with his "things", and he then *transforms his axioms to fit!* His axioms *themselves* are transformed to fit his "things".

"..and then assume all my axioms as relations between these things."

This is not a *simplistic* conceptualization of "relation". I think his perspective here corresponds to that of Quine wherein the latter noted that "total science is like a field of force whose boundary conditions are experience. A conflict with experience at the periphery occasions readjustments in the interior of the field. Truth values have to be redistributed over some of our statements. Reevaluation of some statements entails reevaluation of others, because of their logical interconnections- the logical laws being in turn simply certain further statements of the system, certain further elements of the field. Having reevaluated one statement we must reevaluate some others, which may be statements logically connected with the first or may be the statements of logical connections themselves. But the total field is so underdetermined by its boundary conditions, experience, that

---

[40] But *what kind* of relations? What is the conception of relation itself that Hilbert had in mind?

there is much latitude of choice as to what statements to reevaluate in the light of any single contrary experience. No particular experiences are linked with any particular statements in the interior of the field, except indirectly through considerations of equilibrium affecting the field as a whole....... Furthermore it becomes folly to see a boundary between synthetic statements... and analytic statements...Any statement can be held true come what may, if we make drastic enough adjustments elsewhere in the system... Conversely... no statement is immune to revision... even the logical law of the excluded middle... and what difference is there in principle between such a shift and the shift whereby Kepler superseded Ptolemy, or Einstein Newton, or Darwin Aristotle?"[XXXVIII]

Consider Hilbert's "one only needs to apply a reversible one-one transformation and lay it down that the axioms shall be correspondingly the same for the transformed things." It is his "correspondingly the same" which grabs my attention.

Remember this was the "king of invariants" speaking and I think his meaning was much deeper. "*Correspondingly* the same" would have a very different significance to someone with that background involving complex transformations and invariance in the sense of my "Rosen" and "Kuhn" discussions above. (You might want to think of the Lorentzian transformations here.)

I believe it is the *invariant core*, the context-free sense of the relationality of his axioms that he wanted preserved in the sense of Kuhn's translations of cosmologies or of Quine's relativistic perspective with which we began this journey. This is a much deeper and more radical interpretation of Hilbert's conception than usual, but I think it is justified. I think this is the actual concept of implicit definition of the "young Hilbert".

I believe that Hilbert's was a deeper conception than Shapiro acknowledges, relating to invariance in complex transformations[XXXIX] and to Hilbert's non-simplistic and mathematically nurtured intuitive conception, rather than from his later perspectives drawn from of formalistic logic.

I think this was the actual subject of his initial debate with Frege. Hilbert's conception of implicit definition is *reinterpretation* in its deepest sense, deriving from the larger scope of the principle of duality[41] and complex transformations, and from Hilbert's native, rather than from his formal logic. I believe it has an affinity to Cassirer's perspective in his "Symbolic Forms", (and to Bell's "local mathematics"?). I believe that it is *invariance itself* that was Hilbert's subject.[42] His was, I think, the very first *structuralist* perspective!

Hilbert's original conception was not grounded, as it later came to be, in the formalistic "Byzantian" implementations of logicism and Cantor's set theory.

Shapiro, Mac Lane, and even Hilbert himself eventually became trapped in the abstractive context[XL] implicit in classical logic -e.g. in Shapiro's definition of "an algebraist" and their joint conception of structures as being necessarily "about" some ontological things.

As I read it, Hilbert's original conception, (of the "*young Hilbert*"), was not about ontology: it was not about proof theory; it was about invariance itself. Hilbert's is a world of mathematical conditionality[XLI] *per se,* and it "floats"! It is neither a world of philosophical idealism nor one of Fregean pragmatism. These are the "ladders"[XLII] we must kick away!

Logic as Biology:

Now let us take a radical but, I think, decisive turn, and consider this mathematics from the standpoint of biology. From a purely physicalist and evolutionary standpoint, logic must *itself* be considered as a highly sophisticated but *purely reactive*[XLIII] system for the survival of the entity. (Maturana is surely relevant here.)

---

[41] The algebraic "Principle of Duality" says that if we merely change the reference of each instance of any non-explicitly defined term in an axiom system, that the conclusions drawn from that system apply to and actually define the latter. It is a very deep and profound idea and is the genesis of Hilbert's "implicit definition"

[42] See "Rosen" discussion above

As such logic becomes pure biology[43], and the "concepts" and "percepts" within that logic, (*even those of human mathematics*), become biological objects. Hence logic becomes "bio-logic"! I suggest that this insight might solve many of the deepest issues in the underlying mathematics.

George Lakoff's ICMs, (to be examined later –See Appendix B), are biologically based –on the human organism. Human cognition and human reason consists, for Lakoff, in the application of the best fit of these inbuilt ICM's, (and their respective categories), to a given problem or situation. They constitute an "embodied logic" deriving from the nature of the human organism itself. There is an obvious parallel between Lakoff's "embodied logic" and the more general case I will argue. I will argue that logic is indeed embodied, *but at the primitive level of cellular process!* (See Chapter 4 –"The Specific Case of Biology"). This more general characterization allows the crucial epistemological move,[44] (which Lakoff's does not), beyond the "God's eye view" he disclaims, but nevertheless utilizes throughout!

The distinction is important because at the cellular level of phenomenology biology becomes a *pure form* very much in the sense that I will argue that Maturana's is in Cassirer's sense a "Symbolic Form" and thus compatible with Cassirer's Hertzian premise. This is especially transparent in Maturana and Varela's book, for instance, (see chapter 6), i.e. in its explicit constructiveness and the subsequent purity of its phenomenology.

I think it is relevant to Hilbert's *relative consistency* proofs, Shapiro's problem with "necessarily *assertive statements*", Mac Lane's "existence problems", and the difficulties of Platonism, et al. If logic is actually bio-logic, then we have an actual model *in the human brain itself*,[XLIV] and as such, we can accept its reality and legitimacy in all these perspectives.

---

[43] See my "embodied logic" comment in the Lakoff appendix.
[44] Through what Maturana and Varela call "structural coupling"

Here is another quote from a very recent contemporary source which might make you think.

Raichle:

Compare Raichle:

"Of the virtually unlimited information available in the world around us, the equivalent of 10 billion bits per second arrives on the retina at the back of the eye. Because the optic nerve attached to the retina has only a million output connections, just six million bits per second can leave the retina, and only 10,000 bits per second make it to the visual cortex.

...After further processing, visual information feeds into the brain regions responsible for forming our conscious perception. Surprisingly, *the amount of information constituting that conscious perception is less than 100 bits per second.* Such a thin stream of data probably could not produce a perception if that were all the brain took into account; the intrinsic activity must play a role.

...Yet another indication of the brain's intrinsic processing power comes from counting the number of synapses, the contact points between neurons. In the visual cortex, the number of synapses devoted to incoming visual information is less than 10 percent of those present. Thus, the vast majority must represent internal connections among neurons in that brain region." (This is very much in accord with both Maturana's and W.J. Freeman's conceptions.)

.... Although six million bits are transmitted through the optic nerve, for instance, only 10,000 bits make it to the brain's visual processing area, and only a few hundred are involved in formulating a conscious perception —too little to generate a meaningful perception on their own. *The finding suggested that the brain probably makes constant predictions about the outside environment in anticipation of paltry sensory inputs reaching it from the outside world.*"[45] (My emphasis)

How very similar to Maturana's, W.J.Freeman's and mine is his perspective. But Raichle does not draw the obvious conclusions, as indeed, it seems nobody else seems to. His conclusions are confounded by the epistemological paradox of his own arguments –*his* is a brain also and subject to the same limitations. *His* picture of the world too is built on that same thin data stream of a few hundred bits per second, (DIV 8 ~= bytes per second), imposed on the underlying structure for as many seconds as he has been alive. This stream that we would never allow for even the crudest dial-up connection on our computer modem, (which would normally be minimally about 64 thousand bytes per second), consists, according to Raichle of a mere few *hundreds* of bits per second in which to download reality. And yet he seems to think he has a definite and explicit conception of the world. Whence, then, "the virtually unlimited information available in the world around us, the equivalent of 10 billion bits per second [which] arrives on the retina at the back of the eye." How did he arrive at this world picture?

How much closer is the fit to Maturana's "triggering" of an underlying process than to Raichle's own "informational model" which lies at the bottom of his worldview? His explicit answer has a definite and clear affinity to my own model of an optimization of underlying blind process –to an optimization of strategy rather than of information –or to William James' pragmatism which we will look at in Chapter 12. His implicit and always underlying answer, however, is that of informational naïve realism!

---

[45] Scientific American March 2010 "The Brain's Dark Energy" Marcus Raichle, Washington University School of Medicine in Saint Louis

His formal conclusion does it better:

> *The finding suggested that the brain probably makes constant predictions about the outside environment in anticipation of paltry sensory inputs reaching it from the outside world."* (My emphasis)

The ultimate answers he seeks lie, rather, in the relativism of epistemology I will propose in Chapter 8. Philosophy *does* have a role in science, and most especially in this particular problem –but in support of science, not in pontificating on it. It provides us with new conceptions of possibility! Repeating a relevant quote from Chapter 1 by Cassirer:

> "A glance at the history of physics shows that precisely its most weighty and fundamental achievements stand in closest connection with considerations of a general epistemological nature. Galileo's 'Dialogues on the Two Systems of the World' are filled with such considerations and his Aristotelian opponents could urge against Gallilei that he had devoted more years to the study of philosophy than months to the study of physics. Kepler lays the foundation for his work on the motion of Mars and for his chief work on the harmony of the world in his 'Apology for Tycho', in which he gives a complete methodological account of hypotheses and their various fundamental forms; an account by which he really created the modern concept of physical *theory* and gave it a definite concrete content. Newton also, in the midst of his considerations on the structure of the world, comes back to the most general norms of physical knowledge, to the *regulae philosophandi*
>
> ... But all these great historical examples of the real inner connection between epistemological problems and physical problems are almost outdone by the way in which this connection has been verified in the foundations of the theory of relativity.... Einstein...appeals primarily to an epistemological motive, to which he grants...a decisive significance." (Cassirer: "Einstein's Theory of Relativity",P.353-354)

In short, Raichle's is a clear problem within a framework of epistemological relativity! Cassirer provided a definite picture of such.

Addressing Shapiro again:

Do we really need "assertatory metamathematical statements", or is it only necessary to accept relative consistency proofs? Taking the brain as a machine, then within the bio-logic, I think the latter is the only option. These are *strategies*, not ontologies!

To answer Mac Lane's pointed question in his "Mathematics: Form and Function", (paraphrasing): Why and how does mathematics then work for us? Why and how is it so useful in our pragmatic world?

My answer is that the foundations of mathematics are necessarily just the same as the organizational foundations of brain process. They work just to the best possible extent that the brains of these highly sophisticated organisms are capable of continuing their existence. They exist and they work, to use Maturana's pregnant terminology, just to the extent that these organisms are capable of preserving autopoiesis. But *no more!*[XLV]

Structural Divider: Most of The Remainder of this Book, (Chapters 4 through 13), Constitutes my proposed Specific Solution to the Mind-Brain Problem. The Initial Part was to Enable You to Understand It.

Chapter 4: My First Hypothesis in Detail: (Biology Part One)

1. On Representation: the perspective from biology

Sometimes we tentatively adopt a seemingly absurd or even outrageous hypothesis in the attempt to solve an impossible problem -and see where it leads. Sometimes we discover that its consequences are not so outrageous after all. I agree with Chalmers that the problem of consciousness is, in fact, "the hard problem". I think it is *considerably* harder than even he seems to think it is however.

I think its solution requires new heuristic principles as deep and as profound as, (though different from), the "uncertainty", "complementarity" and (physical) "relativity" that were necessary for the successful advance of physics in the early part of the 20th century. From the preceding chapters, I think you will have some idea of my thoughts on the subject. I think it involves an extension of logic as well. Consideration of those deep cognitive principles: "cognitive closure", (Kant and Maturana), "epistemological relativity", (Cassirer and Quine), and of the extension of logic, (Cassirer, Lakoff, Iglowitz), must await other chapters however.

Sometimes it is necessary to walk around a mountain in order to climb the hill beyond. It is the mountain of "representation", and the cliff, (notion), of "presentation"[1] embedded on its very face, which blocks the way to a solution of the problem of consciousness. This first hypothesis points out the path around the mountain.

Maturana and Varela's "Tree of Knowledge"[1] is a compelling argument based in the mechanics of physical science and biology against even the very possibility of a biological organism's possession of a *representative* model of its

---

[1] For we would surely, then, require some homunculus for it to be presented to!

environment.[2] They and other respected biologists, (W.J.Freeman, Edelman), argue against even "information" itself. They maintain that information as such never passes between the environment and organisms; there is only the "triggering" of structurally determinate organic forms. I believe theirs is the inescapable conclusion of modern science.

    I will now present a specific and constructive counterproposal for a different *kind* of model however: i.e. what I will call the "Schematic Operative Model". Contrary to the case of the representative model, it *does* remain viable within the critical context of modern science. I believe that we, as human organisms, do in fact embody a model. I believe it is the stuff of mind!

2. "The Schematic Model": Definition and Examples.
(Defining What It Means To Be "An Object")

Normally, when we think of "models", we mean reductive or at least parallel models. In the first we think of a structure that contains just some of the properties of what is to be mirrored. When we normally use the term "schematic model", we talk about the preservation of the "schema", or "sense" of what is mirrored. Again it is reductive, however- it is logically reductive. It is, as has been claimed, "just a level of abstraction"[3]. There are other uses for models, however, -those that involve superior organizations! This is the new sense of "schematic model" that I propose to identify.

2.1 The Simplest Case: A Definition by Example

    Even our most simplistic models, the models of even our most simplistic and mundane training seminars, suggest the possibility of another usage for models very different than as representative schemas. They demonstrate the possibility of a

---

[2] See Chapter 6

[3] As a JCS reviewer once tried to characterize my conception

wholly different paradigm whose primary function is *organization* instead.

Look first at the very simplest of models. Consider the models of simplistic training seminars -seminars in a sales organization —even the primitive training seminars of AMWAY©!- for instance. "'Motivation' plus 'technique' yields 'sales'.", we might hear at their sales meeting. Or, (escalating and shifting our ground just a little bit), "'Self-awareness of the masses' informed by 'Marxist-dialectic' produces 'revolution'!", we might hear from our local revolutionary at a Saturday night cell meeting. Visual aids, (models), and diagrams are *ubiquitous* in these presentations!

A lecturer stands at his chalkboard and asks us to accept drawings of triangles, squares, cookies, horseshoes... as meaningful objects -with a "calculus" of relations, (*viz:* an "arithmetic" of signs),[4] between them, (arrows, squiggles, et al). The icons, (objects), of those graphics are stand-ins for concepts or processes as diverse, (escalating and shifting ground just a bit more), as "motivation", "the nuclear threat", "sexuality", "productivity", and "evolution".

Those icons need not stand in place of entities in objective reality, however! What is the *object* which is "*a* productivity" or "*a* sexuality", for instance? What *things* are these?

Consider this: two different lecturers might invoke different symbols, ("objects"), and a different "calculus" to explicate the same topic. In analyzing the French Revolution in a history classroom,[5] let us say, (a classroom is a *kind* of training seminar after all!) , a fascist, a royalist, a democrat might alternatively invoke "the Nietzschean superman", "the divine right of kings", "freedom", ... *as actual "objects"* on his blackboard, (with appropriate symbols).

---

[4] Webster's defines "calculus": "(math) a method of calculation, any process of reasoning by use of symbols". I am using it here in contradistinction to "the calculus", i.e. differential and integral calculus.

[5] I actually attended such a class which dealt with alternative explanations of the French Revolution at the University of Chicago. It was a good school.

He will redistribute certain of the explanatory aspects, (and properties), of a Marxist's entities, (figures) -or reject them as entities altogether.[II] That which is unmistakably explanatory, ("wealth", let us say), in the Marxist's entities, (and so which must be accounted for by *all* of them), might be embodied instead solely within the fascist's "calculus" or in an interaction between his "objects" and his "calculus".

Thus and conversely the Marxist would, (and ordinarily does), reinterpret the royalist's "God"-figure, (and his –the Marxist's- admitted function of that "God" in social interaction[6]), as "a self-serving invention of the ruling class".

It becomes an expression solely of his "calculus" and is not embodied as a distinct symbol, (i.e. object). Their "objects" - *as objects* - need not be compatible! As Edelman noted: "*certain symbols do not match categories in the world.* ... Individuals understand events and categories in more than one way and sometimes the ways are inconsistent."[III]

---

[6] Dennett's term "heterophenomenological" -i.e. with neutral ontological import -is apt here.

Chapter 4 — My First Hypothesis in Detail

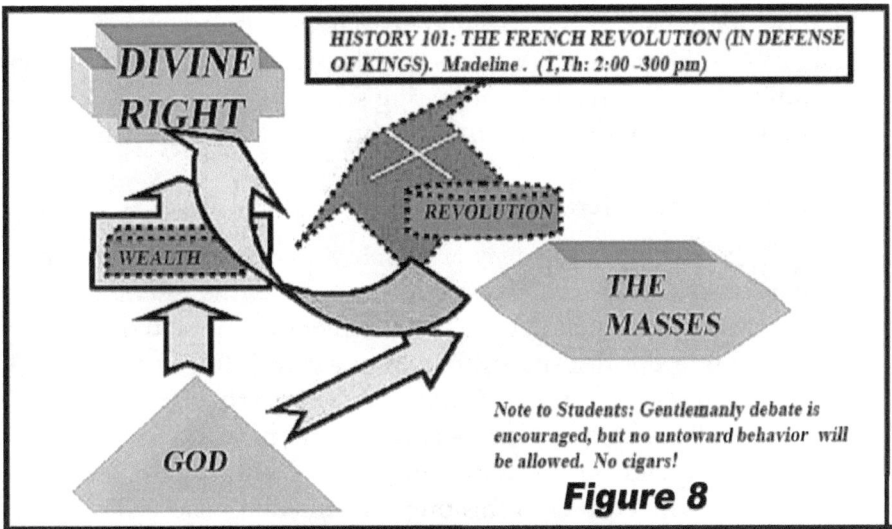

(Madeline's Chalkboard)

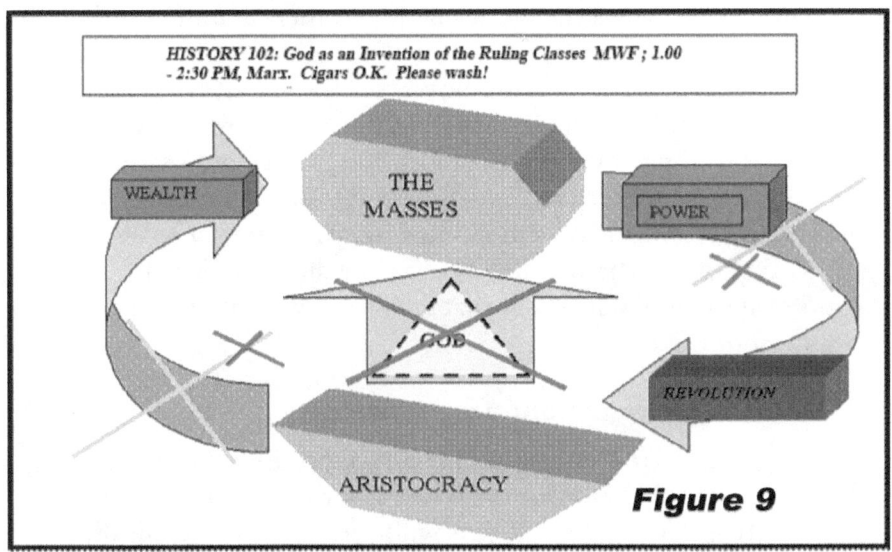

Figure 9, (Marx's Chalkboard)

What is important is that a viable calculus-plus-objects, (a given model), must explain or predict "history" –that is, it must be compatible with *the phenomena*, (in this particular example with the historical phenomena). But the argument applies to a much broader scope. I have argued elsewhere,[IV] (following the strong case of Hertz and Cassirer —see Chapter 7), that the same accounting may be given of competing scientific theories, philosophies, and, indeed, of *any* alternatively viable explanations.

Consider Heinrich Hertz:

> "The [scientific] images of which we are speaking are our ideas of things; they have with things the one essential agreement which lies in the fulfillment of the stated requirement, [of successful consequences], but further agreement with things is not necessary to their purpose. Actually we do not know and have no means of finding out whether our ideas of things accord with them in any other respect than in this one fundamental relation." (Hertz, "Die Prinzipien der Mechanik")

The existence of a multiplicity of alternately viable "calculuses", (sic), and the allowable incommensurability of their "objects"[V] suggests an interpretation of those "objects" contrary to representation or denotation however. It suggests the *converse* possibility that the function and the motivation of the objects themselves of those models, specifically as entities *per se*, (in what I will call these "schematic models"), is *instead* to illustrate, to enable, -to crystallize and simplify *the very calculus of relation* proposed between them! The "objects" of these models, I propose, are manifestations of that structure; the structure is *not* a resolution of the objects.[7]

2.1.1 Reversing our perspective:

I propose and will argue the actual possibility that the boundaries -the demarcations and definitions of these schematic objects, (their "contiguity" if you will) -are formed specifically to

---

[7] You might look at Resnik 1992 for a mathematical "structuralist" viewpoint on the issue.

meet the needs of the operations. I propose that they exist to serve structure- not the converse.^{VI}

The objects of those structures —specifically *as objects per se* - serve to organize process, (i.e. analysis or response). They are not representations of actual objects or actual entities in reality.^{VII} This, I propose, is *why* they are "things" in the first place.[8] *These* "objects" functionally bridge reality in a way that physical objects do not and I suggest that they are, in fact, *metaphors* of analysis or response! The rationale for using them, (as any good "seminarian" would tell you), is clarity, organization and efficiency.

Though set in a plebian context, the "training seminar", (as minimally presented), illustrates and defines the most general and abstract case of schematic non-representative models in that it presumes *no particular agenda*. It is easily generalized: it might as well be a classroom in nuclear physics or mathematics, the boardroom of a multinational corporation, -or a student organizing his love life on a scratchpad!

---

[8] Compare the Benacerraf passage cited earlier: "[Hilbert] argues that what constitutes an object varies from theory to theory, category to category, and that Frege failed to realize this fact....'It [logic] remains the tool applicable to all disciplines and theories, the difference being only *that it is left to the discipline or theory to determine what shall count as an 'object' or 'individual.'*" [Benacerraf, 288]

Lecture Halls

Lovelife?"

2.2 A Case for Schematism More Specific to Our Special Problem: Narrowing the Focus.

(The Engineering Argument)

Engineers' instrumentation and control systems provide an example of the organizational, non-representational use of models and "entities" in another setting. These entities, and the context in which they exist, provide another *kind* of "chalkboard".[9] *Their* "objects" need not mirror objective reality either. A gauge, a readout display, a control device, (the "objects" designed for such systems), need not mimic a single parameter -or an actual physical entity. Indeed, in the monitoring of a complex or dangerous process, it should not. Rather, the readout for instance should represent an efficacious synthesis of just those aspects of the process which are relevant *to effective response*, -and be crystallized *around* those relevant responses!

A warning light or a status indicator, for instance, need not refer to just one parameter. It may refer to electrical overload and/or excessive pressure and/or... Or it may refer to an optimal relationship, (perhaps a complexly functional relationship), between many parameters -to a relationship between temperature, volume, mass, etc. in a chemical process, for instance, or to the urgency of immediate response by a battlefield commander.

The exactly parallel case holds for its control devices. A single control may orchestrate a multiplicity of (possibly disjoint) objective responses. The accelerator pedal in a modern automobile, as a simple example, may integrate fuel injection volumes; spark timing, transmission gearing...

Ideally, (given urgent constraints), instrumentation and control might unify in the *selfsame* "object"! We could then manipulate the very object of the display and it in itself could be the control device as well. Consider the advantages of manipulating a graphic or tactile object which is simultaneously

---

[9] Their designers are the "lecturers", and the instruments they design are the "objects" of their schematic models

both a readout and a control mechanism under urgent or dangerous circumstances.

Now think about this same possibility in relation to our ordinary objects of perception -in relation to the sensory-motor coordination of the brain and the objects of naïve realism in the real world! The brain is a control system, after all, so what should it's "objects" be? The brain is an organ of control and its mechanics must be considered in that perspective. Its function is exceedingly complex and the very continuation of life itself is at stake.[VIII] It is indeed a complex and dangerous world. Might not our naïve world *itself* be such a combined schematic control system?

2.3 The "G.U.I.", the Most Pertinent and Sophisticated Example of a Schematic Model: the Special Case!

The "object" in the graphic user interface, (G.U.I.), of a computer is perhaps the best example of a purely schematic usage currently available. In my simplistic manipulation of the schematic objects of my computer's G.U.I., I am, in fact, effecting and coordinating quite diverse, disparate and unbelievably complex operations at the physical level of the computer. These are operations impossible, (in a practical sense), to accomplish directly.

What a computer object, (icon), represents and what its manipulation does, at the physical level, can be exceedingly complex and disjoint.[10] The disparate voltages and physical locations, (or operations), represented by a single "object", and the (possibly different) ones effected by manipulating it, correlate to a metaphysical object only in this "schematic" sense. Its efficacy

---

[10] In fact, it is totally arbitrary and at the will of the programmer(s) —and any, possibly conflicting, organizational schemes they may have in mind.

lies precisely in the *simplicity* of the "calculus" it enables! It is specifically *the interface itself* that must be simple![11]

Contemporary usage is admittedly primitive. Software designers have *limiting preconceptions* of the "entities" to be manipulated, for the necessity of a preservation of hierarchy, and of the operations to be accomplished in the physical computer by their icons and interface. But I assert that G.U.I.'s and their "objects", (icons), have a deeper potentiality of "free formation". They have the potential to link to *any* selection across a substrate, i.e. they could "cross party lines".[12] They could cross categories of "things in the world", (Lakoff's "objectivist categories"[IX]), and acquire thereby the possibility of organizing on a different and the most pressing issue: i.e. urgency / risk. They need preserve neither parallelism nor hierarchy.

Biology supplies fortuitous examples of the sort of thing I am suggesting for G.U.I.'s –e.g. in the brain's "global mapping" noted by Edelman[X], (I will present Walter Freeman's more explicit case in detail shortly). The *non-*topological connectivity Edelman[13] notes from the brain's "topobiological" maps,[14] and specifically the connectivity, (the "global mapping"), *from* the objects of those maps *to the non-mapped* areas of the brain[15] supplies a concrete illustration the kind of potential I wish to urge for a G.U.I. (Very shortly I will argue a much stronger case based on the researches of the noted neurophysiologist W.J. Freeman.)

Ultimately I will urge it as the rationale for the brain itself. Edelman's global mapping allows "... selectional events", [and, I suggest, *their "objects"* as well], "occurring in its local maps ... to be connected to the animal's motor behavior, to new sensory samplings of the world, and to further successive reentry events."

---

[11] This is clearly related to intentionality, to the facility of implementation.
[12] See Freeman Figure 2 in section 2.4.1 for a physical demonstration drawn from modern biological research.
[13] See Appendix B
[14] from the multiple, topological maps in the cortex
[15] Edelman, 1992 P.89

But this is explicitly a non-topological mapping. This particular mapping, (the global mapping), *does not preserve contiguity*. Nor need it preserve hierarchy.

Here is an actual biological model demonstrating the more abstract possibility of a connection of localized "objects"[16], (i.e. in a G.U.I.), to non-topological (distributed) process -to "non-objectivist categories ", using Lakoff's terminology again. As such, it illustrates the possibility of "schematism" in its broadest sense. Edelman's fundamental rationale is "Neural Darwinism", the ex post facto adaptation of process, not "information", and is thus consistent with such an interpretation. It does not require "information". Nor does it require "representation".[17]

Edelman, (unfortunately), correlates his topobiological maps, (as sensory maps), directly and representatively, (i.e. hierarchically), with "the world". This is a clear inconsistency in his epistemology. It is in direct conflict with his early and continual repudiation of "the God's eye view" upon which he grounds his biologic epistemology.

---

[16] in the brain's spatial maps
[17] See the heading later in this chapter "Turning Our Perspective Around…"

Chapter 4 My First Hypothesis in Detail

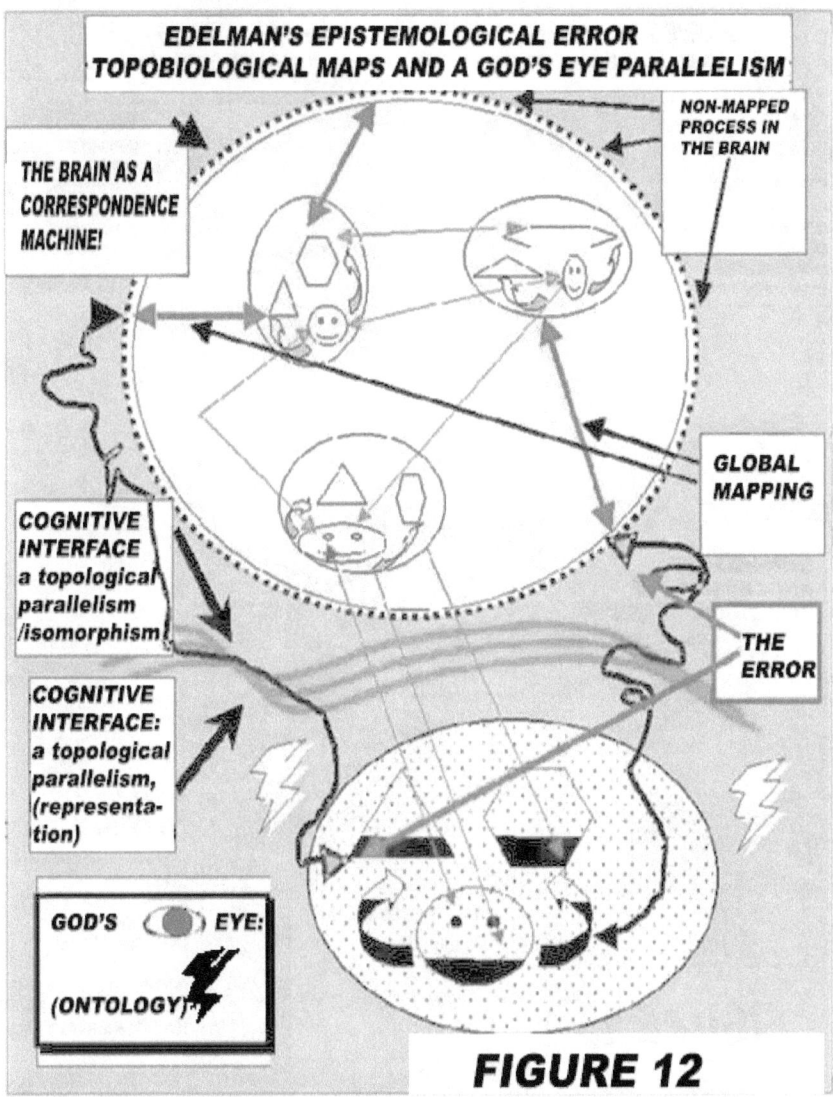

FIGURE 12

A Graphic Rendering of Edelman's Epistemology: (Note: hierarchy and contiguity are implicit in his model!)

But what if we turn Edelman's perspective around however? What if we *blink* the "God's eye" he has himself so strongly and continually objected to, and step back from the prejudice of our human (animal) cognition. What if the maps and their objects *both* were taken as existing to serve blind primitive

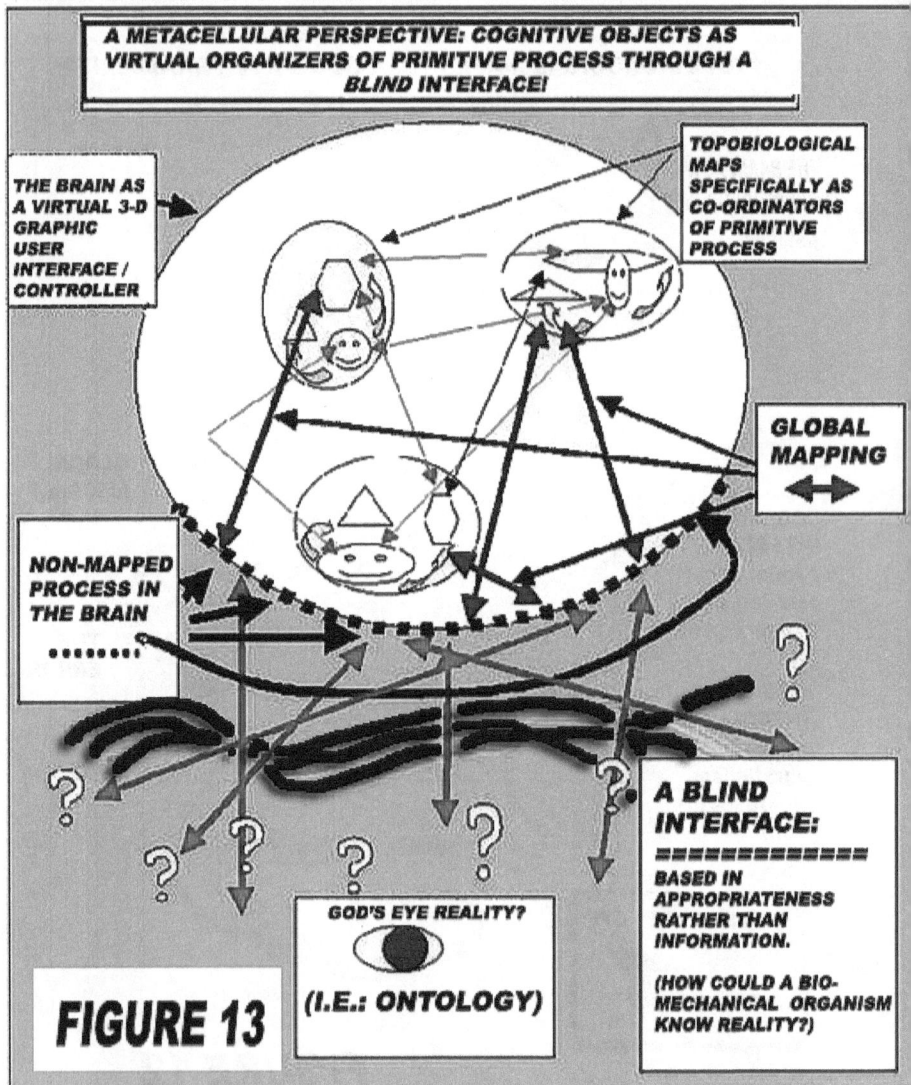

FIGURE 13

process instead of information? (See Figure 13) What if they are *organizational* rather than representative?

Figure 13 is a more consistent rendering of Edelman's epistemology suggesting a new paradigm for G.U.I.'s. (Note: Neither hierarchy nor contiguity are implicit in this model!)

This is the case I wish to suggest as an illustration of the most abstract sense of the G.U.I., (and which I will argue shortly) –i.e. a *non-topological, non-information preserving* correlation! It opens a further fascinating possibility moreover. It suggests that evolution's "good trick", (after P.S. Churchland's usage), was not representation, but was, rather, the organization of primitive process[18] in a topological context. It suggests that the "good trick" was evolution's creation of *the cortex* itself!

2.4 Towards a Better Biological Model

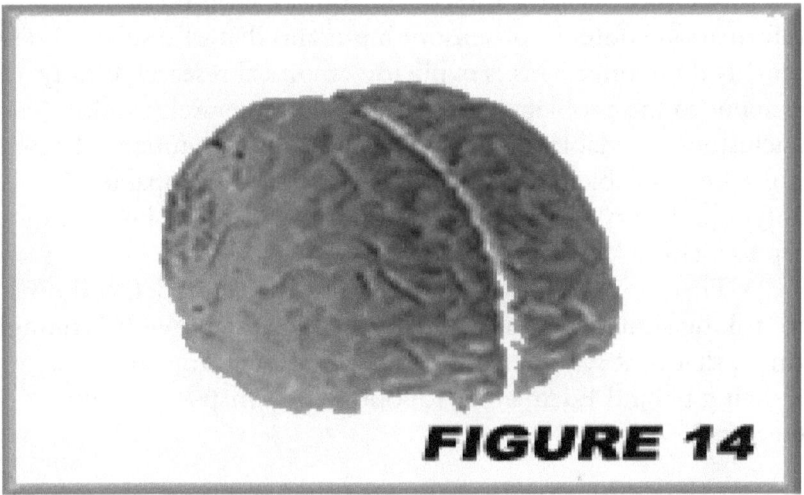

FIGURE 14

2.4.1 Biology, The Real Thing: Freeman's Model

What is needed now is a more explicit model, and a specific research problem to embody the proposal. Edelman's "global mapping" is all very well and good, but it doesn't really do what it has to. It is "too philosophical", too vague, and as Popper would have predictably urged, not falsifiable. A more detailed and quite specific model comes from the work of the noted neurophysiologist, Walter J. Freeman.

---

[18] See Section 3.2: Turning our perspective around –a Model of Process!

Walter J. Freeman

Based on extensive experimental research first with the olfactory cortex, (arguably evolution's first cortex), and then with the visual and other cortices, Freeman argues that the brain does not process information at all —it does other things!

He has approached the problem directly and addressed the crux of the issue: what is the correlation between the "information content" of sensory input and that of resultant brain states? Is there one? This is explicitly empirical research clearly pertinent to the problems of parallelism and hierarchy and, if his conclusions are viable, is totally relevant to my argument. These results are falsifiable![19] But, conversely, they ares capable of falsifying the very premise of the standard paradigm –i.e. that of "representation" itself.

First, however, please look at Freeman's model, and note the striking similarity to my own Figure 13 just above.[20] Strikingly similar, that is, if we interpret his "topographic projections" as following behind Edelman's "topobiological maps". (Feature detectors?)

---

[19] i.e. they satisfy Popper's condition

[20] Please note that figure 13 and figure 15 were generated by myself and W.J.Freeman in total mutual ignorance of the other and in different contexts. It was only later that I discovered this paper —to me it was a blinding coincidence.

Chapter 4   Biology: The Real Thing!

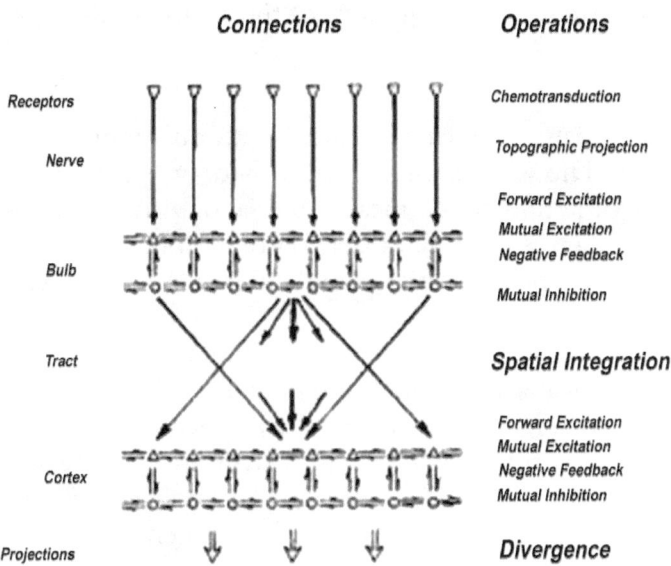

Figure 15, (Freeman's Figure 2)

(Fig. 2 –from Freeman 1983, reproduced by permission.) "The input path from receptors to the bulb has some topographic specificity." [but] "The output path to the prepyriform has broad axonal divergence, which provides a basis for spatial integration[21], (please see important footnote below), of bulbar output and extraction of the "carrier" wave.

"It is based on a striking difference between two types of central path, one that provides *topographic* mapping from an array of transmitting neurons to an array of receiving

---

[21] Understand that Freeman is talking within a context and here "spatial integration" relates to the geometry, (physical space), of the brain, and is not about the structure of the data itself. This ambivalence of perspectives will be resolved in Chapters 7 & 8.

neurons, the other having *divergence* of axons that provides for *spatial*[22] integration of the transmitted activity."[XI]

Now compare Freeman's Figure 2 with my Figure 13 shortly before it. This is an explicit case, truly drawn from biology itself, illustrating the non-topological potential of virtual systems and of models to "cross party lines". It is not a topological mapping, it does *not* preserve hierarchy, and it *does not* preserve information.

This is an actual case emerging from biology itself that demonstrates the ultimate potential of schematic G.U.I.'s for distributing, (or conversely, for centralizing), function into operative "objects" which I had sought. It demonstrates the actual possibility of the "free formation" of schematic models in the real world that I had argued earlier beginning from an entirely independent perspective!

Freeman's model exposes a new paradigm for models. It demonstrates an organizational potential of models far beyond representation. (See overview model in Freeman Appendix in this chapter.)

Freeman begins:

> "This book had its origin ... in an experimental finding....I was tracing the path taken by neural activity that accompanied and followed a sensory stimulus in brains of rabbits. I traced it from the sensory receptors into the cerebral cortex and there found that the activity vanished, just like the rabbit down the rabbit hole in 'Alice in Wonderland'.
>
> What appeared in place of the stimulus-evoked activity was a *new* pattern of cortical activity that was *created* by

---

[22] Again, see usage above.

> the rabbit brain... My students and I first noticed this anomaly in the olfactory system... and in looking elsewhere we found it in the visual, auditory, and somatic cortices too...
>
> In all the systems the traces of stimuli seemed to be replaced by constructions of neural activity, which *lacked invariance with* respect to the stimuli that triggered[23] them. The conclusion seemed compelling. *The only knowledge that the rabbit could have of the world outside itself was what it had made in its own brain.*"[XII]

What does this mean? What does it mean that the new pattern "lacked invariance" in regard to the stimuli? The "invariance" demanded in a referencial conception of the brain would correlate precisely to the "passage of information" –and it could not be found!

> "The visual, auditory, somatic and olfactory cortices generate... waves [that] reveal macroscopic activity ... from millions of neurons. ... These spatial AM patterns are unique to each subject, are *not invariant* with respect to stimuli, and *cannot be derived* from the stimuli by logical operations!'[XIII]

In this paper, ("Chaotic Oscillations..."), Freeman actually makes two cases –one structural and one functional. The structural case is purely physiological and, I think, very strong. It deals with the actual connectivity of nerve tissue and argues against the possibility of maintaining topological integrity –to include hierarchy- of the "data" within the cortex. (The other case is for "Chaos theory" as an explanation of function which I will refer to later.)

---

[23] Please note the use of the word "triggered" and remember Freeman's usage of it here when you come to Chapter 6 and Maturana's conception of "structural coupling".

The former is the case I want to emphasize here as I think it supplies an exact and explicit example of my argument for the non-topological possibilities of schematic models. This model as an ultimate possibility of GUI's "crossing party lines" is what I believe evolution did and how it did it.

The Peripheral Code:

1. Freeman divides nerve physiology into two categories: Those which preserve topological integrity: this is the case for the sensory nerves for instance.

"Sensory neurons exist in large arrays in the skin, inner ear, retina...so that a stimulus is expressed as a spatial[24] pattern...carried in parallel along sensory nerves. Typically only a small fraction of the axons in a nerve is activated...with the others remaining silent" [for isolation] "...so that the 'signal' of the stimulus is said to be 'encoded' in the frequencies of firing of that subset of axons subserving ...the activated...receptors."

"The code of sensory, motor and autonomic parts of the peripheral nervous system is the spatial"[XIV], [topological], "pattern of temporal pulse rates. The same code appears to hold...for the ascending and descending pathways and relays in the brainstem and spinal cord. ...Serious efforts have been made to extend this model to the cerebral cortex with considerable success in characterizing the receptive fields and 'feature detector' properties of cortical neurons in primary sensory areas."[XV]

But he argues that 'feature detection" occurs only early in cortical process. Points on the retina, for instance, are mapped

---

[24] i.e. in "real=informational space" –see my third thesis for a rationale for this seeming paradox.

onto the cortex in a way that preserves the topology of the source and, apparently, feeds the feature detectors *which are just the very beginning of cortical input!*

Cortical Mapping is Very Different, However:

(2) Within the cortex, however, it is a different story. Cortical neurons typically have short dendritic trees on the order of ½ millimeter. They are not, however, typically connected to the neurons physically adjacent to them!

"The main neurons in cortex ...intertwine at unimaginable density, so that each neuron makes contact with 5,000 to 10,000 other neurons within its dendritic and axonal arbors, but those neighbors so contacted are less than one percent of the neurons lying within the radius of contact. The chance of any one pair of cortical neurons being in mutual contact is less than one in a million."[XVI]

"Peripheral neurons", [on the other hand], "seldom interact with other neurons, but offer each a private path from the receptor to the central nervous system. In contrast, each cortical neuron is embedded in a milieu of millions of neurons, and it continually transmits to a subset of several thousand other neurons sparsely distributed among those millions and receives from several thousand others in a different subset."[XVII]

This is reminiscent of Maturana's comment:
> "It is enough to contemplate this structure of the nervous system... to be convinced that the effect of projecting an image on the retina is not like an incoming telephone line. Rather, it is like a voice (perturbation) added to many voices during a hectic family discussion (relations of activity among all incoming convergent connections) in which the consensus of actions reached will not depend on what any particular member of the family says."[XVIII]

And Edelman's:

"... To make matters even more complicated, neurons generally send branches of their axons out in diverging arbors that overlap with those of other neurons, and the same is true of processes called dendrites on recipient neurons .... To put it figuratively, if we 'asked' a neuron which input came from which other neuron contributing to the overlapping set of its dendritic connections, it could not 'know'."[XIX]

Peripheral neurons are relatively isolated, ("private"), within nerve bundles and support a topological (information preserving) case to the point of 'feature detection' at cortex. Within the cortices, however, we are dealing with a *different* sort of connective process.

We are no longer dealing with parallel or hierarchical, (i.e. information preserving), mappings. Because each cortical neuron is embedded in a milieu of millions of neurons, it "continually transmits and receives from several thousand others" and therefore has "continual [non-topological] *background* activity owing to its synaptic interactions with its neighbors". This is a characteristic property of cortical neural populations not shared by peripheral neuron arrays.[25] Cortical process disburses function *spatially*

---

[25] Reconsider and compare Reichle: "Of the virtually unlimited information available in the world around us, the equivalent of 10 billion bits per second arrives on the retina at the back of the eye. Because the optic nerve attached to the retina has only a million output connections, just six million bits per second can leave the retina, and only 10,000 bits per second make it to the visual cortex.

...After further processing, visual information feeds into the brain regions responsible for forming our conscious perception. Surprisingly, the amount of information constituting that conscious perception is less than 100 bits per second. Such a thin stream of data probably could not produce a perception if that were all the brain took into account; the intrinsic activity must play a role.

...Yet another indication of the brain's intrinsic processing power comes from counting the number of synapses, the contact points between neurons. In the visual cortex, the number of synapses devoted to incoming visual information is less than 10 percent of those present. Thus, the vast majority must represent

through the *physical space* of the brain, ("with strong axonal divergence"), through intertwined nerve process -not topologically. It connects point-to-point fitfully within the volumetric and specifically physical space of the brain -not topologically, and not hierarchically —and does not preserve "information"! These cell assemblages act as units which "provide for spatial integration [projection] of the transmitted activity." The cortices generate dendritic potentials…arising from synaptic interactions of *millions* of neurons. They share "a spatially coherent oscillation… by which spatial patterns of amplitude modulation are transmitted in distinctive configurations… The neurons sharing the macroscopic, aperiodic oscillations comprise a local neighborhood that can be viewed as an equivalence class."[xx]

These "equivalence classes" thereby provide a non-contiguous spatial distribution onto the *physical space but not onto an "information space"* of the brain. These spatially

---

internal connections among neurons in that brain region." (This is very much in accord with both Maturana's and W.J. Freeman's conceptions.)
…. Although six million bits are transmitted through the optic nerve, for instance, only 10,000 bits make it to the brain's visual processing area, and only a few hundred are involved in formulating a conscious perception —too little to generate a meaningful perception on their own. *The finding suggested that the brain probably makes constant predictions about the outside environment in anticipation of paltry sensory inputs reaching it from the outside world.*" (My emphasis)
From Scientific American March 2010 "The Brain's Dark Energy"
Marcus Reichle, Washington University School of Medicine in Saint Louis
(Note: Please compare this passage after you have read through to Chapter 13 on d"Espagnat, Chapter 14, and Appendix D on Niels Bohr. I think it confirms my conclusions.)
 How very similar to Maturana's, W.J.Freeman's and mine is his perspective. But Reichle does not draw the obvious conclusions, as indeed, it seems nobody else does. His conclusions are confounded by the epistemological paradox of his own arguments —*his* is a brain also and subject to the same limitations. Those answers lie in the relativism of epistemology I propose. Philosophy *does* have a role in science, and especially in this particular problem —but in support of science, not in uninformed criticisms of it.

extensive and intertwined complexes of cells throughout the cortex achieve the connectivity that mere parallelism, (or hierarchy), cannot. Freeman shows us how a topological mathematical space can be mapped onto the specifically physical space of the brain. But that particular physical space, I argue, is determined by its specific connectivity -by evolution and ontogeny, *not by* representation. Determined by genetics and learning, (ontogeny), it has the ability to connect specific process "ad hoc". It has the ability to self-organize on principles other than topological ones. It can "cross party lines"!

"The local neighborhoods corresponding to cortical columns and hypercolumns seldom have anatomical boundaries of their internal synaptic connections, so that an area of cortex composed of hundreds and even thousands of neighborhoods can act as a coherent element of function in generating a spatially coherent carrier wave. These distributed neural populations are dynamically unstable and are capable of very rapid global state transitions [which can] easily fulfill the most stringent timing requirements encountered in object recognition." (ibid).
(Think of the possibility of these "equivalence classes" functioning as evolutionary "axioms" as I have suggested earlier!)

Freeman concludes:

"The transform effected by the output path defines the self-organized[26] macroscopic activity as the cortical 'signal'…In brief, the central code cannot be the same as the peripheral code."[XXI]

He argues ultimately that the brain is a self-organizing entity, specifically obeying the laws of Chaos theory, ("Chaos can make as well as destroy information!").

I am frankly unqualified to judge this aspect of his argument, nor do I think it is necessary. His physiological case is an actual physical demonstration of the full possibilities and an

---

[26] Please note his usage of "self-organized" which is one of my basic claims.

actual physiological example of my thesis of schematism and of G.U.I.'s that is the thesis of this chapter.

That physiological case: i.e. the connectivity of the CNS, is entirely sufficient in itself to demonstrate the kind of mapping, the broadest logical potential of "schematic G.U.I.'s" and their explicit relevance to cognition. This model actually does "cross party lines"!

That the brain is, in fact, "self-organized" is exactly the case I am making. I argue that it is self-organized specifically for optimal efficiency, (i.e. urgency / risk), not for reference. Freeman's case, I believe, constitutes an actual instance demonstrating the deepest possibilities of the "schematic models" argued earlier. It demonstrates the possibility of a truly useful model organized on non-topological principles, and, as such, demonstrates the deepest capabilities -previously suggested- of a schematic G.U.I. This is *not* just "a level of abstraction."

But where, accepting Freeman's description of the actual brain, do these cell assemblages, (these "equivalence classes"), come from, and what is their function? How do these particular entangled arrays of cells, interconnecting and overarching "the less than one percent of the neurons lying within the radius of contact" arise? I propose that they arise evolutionarily —as internal, organizations of *blind function.* This is exactly what we would expect the organizing principle of a "self-organizing" metacellular entity to be.[xxii] It is also an example of how a machine, in the sense of my figure 3 in Chapter one could arise!

Representation is neither required, nor, accepting Freeman, is it even possible in cortex. This is what we would expect if neural organization were modeled on efficiency over "truth" -and how. Our "percepts", moreover, are what we would expect if we joined the loop of output to input! (See graphic immediately following.)

"In particular, Maurice Merleau-Ponty in "The Phenomenology of Perception" [2] conceived of perception", [itself], as the outcome of the "intentional arc", by which experience derives from the intentional actions of individuals that control sensory input and perception. Action into the world with reaction that changes the self is indivisible in reality, and must be analyzed in terms of 'circular causality' as distinct from the linear causality of events as commonly perceived and analyzed in the physical world." [XXIII] [XXIV]

2.4.2 An Explicit Model of the Mind:

If we turn our perspective around and think of our (input) topographic maps as the looping, *re-entrant* extension of our output, then we can clearly see them, (and the "objects" contained therein), in their specific role as organizing artifacts of cortical function itself. Our "percepts" are just the *combined-in-one icons* previously described in the "engineering" argument! They are the "A-D", ("analog/digital", or, better yet, the hierarchical/chaotic), converters[27], so to speak, of the reentrant loop of process.

This is what we would expect from my prior arguments taking "percepts" as expressly schematic objects of process. That is, these are what we would expect "*to see*"! (See Figure 16) I propose that our cognitive interface lays precisely in the topobiological models themselves, mediating between an unknowable externality and the optimized functionality of the cortex. I claim that this constitutes an explicit and non-representational model for the mind. [XXV] (See graphic model immediately following.)

---

[27] This is, at best, a crude metaphor —but it crystallizes the idea nicely. A more apt characterization would be "topological / non-topological" converters.

Chapter 4   Biology: The Real Thing!

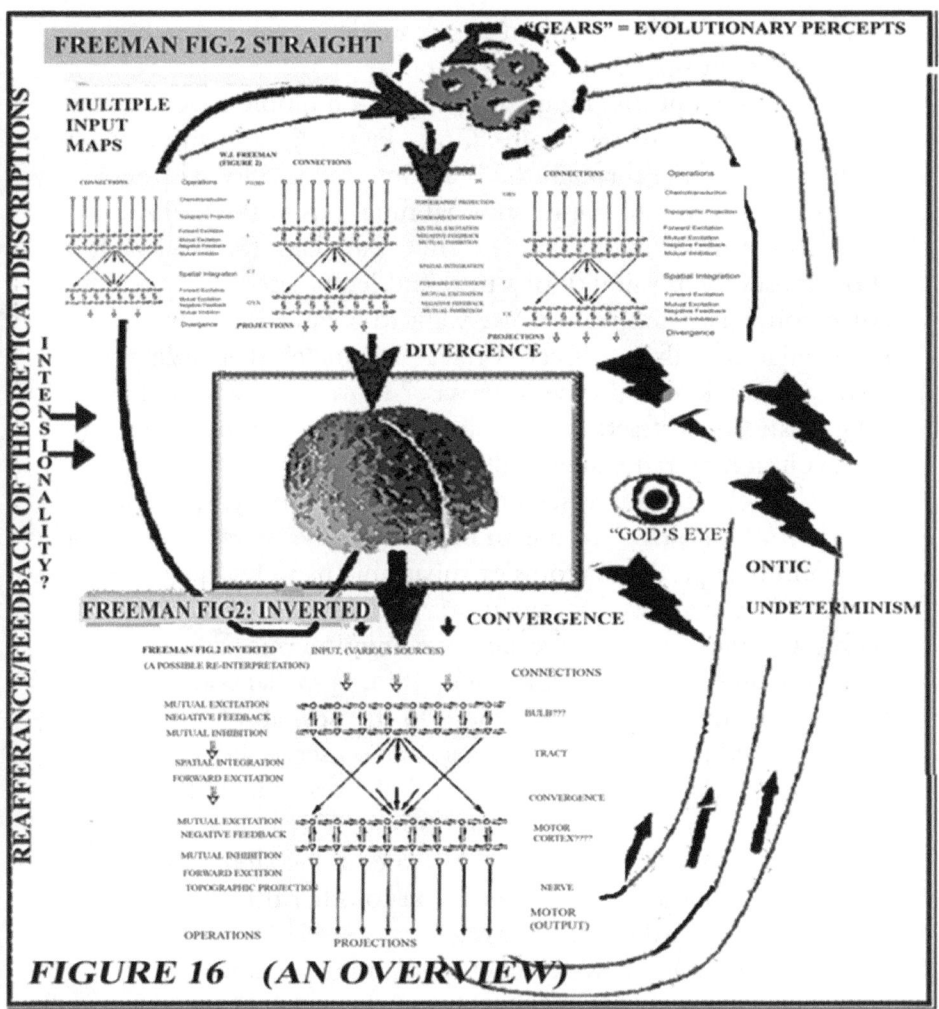

FIGURE 16 (AN OVERVIEW)

GOD'S EYE?

(Edelman -to Freeman -to Edelman!)

    Note: To repeat a prior reference, some of my illustrations require a higher resolution that is possible within this book format. I am therefore supplying them in a hi-resolution form at: http://foothill.net/~jerryi/Illustrations. You are free to examine or print them, but are required to acknowledge their source in any subsequent usage.

Simply put, the figure above shows multiple inputs utilizing W.J. Freeman's Figure 2, (my Figure 6), and an inversion of the same passing through the brain into externality with re-afferent feedback into the intentional functions of the brain.

Freeman's model exposes a new paradigm for models. It also exposes the possibility of a new and legitimate correspondence with reality. We want to believe that our knowledge of reality is direct —or at least parallels that reality. How, we think, could it be otherwise? How could a model be other than "an abstraction" and still be useful? Moreover, what is the evolutionary rationale for all of this?

Much of modern science says that what truly is, absolute reality, (or "ontology" to use an old but precise word), consists of some ultimate particles: atoms or subatomic particles, quarks, etc.

We are allowed to retain our normal view of reality within this view however because we envision our ordinary objects, (baseballs, you, me, the sun, etc.), as spatial containers, (and logical, theoretical hierarchies), in the new absolute reality we are forced to believe in. We may still preserve the sense of our ordinary objects as physical and logical clusters, (hierarchies), of those deeper existences. I can think of myself as a cluster of atomic particles and fields shaped like me, doing all the things I do, and positioned in ontic reality next to other things and persons just as I ordinarily see myself.

There is a necessary belief in a continuity, and a contiguity, ("next-to-ness"), in this belief system. This is the "hierarchy" or "logical containment" implicit in the Newtonian World and it is mirrored in the hierarchies of contemporary mathematics and of logic. *Truly modern* science says otherwise, however. Quantum theory[28] and Relativity say that the world, (reality), is an even stranger place. Freeman's conclusions,

---

[28] See Chapter 13, (d'Espagnat)

moreover do not allow it at all. If we live anywhere, we live in cortex!

On P.S.Churchland:

"At some point in evolutionary history, nature performed a "good trick". It allowed for an internal representation of environment.... and this allowed competence in the larger world." (P.S. Churchland, paraphrase)
I suggest that the "good trick" was evolution's invention of the schematic model, and specifically in the GUI enabled in cortex!
Unless, of course, we were to posit a "pre-established harmony". This, however, would be mysticism, not science. This is *our* world, not God's. We do not and *cannot have* a God's eye view!

3. The formal and abstract problem:

3.1 The formal argument

Consider, finally, the formal and abstract problem. Consider the actual problem that evolution was faced with. Consider the problem of designing instrumentation for the efficient control of both especially complex *and* especially dangerous processes. In the general case, (imagining yourself the "evolutionary engineer"), what kind of information would you want to pass along and how would you best represent it? How would you design your display and control system?
It would be impossible, obviously, to represent all information about the objective physical reality of a, (any), process or its physical components, (objects). Where would you stop? Is the color of the building in which it is housed, the specific materials of which it is fabricated, that it is effected with gears rather than levers, -or its location in the galaxy- necessarily relevant information? (Contrarily, even its designer's middle name might be relevant if it involved a computer program and you were considering the possibility of a hacker's "back door"!) It

would be counterproductive even if you could as relevant data would be overwhelmed and the consequent "calculus", (having to process all that information),[XXVI] would become too complex and inefficient for rapid and effective response.

Even the use of realistic abstractions could produce enormous difficulties in that you might be interested in many differing, (and, typically, conflicting), significant abstractions and/or their interrelations.[XXVII] This would produce severe difficulties in generating an intuitive and efficient "calculus" *geared towards optimal response.*

For such a complex and dangerous process, the "entities" you create must, (1) necessarily, of course, be viable in relation to both data and control -i.e. they must be adequate in their function.[XXVIII]

But they would also, (2) need to be constructed with a primary intent towards efficiency of response, (rather than realism), as well -the process is, by stipulation, dangerous! The entities you create would need to be specifically fashioned to optimize the "calculus" while still fulfilling their (perhaps consequently distributed) operative role!

Quoting from my arguments in Chapter 12:   In the terminology of computer languages, "danger" may necessitate a "Go To" command which can absolutely violate the "structure"/hierarchy of a program to go elsewhere —even outside the program itself by reason of urgent necessity!

But your "entities" would need to be primarily fabricated in such a way as to intrinsically define a *simplistic* operative calculus of relationality between them -analogous to the situation in our generic training seminar. Maximal efficiency, (and safety), therefore, would demand crystallization into schematic *virtual* "entities" -a "G.U.I."- which could resolve both demands at a single stroke. (This, I think, is the ultimate import of Freeman's discoveries.)

Your "objects" could then distribute function, (in a global / cortical mapping), so as to *concentrate and simplify control,* (operation), via an elementary, intuitive calculus. I think they serve the intentional functions of the brain.

These virtual entities need not necessarily be in a simple (or hierarchical -i.e. via abstraction) correlation with the objects of physical reality however.[29] [XXIX]

But they would most definitely need to allow rapid and effective control of a process which, considered objectively, might not be simple at all. It is clearly the optimization of the process of *response* itself –i.e. the existance of a simplistic "calculus"- that is crucial here, not literal representation. We, in fact, *do not care* that the operator *knows* what function(s) he is *actually* fulfilling, only that he does it (them) well!

3.2 The Specific Case of Biology

Biological survival is exactly such a problem! It is both especially complex *and* especially dangerous. It is the penultimate case of complexity and embodies a moment-by-moment confrontation with disaster. It is therefore a schematic model in just this sense that I argue evolution constructed, and I propose it is the basis for both the "percept" and the "mind".

Turning our Perspective Around –a Model of Process!

But it is just the *converse* of the argument made above that I propose for evolution however. It is not the *distribution* of function, but rather the *centralization* of disparate "atomic" -*but blind* -biological function into efficacious schematic -and virtual- objects that evolution effected while compositing the complex metacellular organism. (These are clearly just the complementary perspectives on the same issue.)[XXX]

But let's talk about the "atomic" in the "atomic biological function" of the previous statement. There is another, and crucial step in the argument to be taken at the level of biology. The "engineering" argument, (made above), deals specifically with the schematic manipulation of "data".

---

[29] See Chapter 13 re d'Espagnat on "multitudinism"!

At the level of primitive evolution, however, it is modular (reactive) process that is significant to an organism, not data functions. A given genetic accident corresponds to the addition or modification of a given (behavioral/reactive) process which, for a primitive organism, is clearly and simply merely beneficial or not. The process itself is *informationally indeterminate* to the organism however –i.e. the process is a modular whole.[30]

No one can presume that a particular, genetically determined response is informationally, (rather than reactively), significant to a Paramecium or an Escherichia coli, for example, (though *we* may consider it so). It is significant, rather, solely as a modular unit which either increases survivability or not.

Let me therefore extend the prior argument to deal with the schematic organization of atomic, (modular), process, rather than of primitive, (i.e. absolute), data. It is my contention that the cognitive model, and cognition itself, is solely constituted as an organization of that atomic modular process, an organization designed solely for computational and operational efficiency within a simplistic and optimized schematic calculus. The atomic processes themselves remain, and *will forever remain*, informationally indeterminate to the organism!

The evolutionary purpose of the model was *computational* simplicity itself! The calculational facility potentiated by a schematic and virtual object constitutes a clear and powerful evolutionary rationale for dealing with a multifarious environment. Such a model, (*the "objects" and their "calculus"*), allows rapid and efficient response to what cannot be assumed, a priori, to be a simplistic environment.

---

[30] See Maturana's "structural coupling" in Chapter 6 for a rationale.

But How can these Brain "Axioms" be Physically Conceived?

Think of these "axioms" of the brain as the most primitive *rules* physically operational in that brain —as the most primitive, basic and "atomic" units of functioning!

"Axioms"

To reiterate a prior comment, it does not seem difficult to conceive of "axioms" *as rules per se* –even within pure mathematics, so the conception of the most primitive operative units, (rules), in the brain in the same light —via Maturana's "structural coupling"[31] *as rules-* does not seem problematical. Then, expanding our scope to include Hilbert's "implicit definition"[32] of an overall *system* of purely mathematical axioms, (rules), to what I propose is an analogous case for the self-organization of the brain seems to me to be straightforward.[33] It is accomplished, I propose, by evolution as an optimizing self-organization of pure process.

My expansion of Cassirer's "Mathematical Concept of Function"[34] still further into what I call "the concept of implicit definition", (C.I.D.), -enlarging his idea on the basis of Hilbert's, allows *percepts themselves* to be embedded in this context!

Revisiting Cassirer:

Cassirer visualized "concepts" themselves specifically *as rules* within his new "mathematical concept of function". But for Cassirer the *rule* of the concept was *not* implicit in the presentations themselves, (i.e. in extensionality), but, contrary to the interpretation within the Aristotelian concept, (wherein it was

---

[31] See Chapter 6

[32] See Chapter 2

[33] Hilbert's initial axioms of Euclidean Geometry might provide the beginnings of a prototype model, (cerebellum?) —as supplemented by the axioms? of Quantum Mechanics -probability and intentionality and their subsequent formalism.

[34] See Chapters 3 & 5

a product of presentation/attention/abstraction), it came from the mind itself! It was, he said, "a new form of consciousness". It reflected the classical distinction between the *content* of a series and the *form* of the series.

He concluded, (incorrectly I will argue), that this concept, *this rule*, is not to be hypostacized, however, because he was still unable to conceive of percepts themselves, ("the phenomena"), outside of the classical context of "perceiver"/"perceived"!

My extension of his brilliant beginning from "the mathematical concept of function" into my own "concept of implicit definition", (C.I.D.) –which incorporates Hilbert's profound new perspective on the axiom systems of mathematics into a *further* redefinition of the concept itself. It allows *percepts themselves* to the conceived as "positions in a structure", (i.e. as "concepts") –as *virtual* expressions of the biological "axioms" themselves!

The brain, I argue, is specifically a *conceptual* organ rather than an informational one! It translates and refines intentional concepts into subsequent *later* intentional strategic concepts through Merleau-Ponty's feedback loop. But they will always remain intentional concepts! I propose, then, a *strategic*, not an *informational* brain.[35]

A Unicellular Perspective

From the viewpoint of the seventy trillion or so individual cells that constitute the human cooperative enterprise, the assumption of environmental simplicity is implausible in the extreme!

But theirs, (i.e. *that* perspective), is the most natural initial perspective from which to consider the problem. For five-sixths of evolutionary history, (three billion years), it was the one-celled organism which ruled alone. As Stephen Gould puts it, metacellular organisms represent only occasional and unstable

---

[35] Note: I think this passage corresponds very closely with d'Espagnat's "what we would expect to observe" as his fundamental rationale of Quantum Physics!

spikes from the stable "left wall", (the unicellulars), of evolutionary history.

> "Progress does not rule, (and is not even a primary thrust of) the evolutionary process. For reasons of chemistry and physics, life arises next to the 'left wall' of its simplest conceivable and preservable complexity. This style of life (bacterial) has remained most common and most successful. A few creatures occasionally move to the right... "

> "...Such a view of life's history is highly contrary both to conventional deterministic models of Western science and to the deepest social traditions and psychological hopes of Western culture for a history culminating in humans as life's highest expression and intended planetary steward."(Gould, 1994)

3.3 Retrodictive Confirmation

Do you not find it strange that the fundamental laws of the sciences, (or of logic), are *so few*? Or that our (purportedly) accidentally and evolutionarily acquired logic works *so well* to manipulate the objects of our environment?

A Profound Teleological Consequence

From the standpoint of contemporary science, this is a subject of wonder -or at least it should be! (cf contra: Minsky, 1985) It is, in fact, a miracle![xxxi] From the standpoint of the schematic model, however, it is a trivial, (obvious), and necessary consequence. It is precisely the purpose of the model itself! This is a profound teleological simplification!

3.4 Conclusion, (section 3)

Evolution, in constructing a profoundly complex metacellular organism such as ours, was confronted with the problem of coordinating the physical structure of its thousands of millions of individual cells. It also faced the problem of coordinating *the response* of this colossus, this "Aunt Hillary", (Hofstadter's "sentient" ant colony).[36] It had to coordinate their functional interaction with their environment, raising an organizational problem of profound proportions!

Evolution was forced to deal with exactly the problem detailed above. The brain, moreover, is universally accepted as an evolutionary organ of response, (taken broadly[XXXII]). I propose that a *schematic* entity, (and its corresponding *schematic model*), is by far the most credible possibility here.

It can efficiently orchestrate the coordination of the ten million sensory neurons with the one million motor neurons,[XXXIII] -and with the profound milieu beyond. A realistic, (i.e. representational /informational), "entity", on the other hand, would demand a concomitant "calculus" embodying the very complexity of the objective reality in which the organism exists, and this, I argue, is overwhelmingly implausible.[XXXIV]

---

[36] cf Hofstadter, 1979. His is a very nice metaphor for picturing metacellular existence.

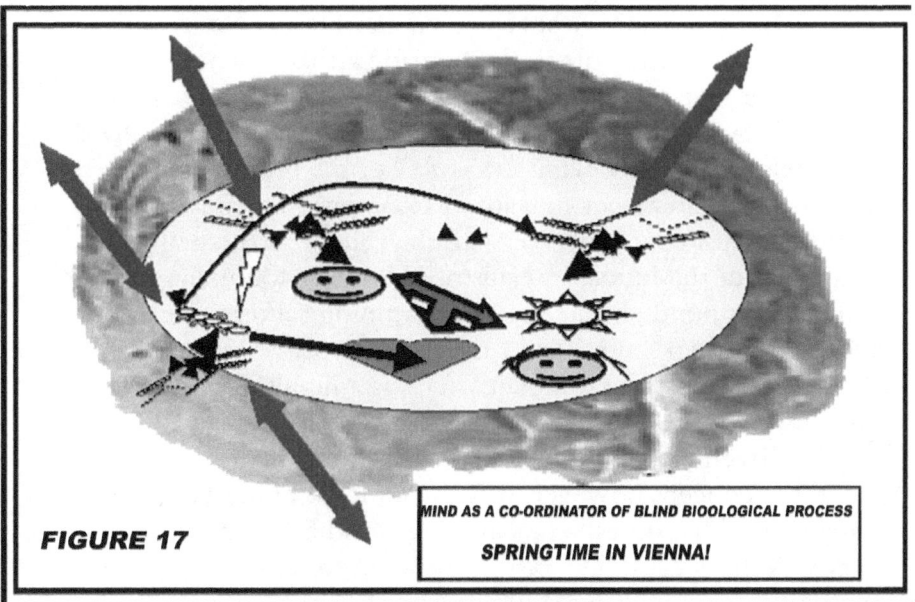

FIGURE 17

MIND AS A CO-ORDINATOR OF BLIND BIOOLOGICAL PROCESS

SPRINGTIME IN VIENNA!

"Lovelife"?

4. The Concordance: Biology's Proper Conclusion

Now I will move to what I think is the most important purely scientific implication of the combination of this and the "implicit definition" briefly expounded in chapter 1, and treated again in Chapter 5, (to follow), where will I formally state it as my second hypothesis.[37] I call it "the concordance".

In those chapters I have argued that the "objects" of mind are solely virtual. I argued that they are logically and implicitly defined by the "axioms" of brain function. I believe this line is profoundly explanatory for the deepest dilemmas of mind as we normally conceive it.

---

[37] I have always had a problem deciding which of these two hypotheses should be stated first. After long consideration, I think this is the way it should be.

In this chapter, I have argued another course -that the objects of mind are biological schematic artifacts. They are optimizing metaphors, artifacts integrating primitive brain process.

Now I propose the biological argument which *relates* the two themes. By *identifying* the "rule" of the brain, (which, accepting Cassirer's conclusions of Chapters 3 and 5 specifies a distinct logical concept[xxxv]), with the rule of "structural coupling" of the human organism, (following Maturana and Varela's profound characterization of biological response –see Chapter 6), then "mind" may now reasonably be defined as the "concept", (/rule), of the brain. This is a highly significant conclusion!

Given that the rule is of the specific structure of my extended concept however, (i.e. the concept of implicit definition - my second hypothesis-see Chapters 3 and 5), then mind becomes the specifically *constitutive* concept of the brain in the sense of Immanuel Kant, and not an ordinary concept. It is a concept necessary to -inbuilt into- our cognition in the exact sense that Kant used the word, and not one imposed upon it.

It is not something *with which* we conceive; it is, rather, the "we" *which conceives*! Following the arguments of the earlier chapters, it implicitly defines and therefore *knows* its "objects".

Combining the results of the two perspectives, I now assert a concordance. I claim that their conclusions are commensurable. The earlier and subsequent chapters make the case that it is only by considering our mental objects as operative *logical* objects, as objects *implicitly defined* by the system, that the wholeness and the logical autonomy of sentiency becomes possible. Referential objects do not allow the same possibility.

The present chapter has made the case that it is only as *virtual and metaphorical* objects, artifacts of the system of control, that the profound difficulties of the integration of megacellular response may be overcome. Again, referential objects do not convey the same possibility. The "objects" of each thesis are thus solely objects of their systems! The objects of the earlier, purely logical and cognitive thesis are thus commensurable with the objects of the second, purely biological and operative thesis. The discovery of such correspondences has always been crucial in the history of science.

But the perspective of modern biology affirms the correlation. Modern day biology must ultimately and *necessarily* reduce logic itself! From an evolutionary perspective, human logic must *itself* now be taken as a strictly biological and evolutionarily derived rule of response![xxxvi]

So too must the "concepts" and "categories" embodied within it. Logic *per se* can no longer be taken as "God-given", or "God-knowledgeable". Such mysticism is not *compatible* with the perspective of modern science.

It is more than plausible, therefore, for biology to identify that human "logic", (that *bio*-logic[38] -and the "implicit definition" resident within it), with the *rules* governing the "objects" of the cognitive G.U.I. of this chapter. "Mind", as the constitutive concept of that bio-logic, (in Kant's and Cassirer's sense), then, is the *biological interface:* the constitutive, holistic, and logical, (i.e. *bio*-logical), expression of the human organism's organization of response.

This conclusion restores "mind" as we normally conceive it to biology and enables a science of mind. This, the biological perspective of the concordance, I maintain, is the logical and proper biological perspective on the whole of the mind-brain problem. It is where biology must ultimately come to stand.

The special significance of the "concordance" for neuroscience is that it finally enables a viable perspective within which biological and specifically neural process might be scientifically correlated with the actual specifics of the mind under evolutionary and operational paradigms. The latter, however, remain the most productive heuristic principles in contemporary biology. It opens, moreover, the prospect of a physical description of mind itself!

---

[38] See the discussion in Appendix B: "Lakoff's ICM's" comparing it to my own "bio-logic"

Our perceptual objects are not objects in reality; they are the implicitly defined logical objects, (alternatively, clearly now, operative objects), of this constitutive logic. They are objects of process.

5. Plain Talk:

Let's talk loosely for a bit. We do not start with absolutes anywhere in our logical and scientific endeavors. Somewhere we start with beliefs. I, for one, believe that I have a mind and a consciousness in the naïve senses of those words. I think most of you believe that you do too.

By this we do not just mean that our bodies mechanically and robotically produce words and actions which "cover the territory" -which merely simulate, (substitute for), sentiency in our naïve sense of it, but that there is some universal and unified existence which is aware. But how?

The solution I propose lies in the combination of the concepts of implicit definition, virtual existence -and logic as biology. This is the only model within our intellectual horizons that seems to hold even any promise for sentiency in our ordinary sense of it. It suggests the only scientifically plausible solution to "the mind's eye" and the "Cartesian theatre" and the only non-eliminativist answer, (for "mind" as such), to the homunculus problem. But these are answers which *must* exist if mind in our ordinary sense is to be real. The "Implicit definition" of my second hypothesis will permit knowing, (as a whole), what are, in some real sense, our distinct and separate parts precisely because those parts, (objects), are in fact non-localized and virtual (logical) expressions specifically *of* the whole! It will open the first genuine possibility, therefore, for a resolution of this essential requirement of "naïve" consciousness.

But that pathway, (implicit definition), does not make sense from the standpoint of representation! Implicit definition solves the problem *logically* -from the standpoint of constitutive logic -and speaks to nothing other than its own internal structure. "Objects", (under this thesis), are known to a system, (i.e. universally/globally), *only* because they are specifically expressions *of* the system.

It becomes a viable and natural solution to the problem of awareness, therefore, only when the objects of consciousness themselves are conceived operationally and schematically, (and specifically, logically), rather than representatively. When our "objects" are taken as specifically schematic representations of process however, (as per the present chapter), the solution becomes both natural and plausible. The logical problem of sentiency is resolved.

How could evolution organize -as it had to organize- the reactive function of this colossus of seventy trillion cells? Even this formulation of the question disregards the yet more profound complexity of the reactivity of the individual cells -also organisms- themselves! It was the overwhelmingly crucial issue in the evolution of complex metacellulars.

My thesis of schematism is both viable and plausible in this context. But what does this evolutionary development and organization of the reactive process of complex metacellulars have to do with "information"? It is an organization of process, not an organization of "data"!

That the progressive evolutionary reactivity of this megacollosus occurred under the bounds of real necessity is, of course, a given. It is the basic axiom of Darwinian "survival". But that it could *match* that possibility[39] -i.e. that it could achieve a (reactive) parallelism to that bound -i.e. "information!" -is a hypothesis of quite another order and teleologically distinct. It is, I assert moreover, mathematically immature.

Objective reality is a *bound* to the evolutionary possibility of organisms, (in Quine's words of my Preface: "the boundary condition"), but under that bound infinitely diverse possibilities remain. I may, as a crude metaphor for instance, posit an infinity of functions under the arbitrary bound $Y = 64,000,000$. I may cite semi-circles, many of the trigonometric functions, curves, lines ... ad infinitum. Only one of these matches the bound, and

---

[39] See Chapter 6: Maturana and my arguments on the specific issue of "*congruent* structural coupling" versus raw "structural coupling".

only a specific subset, (the horizontal lines Y = a, a <= 64,000,000), parallels it. It is a question of the distinction between a bound and a limit. (See Figure 18 following shortly.)

The reactive evolutionary actuality of an organism certainly exists within, (and embodies), a lower bound of biologically possibility. But that some such, (*any* such), organism, (–to include the human organism!), embodies a *greatest* lower bound –i.e. that it, (or its reactivity), matches and meets, (or parallels, i.e. knows!), the real world does not follow.

It is incommensurate with the fundamental premise of "natural selection" and stands as the "parallel postulate" of evolutionary theory. Organisms do not know; organisms do! Organisms survive!

How much more plausible, is it not, that the primary and crucial thrust of evolution was coordination itself, and specifically a coordination of allowable or appropriate, (rather than "informed"), reactive response? I submit that from a biological perspective the schematic object is far more plausible than the representative one. It involves no "magic", and is totally consistent with our deepest conceptions of biology.

I submit that no other viable, (i.e. non-eliminative or non-dualistic), explanation, –an *actual* explanation rather than a prevarication– has ever even been offered for mind and consciousness as understood in our ordinary sense. The argument, then, is one of demonstration. If no truly viable alternative can be offered, then this one must be considered seriously.

The operational process of brain, (and its evolutionarily determined structural optimization), I argue, implicitly defines its "objects", its "entities" in the same sense and in the same manner that the "process" of an axiom system implicitly defines its "objects". They are "positions in a structure"! The "objects of perception" are "intellectual objects". They are (constitutive) conceptual objects. But *those*, in turn, are schematic objects, (alternatively, "operational objects"), *only*, in no necessarily simple correspondence with objective reality. They are *metaphors* of response.

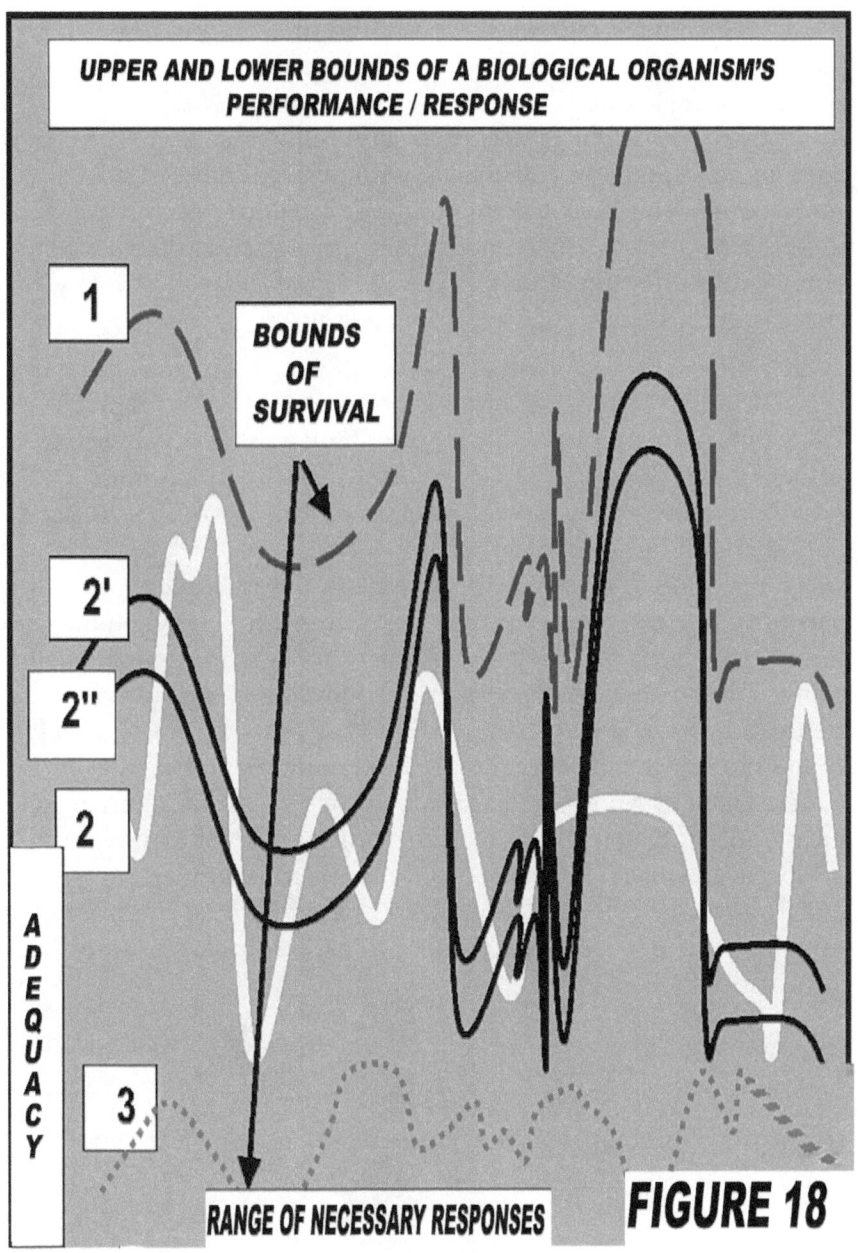

Bounds and Limits

(1) and (3) represent the best and the least possible performance for an organism over the domain of its behavior in absolute (ontic) reality. Less than (3) results in lessened survivability or death; greater than (1) is impossible as it is envisioned as perfect performance with perfect knowledge in actual reality. Between the two bounds, "adequate performance", (... (2), (2'), (2"), ...), need not match, *nor even parallel* these outer bounds. [Note: 2' and 2" parallel 1, but 2 does not!] Any curve within them is consistent with evolution.[40]

"Evolutionary niches" are the obvious biological explanation and confirmation of my "bounds & limits" figure. Evolution must provide some answer, (*any* workable answer), to any particular crucial difficulty, but it is not true that it must provide an answer based on parallelism or "information". It need only work! This is an operationalist answer.

Edelman, for instance, talks about the multiple, non-derivative antibody responses to a given antigen. The same must surely apply to cognition itself, another "recognition system", (using Edelman's characterization of cognition). Cognition and response must be adequate, but it is not obvious that there is only *one* way -a mirroring way. Nor is it inherent that all ways be commensurate! An organism's performance in its environment is measured, fundamentally, not in perfection or in rationality, but in simple adequacy. It is very easy to envision multiple, noncommensurate, blind-though-adequate responses to a given situation. It is not easy to envision rational responses *informed* by information!

---

[40] "Evolutionary niches" are the obvious biological explanation and confirmation of my "bounds & limits" figure. Evolution must provide some answer, (*any* workable answer), to any particular crucial difficulty, but it is *not* true that it must provide an answer based on parallelism or "information". *It need only work*! This is an *operationalist* answer.

Appendix, (Freeman & Automorphism)

An aside: a fascinating quote from Freeman, (it rings strong "bells" in my head)!

> "Some people turn to chemicals as a way to deepen the privacy within solipsistic chasms, and in order to retreat from social stress into inner space. A few have induced these states so as to peer through the solipsistic bars and dirty windows in order to see what is 'really there', although, as minds disintegrate, what comes are swirls and tinglings, *and ultimately the points of receptor inputs* like stars, flies or grains of sand." (Freeman, 1995, my emphasis)

Freeman and I have the same problem –in our innate resistance to the consequences of our own nonrepresentationalism. I too have wrestled with the "points" of sensory input –"like stars, flies or grains of sand". The conclusion I have reached however is that our "points" are, in fact, primitive, "atomic", (unspecified) process, not information. From the simpler perspective of ordinary biology, this is more obvious. These processes, (i.e. pragmatic and adequate, but *not* informational processes), are the necessary basic building blocks of biological cognition. These are our "points'. The difficulty lies in the automorphism we presume in cognition itself, and this is not an easy problem.

How can science continue to make new, profound discoveries? How can the level of verifiable intricacy continue to multiply, seemingly without bounds within the legitimate confines of science? How can the various branches of science continue to integrate and resolve themselves within one comprehensive picture? How could, and why does statistics in fact work? These are the real and crucial questions that a non-representational conception of mind must address.

The fact that the overall picture is getting better –that it is completing itself– does not in itself invalidate the hypothesis that it is non-representational however. Nor does its overwhelming level of intricacy.

To answer the objection, let me reiterate a *counter* question: Is it not possible that we, like a swarm of bees, are merely building, (completing), a *"hive"*, (our worldview)?

We may be completing our interface with externality, but it does not follow at all that that interface is representational. What does follow is that it is the most efficient one possible within our context. This, I believe, is a system with (mathematical) closure[41] –it never escapes itself.

We presume that our science maps back, (automorphically), onto the very model we visualize. But the *path* of the automorphism we seek, I propose, lies through the very "gears and levers" of the *original* evolutionarily derived topobiological cognitive model itself, (*re-using* its "objects") - through another iteration –in another re-entrant mapping which supplies the mechanics and the transformation (back into Freeman's non-topological dispersive mapping into the overall brain) that we seek.[42]

I propose that *reafferance* within the loop of brain function combines with input from outside the loop, (passing through the environment), to yield a consistent, compound map which either does, or does not confirm our theoretical constructs. Nowhere does this conception demand the absolute (ontic) reality of the objects of those constructs, however. It is a reuse of our evolutionarily pragmatic (cortical) objects, (like Rosch's prototypes perhaps), saying nothing whatsoever about the real (external) world in which we live.

Why is this an important advance in our perspective? Because it allows the use of my second hypothesis of "implicit definition" in a legitimate scientific context. (See Chapter 5[43]). That second thesis will enable, for the very first time, legitimate scientific conceptions of the most fundamental aspects we demand

---

[41] See Chapter 6 on Maturana's concept of biological "closure"

[42] I think that Chapter 13, (d'Espagnat on Quantum Physics), and Appendix D, (on Niels Bohr's epistemological origins), validate this conclusion based on modern physics.

[43] -and remember Chapter 2!

for "mind" itself: i.e. a "Cartesian Theatre", the elimination of the problem of the "homunculus", and "knowing" per se. These are not trivial consequences.

Thus, I propose, microscopy, anatomy, biology, physics ... is fed through the same interface to yield an image --of the body of another being or of our own, for instance, or the nature of our environment. But the "objects" are functions of the interface itself, not of an external ontology. This, I believe, is the mechanics of the automorphism we seek --i.e. the one processed by the brain, using its own transformation and mapping back onto its own map *reusing* the (stable) "objects" of that map! It is Edelman plus Freeman plus Merleau-Ponty and back to Edelman. It already exists. (The automorphism can be skewed by the *intent* of the model however --i.e. it can be processed to a different purpose.)

(The whole of this discussion is nonsense, of course, in the absolute form within which it is stated.) Does our feedback really preserve parallelism in the absolute form I have proposed?

It is a valid statement *within a context*, but in an absolute ontological sense these are things we can never truly know. A proper formulation must await the introduction of a completely new philosophical perspective -i.e. that of Cassirer's Philosophy of Symbolic Forms which I will detail in Chapter 7. This supplies the rigorous, (and biologically necessary), scientific epistemological relativism required by the parameters of the problem and is validated in the "real world" from the perspective of modern physics in Chapters 13, 14 and Appendix D, (Niels Bohr).

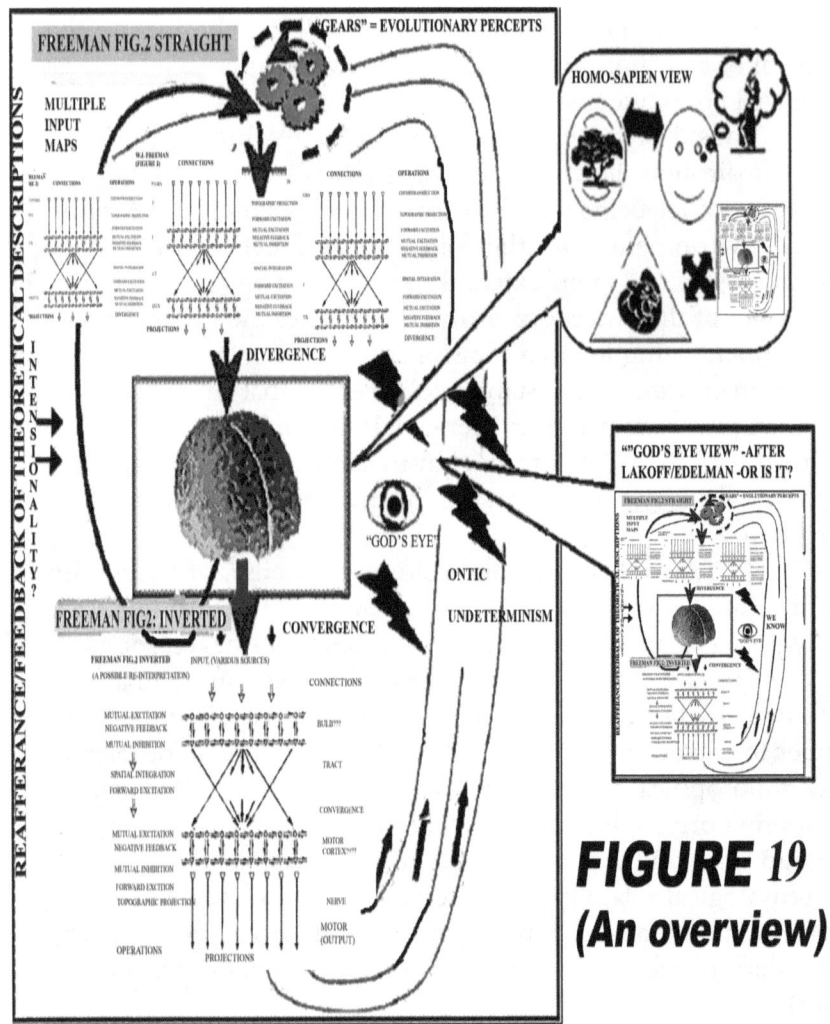

FIGURE 19 (An overview)

GOD'S EYE?

Edelman to Freeman to Edelman[xxxvii]

----------DIV-------------- = Epistemological
(DIV Merleau-Ponty)   Relativism!

Repeating a citation from W.J. Freeman:

> "To explain how stimuli cause consciousness, we have to explain causality. [But] We can't trace linear causal chains from receptors after the first cortical synapse, so we use circular causality to explain neural pattern formation by self-organizing dynamics. But an aspect [a *key* aspect] of intentional action is causality, which we extrapolate to material objects in the world. Thus causality [as far as humans are concerned] is a property of mind, not matter." (Freeman, 1999)

Where is the world outside? *What* is the world outside? Freeman describes his stance as "epistemological solipsism". I understand his rationale, but let me suggest something else. As realists, we necessarily accept[44] the actual existence of an external reality, (as does Freeman), but the fact is we can never know it. Instead of epistemological solipsism, (which is circular ontological language at best), let me suggest another characterization: i.e. ontic indeterminism. (I think Maturana came closest to a realization of this characterization: See Chapter 6).

We must accept the existence of externality, but, as biological organisms, there is not even a possibility that we may ever know it. We can never attain a "God's eye view".

There is a good side to this, however. If we accept the existence of other beings as well, (as I think both you and I do as intentional *beliefs*, then we are not limited to enclosing them hierarchically. We are not obliged to limit them to their "properties". Who is old or young? Who is white or black? Who is crippled or sound? Who is beautiful or ugly? What is the possibility and the "soul" of man?

I have made a point earlier that I think is worth repeating here. I argued that it is not important that the "operator" of such a complicated process knows what it is, (specifically), that he is doing. It is important *only that he does it well*! It is crucially important that he does it diligently, however. It is imperative that

---

[44] It is a key aspect of our specifically *intentional* realist perspective!

he be locked into the loop of his virtual reality -that he "pay attention".

This introduces the necessity of an *inbuilt realistic imperative* -i.e. a mechanical guarantee of his dedication.[xxxviii]. The universal and dogmatic belief in the simple reality of our natural naïve world is thus itself a consequence of my thesis -and the greatest obstacle to its acceptance!

D'Espagnat contributes another perspective:[45]

The Realism of the Accidents

D'Espagnat's discussions about the power and the central role of the "realism of the accidents" argument as perhaps the deepest underlayment to "Scientific Realism" is entirely relevant here, (see Chapter 13). But, surprisingly and conversely, "the realism of the accidents" might be taken as a specific argument I will make for my own "schematic model"!

This philosophical argument, ("the realism of the accidents"), goes roughly like this:

I observe some commonplace fact, e.g. the position of a book on my table, and then leave to have my breakfast at a local cafe, let's say. When I return, I expect to see that book exactly where I left it *unless*, for instance, someone has cleaned and re-arranged the room, there had been a fire which burned my house down, a meteor had fallen to earth directly on my house, ... The point is that the macroscopic object, "the book", is an *enduring* entity, whose modification is entirely subject *only* to the type of qualifiers I just mentioned —to the natural laws of our naïve calculus and their "natural" extensions! *These* "objects" *are* our "decoherent objects"![46]

---

[45] See Chapter 13: "Introducing Bernard d'Espagnat...."
[46] See Chapter 13 for a discussion of decoherence

But this scenario matches in detail my assertion of the *permanence* of our fixed evolutionary "objects"[47], -these metaphors and nexuses -and the concomitant naïve (schematic) "calculus" to which they are subject.

It is almost a precise restatement of the argument I have made for my schematic model and supplies the rationale for the biological commitment to naïve realism I argue. It shows that we require an *absolute dedication to the algorithm* for our continuing existence! It supplies a "higher level language" to enable the use of our intentional faculties.

As an aside on another subject -that of "falsifyability", (and Popper's criteria), by the way, consider Dennett's "color phi" argument[48] from our new perspective. Here is a case where the mental content *would be* falsifiable under the ordinary interpretation of "observability" And yet this particular "mental content" exists -*it has been experimentally confirmed repeatedly*, (observers actually *see* the color change). Following Popper then, Dennett's argument thereby *has itself been falsified!* What *else* follows? Phantom limbs, blindsight? Are these not clear examples, falsifying the standard paradigm, (i.e. representationalism), and easily incorporated into the converse picture of a *virtual* mind?

Chapter Conclusions and a Foreshadowing:

This (specific) chapter, by itself, does not answer all the questions of consciousness. I do claim it as a valid, but specifically biological perspective and *part of* the solution however. It is important at this early stage because it actually enables my *other* crucial hypothesis: i.e. that of "implicit definition".[49] That

---

[47] These "A/D /hierarchical/non-hierarchical" converters in the cortex seen in the context of W.J. Freeman's diagram —See Chapter 1, Figures 6 & 7, and Chapter 4, Freeman Appendix

[48] See Appendix A, (the Dennett Appendix)

[49] See Chapters 2 & 5

second hypothesis will finally offer an explanation of the deepest and most profound problems of mind, *per se*. It will finally elucidate Leibniz' profound problem: "How is it possible for the one to know the many?" It will answer it by finding that "the many" are, in fact, *part of* "the one". The (biological) logic[50] of brain *implicitly defines* our objects because, (*but if and only if*), they are *operational objects* as seen under my previously presented "concept of implicit definition". This is how we are able to know them! *This* is the ground of the "Cartesian Theatre" and finally lays the "homunculus" to rest. But implicit definition as a solution to these problems makes sense only in an operational system, not an informational one.

But still we will not be at the end of our quest. There still remain two more critical steps. The first is an examination of what *any* kind of knowledge *per se* could possibly be. Ernst Cassirer proposed that all knowledge is axiomatic. Otherwise stated, it is all hypothesis and organization, (commensurate, of course, with experience).

His brilliant conclusion[51] was to realize that there could be many beginnings, many organizations, and that the comprehensiveness of a one given theory did not preclude the comprehensiveness of another.[52] What it leads to is a conclusion of the indeterminacy of our absolute understanding of the world around us, (ontic indeterminacy). But this is just what we would expect of the biological organisms we, you and I *both,* understand ourselves to be!

This frustrating conclusion actually leads to the proper ground for an understanding of "mind" however. That ground lies in the realization of our basic realist –*and intentional*- posture itself – of our belief system itself. It is what we, *as realists*, absolutely refuse to give up and which is innately incorporated in

---

[50] See again my comments in Appendix B: "Lakoff's ICM's"
[51] In his "Theory of Symbolic Forms"
[52] Try to think within the context of my citation of Quine's profound overview of knowledge with which I began this book. The parallel is very clear. I will develop the consequences further in later chapters.

any theory we will countenance. But these remain, (and, I argue, will always remain), specifically *intentional* postulates!

Putnam, Lakoff and Edelman, (and Kant himself), propose three basic tenets of scientific realism. They are:

(1) "A commitment to the existence of a real world external to human beings

(2) a link between conceptual schemes and the world via real human experience; experience is not purely internal, but is constrained at every instant by the real world of which we are an inextricable part. A concept of truth that is based not only on internal coherence and "rational acceptability", but, most important, on coherence with our constant real experience

(3) a commitment to the possibility of real human knowledge of the world." (I differ with this last postulate for what should now be obvious reasons.)

(Note: Points 1 and 2 might almost be a restatement of my claims I will make regarding and framing the "interface: in Chapters 9 and 10!)

But I propose a further intentional postulate of scientific realism, (elaborating on the sense of postulate two above to become the new (3') –replacing (3) above). I will propose the actual *ontic existence* of an "interface" *between* the "real world", (1), and "experience", (2), as well –but one consistent with Freeman's conclusions.

It is this postulate –of the actual ontological existence of this "interface" –that I will propose is the actual "substance"[53] of the mind. (Cassirer will place strong limitations on our description of this interface however –it will have to be a *context-free* description[54].)

My third hypothesis, (still foreshadowing), which I will state in Chapters 10 and 11, will be to assume that this "interface" is *structured* in the same way as I have postulated for the brain in

---

[53] Whatever that word, "substance", may possibly mean in this context!
[54] Note: 12-2011: You might want to take a peek at Appendix D, (Niels Bohr) here.

this chapter and will postulate for "experience" itself as well in Chapter 5, (together my first and second hypotheses). All the other substantive problems will have been answered in my first and second hypotheses. Thus it will follow that we are, (this interface is), "live", we are, (this interface is) "conscious", and we, (as minds), *do exist!*

Chapter 5: my Second Hypothesis –a Short Sketch

Note: This second thesis is already better rendered at your level of preliminary understanding in the beginning stages: chapters 1, 2 & 3 of the present paper so here I will only sketch an overview and then proceed to an elaboration as a "snippet" drawn from the first edition of my original book which corresponds to that opening material. That original chapter of the first edition elaborates further the philosophical argument regarding the "Concept of Implicit Definition" and I have added to it a bit. It was mostly conceptually original with me, contrary to the case with Chapter 2 of this current writing, but with which I believe it is totally consistent. It derived from my early understanding of mathematics.[1]

I always hated High school mathematics and had an extremely difficult time with it. The reason lay in the fact that on days 1 and 2 of most of these courses, the beginnings were both plausible and highly interesting to me. But then on day 3 –and it was *universally* so- the instructors began their dialogue with the words "and therefore", and jumped to conclusions which totally confounded me, and which, it turns out, were totally unwarranted.

It was only in my autodidactic digestion of Mac Lane's book that I realized that there were months of intense work between days 2 and 3, and it made me furious for being so deceived. I never forgave them. Even the simplest and most primitive of mathematical operations involved laborious computations and intermediate theorems derived solely from the axioms and definitions, (which primed me for Hilbert's "Implicit Definition"), and had absolutely nothing to do with the "permissive" and totally blind, (and never specified), "objects" themselves.

---

[1] Important Note:: you might want to jump ahead to read Appendix E: "Cassirer Speaks to Shapiro and Mac Lane" before beginning this particular chapter –the two writings are absolutely linked and the latter is an argument for the mathematical and logical *validity* of the conception embodied herein!

Contrary to Dennett, Hofstadter, Churchland, et al, this, my second hypothesis, asserts that the problems of sentiency —of consciousness: the "homunculus" problem, the "mind's eye", "the Cartesian theatre", ... actually *are* capable of solution within the physical world, (and I have proposed an explicit solution).[1] Indeed they *must* be solvable if mind in our ordinary sense of the term is to exist at all. (Dualism is a non-answer. It is a philosophical "cop-out"!) But these problems are not solvable within the confines of classical Aristotelian logic or its modern embodiments. Current logic, still based essentially in the Aristotelian, (i.e. "generic" and hierarchical, set-theoretic), formal concept, is inadequate, I maintain, for the *specifically logical* problems implicit in the mind-brain problem.

An Aside for Clarification:

Let me introduce two diagrams which I will replicate again in Chapter 12. These are fundamentally just input-output loops, (sensors/motor nerves), -but with feedback!

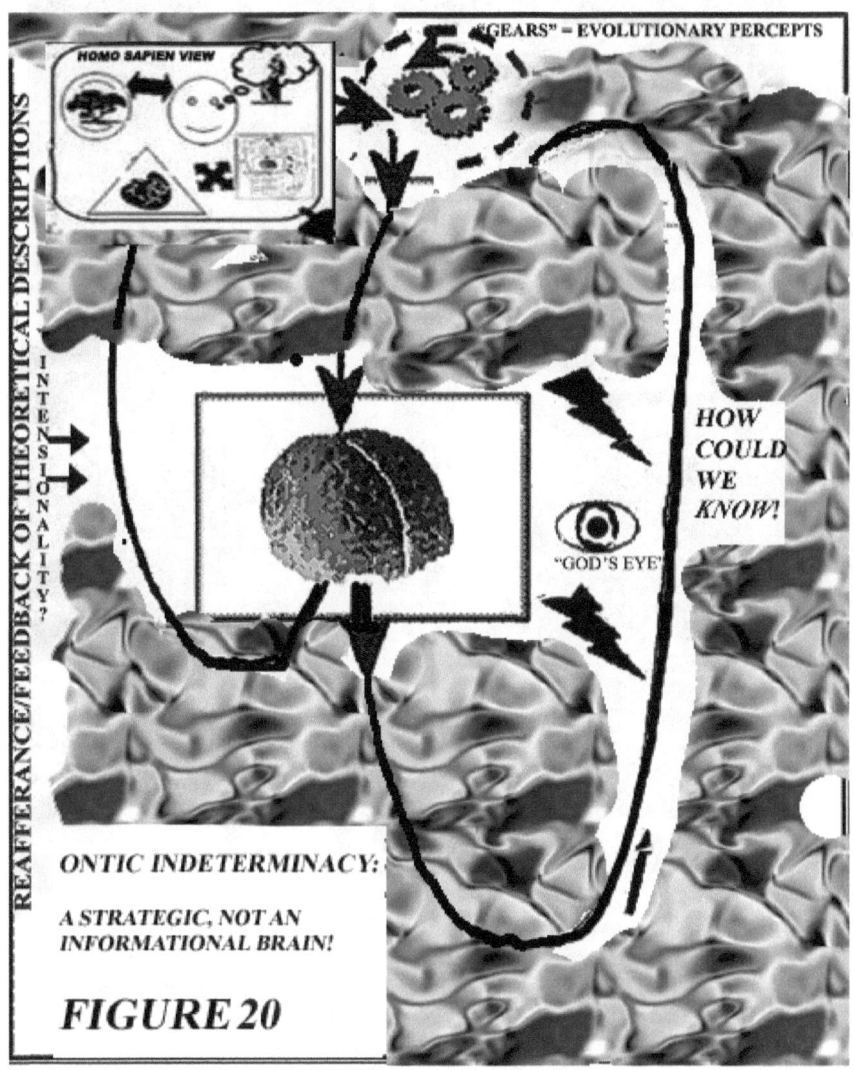

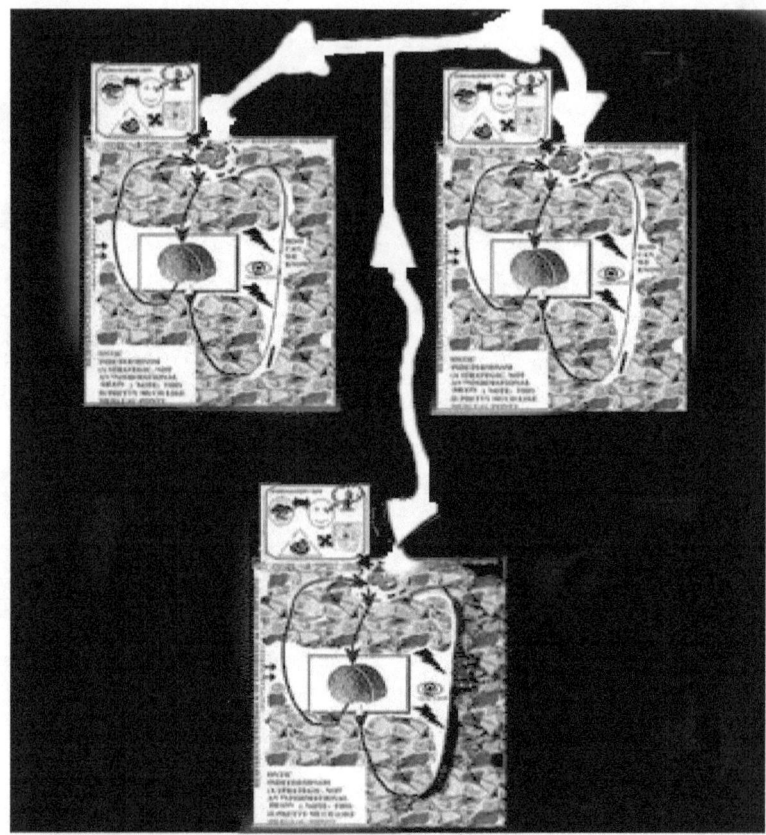

FIGURE 21

This is the model I propose for human reality, but it is lived through the "gears and levers" of our evolutionary artifacts, the latter being understood from the perspective of Biology, itself being just one of Cassirer's multitudinous "Symbolic Forms". This, I believe, is the homo sapien cognitive reality!

Any physical description of a mind, (or of a community of such minds), as mechanisms/organisms, must meet the minimal necessities of these diagrams. They must embody action into the world and self-correcting feedback in the sense of Merleau-Ponty as cited in the previous chapter..

Repeating the Freeman quote yet again:[2]

> "In particular, Maurice Merleau-Ponty in "The Phenomenology of Perception" [2] conceived of perception", [*itself*], as the outcome of the "intentional arc", by which experience derives from the intentional actions of individuals that control sensory input and perception. Action into the world with reaction that changes the self is indivisible in reality, and must be analyzed in terms of 'circular causality' as distinct from the linear causality of events as commonly perceived and analyzed in the physical world." [II]   W.J. Freeman, 1997

What is the "space" that a *machine per se* must necessarily live in? It is a space of total unknowns. It is a space of ontic indeterminacy and the machine really doesn't "care" or "know". Machines only "do". They do it on the surface of the Moon, or in the Pope's living room. It is simply a question of whether or not it works and survives intact!

A machine/mechanism cannot "know"; it can only react. How then could there be "knowing" other than in its contemporaneous physically reductive sense, (Dennett), as mere pertinent mechanical reaction? How could there be a Cartesian theatre, and what of the homunculus? How could a whole know its parts in Leibniz's sense? These questions, moreover, embody pretty much the whole of the very essence of our demands for "mind" in our ordinary intuitive sense of the word.

This was the first formulation of the mind-brain problem I conceived about 50 years ago and I think it was precisely on target. The sole possibility of a solution I saw then, and still the only one I can see now lay in an adaptation and incorporation of

---

[2] I cannot expect that every reader will read this book in context. I will therefore repeat the critical citations where I think it is necessary.

something very strange. It was Hilbert's "concept of implicit definition" which seemed to offer the only conceivable answer to the dilemma other than a circumvention. Hilbert's conception, taken operatively, wherein the "axioms" are re-interpreted as the physical building blocks of the brain,[3] allowed "live", (but specifically *virtual*), objects to physically exist even within the sense of a pure mechanism.

To repeat the Schlick citation:

> "The [Hilbert's] revolution lay in the stipulation that the basic or primitive concepts are *to be defined*[4] just by the fact that they satisfy the axioms.
>
> [They] "acquire meaning only by virtue of the axiom system, *and possess only the content that it bestows upon them*. They stand for entities *whose whole being* is to be bearers of the relations laid down by the system", (my emphasis)[5]

The hard conclusion followed, however, that those "objects" would necessarily have to be "implicitly defined" within the very mechanics of the system itself —i.e. they would have to be operative, internal and logical objects and not referential ones except to the system itself. And how could this be?

This route led me through the development of my first thesis, (Chapter 4), and eventually through my third thesis which, together, will answer these profound difficulties.

---

[3] Alternatively, the primitive embodied rules of the brain —see Chapter 6 and Maturana's "structural coupling"

[4] It is crucial to understand that "defined" is used in a very different sense in mathematics than in the sense of ordinary "dictionary definition". It specifies the *actual*, the whole and exclusive *existence* -for mathematics- of the entity defined. Mathematics students are ingrained in this as the very first step towards "mathematical maturity".

[5] Please note the close parallel to the argument I made in the "training seminar" of Chapter 4

Chapter 5   The Concept of Implicit Definition

Building on Ernst Cassirer's innovative *rule-based*, (rather than property-based), reformulation of the classical concept itself, (his "functional concept of mathematics"), and a new application of David Hilbert's brilliant logical reorientation of mathematics onto purely axiomatic grounds: "implicit definition", [as strongly distinguished from his later "Formalism"[6] –the current paper probably explains the foundations for this thesis best], I propose a further extension of Cassirer's formal and technical "Concept", (and its subsequent logic), to a new and *largest* formal "Concept": i.e. "The Concept of Implicit Definition", (C.I.D.). It is largely equivalent to the complex *logical* rule of the whole of an axiom system –seeing it through Hilbert's eyes as discussed in Chapter 2.[7]

The Concept of Implicit Definition

Following and extending Cassirer's cogent arguments, (see Chapter 3 and especially the "snippet" to follow shortly), dualism and opposition, (innate in classical logic and themselves the actual genesis of the "homunculus", I argue), are then, (after Hilbert), no longer innate in this new Concept. Just as Cassirer argued for his own "Functional Concept of Mathematics", I argue that C.I.D., (my Concept of Implicit Definition), no longer derives from presentation vs. attention and abstraction in cognition- which latter is generally accepted as the theoretical basis of the classical Concept, but rather is unary[8] and *internally*, (i.e. *logically*), resolving of its objects in the sense of Hilbert and modern mathematics.

The extended Concept, (CID), is no longer confined to *intellectual* cognition, (i.e. logic and concepts), however, but is adequate to *perceptual cognition*, (i.e. to *perceptual* "objects"), as

---

[6] Which some still persist in confusing –See Shapiro discussion of Chapter 3.
[7] See Chapter 3 for a full elaboration of this whole concept.
[8] In the sense of Hilbert's "from the *whole* of the axiom system" quoted earlier

well. (From the physicalist perspective[9], I argue that our primitive perceptual "objects" are evolutionary optimizations of process.) C.I.D. is a *constitutive* logic in the sense envisaged by Kant —or it may itself *actually be* that constitutive logic that he envisioned! In concert with the first hypothesis, (non-*re*presentation = "not presentation"), it allows a solution of the logical problem by permitting cognition and "objects" *without* presentation and without the latter's implicit oppositional "cognator" –i.e. without a homunculus. Reconceiving brain function as *organization* rather than representation, (Chapter 4), allows mind and cognition in our ordinary, unified sense.

    A significant corollary of this hypothesis is that it allows mind to be productively defined as the biologically logical, i.e. the *operative* "concept" of the brain. It allows it to be seen as an expression of the *behavioral rule* of the brain, (taken in Cassirer's sense of "the rule of the Concept"). (But here "logical" itself and "concept" itself are expressed in a reductively materialist sense. My third thesis, to be developed shortly in Chapters 7 through 9 will rectify this.) This present conclusion, however, is an important result since I will argue that it is only in taking our objects as *specifically logical* objects that the homunculus problem can be solved, and it shows the relevance of that conclusion to the fundamental biological problem begun in the previous chapter. But the "logic" just mentioned is *biological* logic in the sense of the first hypothesis. It is the "calculus" of our biological "schematic model".

---

[9] Itself taken as a legitimate though relativized "symbolic form" in the sense of Chapter 8

From Appendix C: Cassirer Again
(An extract from Iglowitz 1998: Chapter 2 to Expand the Cassirer and Concept of Implicit Definition Dialogues Incorporated Heretofore.) [10]

How? The Logical Problem of Consciousness
(Cassirer- Hilbert- Maturana: an Archimedean Fulcrum)

"... Every attempt to transform logic must concentrate above all upon this one point: all criticism of formal logic is comprised in criticism of the general doctrine of the construction of concepts." (Ernst Cassirer) [11] [12]

---

[10] Note: This is an excerpt from my original Chapter 2 from my initial MS. It is an expansion of my compacted versions rendered earlier in this book as I thought it might have interrupted the flow of the argument. I think it is a reasonably good overall presentation of Cassirer's perspective and of my expansion of it so I incorporate it here. I think it is worth reading for depth.) Please forgive the repetition of parts of this text, but I want to present it as a whole and completed body. You might want to examine the original book. Note: this was written prior to my acquaintance with the modern "structuralism" / "category theory" dialogue.

[11] Compare also Lakoff: 1987, p.353. "Most of the subject matter of classical logic is categorization."

[12] Cassirer 1923 pps.3-4

He continues: "The Aristotelian logic, in its general principles, is a true expression and mirror of the Aristotelian metaphysics. Only in connection with the belief upon which the latter rests, can it be understood in its peculiar motives.

The problem of "consciousness" and the profoundest paradoxes of the mind-body problem: the "Cartesian theater", the "mind's eye", and the "homunculus" are *logical* problems. They are problems of logical *possibility*!

How could cognition, *how* could mind, ordinarily taken, even *exist*? It is not so much a problem of what it is that they *actually are*, but rather a problem of how it is it even possible to conceive that they *could be*!

How, as Leibniz framed it, *could* "the many be expressed in the one"? How could we *know*? In the context of ordinary realism, ordinary logic allows not even a possibility –other than an eliminative reduction, (a denial), of the problem –which entails a denial of sentiency itself.

The "schematic model" of my first hypothesis cuts to the core of these problems. Coupled with Ernst Cassirer's extension of traditional logic, (his "Functional Concept of Mathematics"), itself extended again by myself in light of the expansion of logical possibility innate in David Hilbert's "implicit definition"[13] for the

---

The conception of the nature and divisions of being predetermines the conception of the fundamental forms of thought. In the further development of logic, however, its connections with the Aristotelian ontology in its special form begin to loosen; still its connection with the basic doctrine of the latter persists, and clearly reappears at definite turning points of historical evolution. Indeed, the basic significance, which is ascribed to the theory of the *concept* in the structure of logic, points to this connection. ..."

[But] "... The work of centuries in the formulation of fundamental doctrines seems more and more to crumble away; while on the other hand, great new groups of problems, resulting from the general mathematical theory of the manifold, now press to the foreground. This theory appears increasingly as the common goal toward which the various logical problems, that were formerly investigated separately, tend and through which they receive their ideal unity."

It is just this "general mathematical theory of the manifold" to which he refers at the end which, I will argue, forces an even further extension of Cassirer's own arguments.

[13] as strongly distinguished from his "Formalism" which is quite a different issue

axiom systems of pure mathematics, it illuminates them and demonstrates a specific "how" for the first time.

The answer turns on a further extension, (again), of the formal logical Concept[14] and with it, of logic itself. Surprisingly that answer will allow us to retain our normal, ("folk"), conception of mind as well.

Cassirer and Classical Logic:

2. Cassirer argued that "the object" of modern mathematics, and "the object of mathematical physics"[15] as well, (their "ideal" objects), are *conceptual* objects (only). He maintained that the Concept they actually embody in modern science is not the classical (Aristotelian) "generic Concept" however, but is rather a new "Functional Concept of Mathematics", (Cassirer's *reformulated* "Concept"). He argued that modern mathematics and modern physics have *already* reconceived the formal logical "Concept" itself, albeit tacitly.[16] [17]

[Repeating just a bit:]

Concept vs. Presentation:

Cassirer's new formal Concept is no longer logically derivable *from its extension* however:

---

[14] I will be employing a convention of capitalizing the word "concept" when it denotes the formal, technical notion of the concept to avoid such verbiage as "the concept of the concept", etc.
[15] See the Heinrich Hertz citation in Chapter 8.
[16] ibid. Also see his "Einstein's Theory of Relativity"
[17] Note, 2011: I did a much more expansive examination of Cassirer's conceptual structure in an appendix to an earlier writing dealing with George Lakoff and Gerald Edelman's books. I have added that writing to this as Appendix B: Lakoff and Edelman. You should probably examine it in conjunction with this argument as it is very "clean", and, I think perhaps, more understandable. Start there with "Cassirer and Lakoff's Logic"

> "The meaning of the *law* that connects the individual members is not to be exhausted by the enumeration of any number of instances of the law; for such enumeration lacks the generating *principle* that enables us to connect the individual members into a functional whole."[18]

If we know the relation by which a b c . . . are ordered, we can deduce them by reflection and isolate them as objects of thought. "It is impossible, on the other hand, to discover the special character *of the connecting relation* from the mere juxtaposition of a,b,c *in presentation*."[19] [20]

> "That which binds the elements of the series a,b,c,... together is not itself a new element, that was factually blended with them, but it is the rule of progression, which remains the same, no matter in which member it is represented. The function F(a,b), F(b,c),..., which determines the sort of dependence between the successive members, is obviously not to be pointed out *as itself a member of the series*, which exists and develops according to it."[21]

This latter is the *definitive* argument against "abstraction" as the *general* case and "presentation" as an ultimate foundation for logic. The association of the members of a series by the possession of a common "property" is only a *special case* of logically possible connections in general. But the connection of the members "*is in every case* produced by some general law of

---

[18] ibid P.26

[19] ibid P.26, my emphasis

[20] cf. Stewart, 1995, "Fibonacci Forgeries". Stewart's article illustrates the case. The "insufficiency of small numbers" leads to an indeterminability of any finite series.

[21] ibid P.17, my emphasis

arrangement [order] through which a thorough-going rule of succession is established."²²

Contra the Theory of Attention:

The "*theory of attention*"²³ therefore "loses all application in a deeper phenomenology of the pure thought processes", (i.e. cognition). The similarity of certain elements, (under the classical view), can only be (conceptually) meaningful when a certain point of view has *already* been established²⁴ from which the elements can be distinguished as like or unlike. This identity of reference under which the comparison takes place is, however, "*something distinctive and new as regards the compared contents themselves.*"^III

The distinction between the concept and its extension, therefore, *is categorical*²⁵ and "belongs to the 'form of consciousness'".^IV It is "a new expression of the characteristic contrast between the member of the series and the form of the series".^V

Cassirer argued that it is the equivalent of his "Functional Concept of Mathematics", rather than the generic "classical" concept, that is the *actual* "Concept" which has been employed

---

²² ibid P.17, my emphasis

²³ It is "presentation" vs. "attention" which is at the basis of the oppositional orientation of classical logic, and which is ultimately, I will argue, the origin of the problem of the homunculus.

²⁴ Compare Lakoff: "Category cue validity defined for such psychological (or interactional) attributes might *correlate*", (his emphasis), "with basic-level categorization, but it would not *pick out* basic-level categories; *they would already have to have been picked out* in order to apply the definition of category of category cue validity so that there was such a correlation." (Lakoff: P.54, my emphasis) See Afterword: Lakoff / Edelman. This is surely directly relevant to the context problem as well, (i.e. "the frame problem"), in Artificial Intelligence research. (cf. Dreyfus, 1992)

²⁵ But see my discussion later.

throughout the history of modern science.[26] He offered a convincing co-thesis, furthermore, that the objects of mathematics and science are "implicitly defined", (in Hilbert's sense), specifically.[27] The "functional concepts", (their primitive laws), implicitly define their conceptual "objects" –and these are the actual working objects of science.[28]

Major Consequences:

Cassirer's "Functional Concept" marks a profound advance to understanding, (and our specific problem), in two respects:
(1) it redefines the formal Concept, *fundamentally*, as a "functional rule" and,
(2), it isolates the (new) "Concept" as (logically) separate from, –as from a "different world" than –the "objects" it "orders". The Concept is no longer inherent in the elements it orders, (e.g. of "perception"), nor is it (logically) derived from them.
It is:

> "a new 'object' ... whose *total content* is expressed in the relations established between the individual elements *by the act of unification.*"[VI]

---

[26] "...the concept of function constitutes the general schema and model according to which the modern concept of nature has been molded in its progressive historical development." (ibid, P.21) See also especially: *Einstein's Theory of Relativity*, Cassirer 1923

[27] Discussing Hilbert, Cassirer says: "The procedure of mathematics here", (implicit definition), "points to the analogous procedure of theoretical natural science, for which it contains the key and justification." ibid p.94 See also Appendix E of this book for an elaboration.

[28] Heinrich Hertz is relevant here:

Re Presentation:

The Concept is a purely intellectual -and original- entity, a "peculiar form of consciousness, *such as cannot be reduced to the consciousness of sensation or perception.*"[VII] It is neither a copy of nor an abstraction from its extension. It is an *independent* and "mathematically" functional "ordering" –*an act of unification!* It is a rule not logically derivable[29] from presentation. *That* rule, I will argue, is provided by biology, *not* by revelation.[30]

Cassirer has removed logic, (in his critique of the formal Concept), from the simple abstraction from perceptual objects, (i.e. from presentation). It becomes instead an *internal* function of the mind, (and hence, I will argue, of biology). He calls it "a new form of consciousness"!

I will now proceed to argue a very natural extension (and, I think, a completion) of Cassirer's thesis: "the Concept of Implicit Definition", (C.I.D.). This Concept[31], part of that *same* "new form of consciousness" I will argue, is also internal and logically independent from perceptual presentation as well. I will argue, in fact, that it *creates* its very "objects" – its "extension" - *within the same* free act of unification. Even our very "perceptual objects", (as well as our "intellectual objects"), I will argue, are resolved within the same internal (biological) act.

This will remove, (in agreement with Maturana, Walter Freeman, Edelman and myself), the need for "presentation", (*metaphysically taken*), altogether. It is the (presented) "perceptual object", I will argue in specific disagreement with Cassirer however, which has been hypostasized! This further expansion and reformulation of the Concept itself, and its subsequent logic will finally allow the resolution of the logical paradoxes of sentiency.

---

[29] i.e. under classical logic
[30] i.e. it is not transcendent –nor does it provide a "God's eye view"!
[31] i.e. the concept of implicit definition!

Cassirer's Concept, (the Functional Concept of Mathematics), is unique in that its arguments show that the fundamental logical Concept is *not* derived from presentation or perception. It is a free and independent act (of unification). It is a "new form of consciousness" according to Cassirer and *not* dependent on them.

But if his arguments are believed, (and I think they are *very* strong), then there is a very *natural* extension of Cassirer's Concept wherein the rule, (which determines the concept), can be likened to the conjunction of the axioms in a mathematical axiom system and its objects, therefore, to the objects of Hilbert's "implicit definition".[32]

This is the import of my prior arguments in Chapter 3 regarding the Cantor diagonal argument and its implications for structuralism. Putting this conception within the framework of biology, it opens a new possibility —it potentiates the possibility *that objects as well*, (and not just intellectual concepts), can be free creations, acts of unification of that same "new consciousness" and therefore not dependent on presentation or perception either!

It is clearly in *"presentation" itself* that the paradoxes of the homunculus and the Cartesian Theater arise, after all, and these are specifically paradoxes of presentation. If our perceptions were *presented to* us,[33] -if mind, consciousness and perception were presentational and dualistic, (which is implicit in the presentation/attention → abstraction of classical logic[34]) -then the paradoxes of sentiency would be innate and irresolvable.

But if those perceptions arose *within* us, and if consciousness arose as a whole, (as the unified rule of "ontogenic coupling", after Maturana, as I will argue), then sufficient grounds for a complete resolution of the problem would be established. This is not an answer from solipsism, dualism or idealism

---

[32] See also Appendix E for a further elaboration of Hilbert's ideas

[33] as is assumed under the classical view

[34] See Chapter 3: Cassirer and Logic

however, but an answer from *realism* sans information and presentation.

The Concept of Implicit Definition:

(a natural extension of Cassirer's "Functional Concept of Mathematics")

3. Cassirer's "Functional Concept of Mathematics" does not exhaust the possibilities however –not even for mathematics. Hilbert's "implicit definition" of axiomatic mathematics has specific and *converse* consequences for the formal Concept. Since, (following Cassirer), an actual concept is now defined by any (definite and consistent) conceptual rule, I propose that the rule of a mathematical axiom system, [in the sense of structuralism], itself generates a perfectly good Concept in Cassirer's sense. Axiom systems embody more profound rules than Cassirer considered however, and I propose that they define the *ultimate* concepts.

Here it is a logically complex, (and typically non-serial), rule which defines the concept, (i.e. the conjunction of the axioms[35]), and conversely. Significantly, following Hilbert and modern mathematics, it is a definite, logically precise and consistent rule of generation of its "extension" –i.e., of its implicitly defined "elements" as well.[36] But axiom systems are not logically "dimensional", (strictly implied in Cassirer's more limited $F(x,y,z...)$), nor do they normally define a "series". They

---

[35] see chapter 2 re: Hilbert's "only the *whole* of the axiom system" constitutes a definition of the "objects" –as seen through a structuralist perspective.
[36] I am concerned here with the object of implicit definition *only insofar* as it is a logical object, only *insofar* as it is a mathematical object. This is the actual object of implicit definition. I am not concerned with the (different) objects of models with which it may be made to correspond, i.e. with the objects of its possible realizations. This is quite a different case and quite a different object. It is the logical object *per se*, I will argue, that solves the homunculus. This is the significance of my objection to Shapiro's critique of Hilbert in Chapters 2 and 3.

define the raw (broadest) manifolds of their domains themselves!"[37] [Note 2010: That is, they define the permissive and virtual "objects" implicitly defined by their axioms. They define "positions in a structure"!]

There is no *a priori* presumption of dimensionality *per se* in the domain of an abstract axiom system. Nor can the elements of the mathematical manifold be characterized a priori, (dimensionally), *as functional values* of the individual axioms. Their "objects" are *not* "objects" of the sort: (a1(x), a2(y), a3(z), ...).

Axioms do not interact *dimensionally*, they interact *operationally* [at the fundamental level of meaning!][38] The combination of axioms, and their rule of generation, (corresponding to Cassirer's "continuous transformation"), is purely, profoundly and complexly logical. A mathematical axiom system need not characterize a "series" or a "series of series" moreover.[39] Indeed, this is the exception rather than the rule.

---

[37] I.e. the abstract set taken in its broadest, most general mathematical sense

[38] Note 2011: But "meaning" *per se* is a function of a given simplistic schematic model as discussed in Chapter 4 of the current writing. It is a function of its operative rules.

[39] Cassirer, like Kant before him, considered the "series", (or a series of series), as the ultimate possible mode of logical and conceptual organization. He saw it as the ultimate expression, and the only *possible* principle, (rule), for a logical function, (i.e. a logical principle which specifies its extension), other than identity. He based his new formal concept, ("the Functional Concept of Mathematics"), upon that belief.
But that conception is inadequate and inaccurate for the case of modern mathematics which was forced to deal with the continuum. Axiom systems exactly describe, (specify), elements, (their extension), that are not generally, (i.e. not a priori), organizable on a series principle. Axiom systems embody a larger and broader logical principle, (a rule which specifies its instances), and a broader logical concept, (as demonstrated, I suspect, by Gödel). The elements of a mathematical domain are fully prescribed, ("functionally" in Cassirer's sense), by their axioms, (their rule), but this rule is not "series". It is a complex logical rule - not referring to, but internally generating its extension as a virtual expression of its own innate ordering. It is the rule of implicit definition. This rule, following Cassirer, (I will argue), defines a new concept, the "Concept of Implicit

What it must and does embody, however, is the raw manifold itself, (its domain).[40] It embodies the "logical continuum" generated by its axioms [which is the lesson we learn from my Cantor argument in Chapter 3 and from modern structuralism.]. It embodies an "order" of a higher degree of freedom.

The instances of Cassirer's "Functional Concept of Mathematics", (the objects of its extension), are the continuous generation of its rule. The instances of the implicit definition of mathematical axiom systems, the implicitly defined "elements" of their manifolds, are logically continuous as well -they are the continuous generation of a more profound rule which, *by definition*, exhausts, (and defines), its extension in Hilbert's original sense as Schlick[41] was quick to perceive.[42]

The "elements" of the mathematical domain are precisely *all and only* those "values" implicitly defined by, (logically generated by), a particular system of axioms -in a sense *precisely parallel* to Cassirer's.

They are the pure embodiment, (crystallization), of the "order" of its rule —but this is taking "ordering" in a vastly expanded, and, I argue, its *largest* sense —in its structuralist sense! Its elements are *virtual* elements expressing that innate order. The whole of their meaning and the whole of their being, (*mathematically*), is solely such."[43] "The manifold, (domain), represents the functional and conceptual, (i.e. structuralist), "values" of its system of "generating relations". Its elements are logical elements.

The "elements", (mathematically conceived), of axiom systems are not "objects" *upon which* a system of "generating

---

Definition". And, following Hilbert, it specifies meaning itself! This is a new thing!

[40] which is not, a priori, *implicitly* dimensional.

[41] See Chapter 1 for the reference to Schlick's reaction.

[42] This is the import of Hilbert's reference to the *whole of the axiom system* in Chapters 1 & 2.

[43] Note 2010 —see Hilbert, Chapter 2 of the current writing

relations" acts, however, or to which it relates. They are *products* of it. There is no a priori presumption of their distinct and separate existence. Wilder, pertinently, characterizes the "existence" terms of axiom systems as "presumptive" and "permissive" only.[VIII] Axiomatic "existence" is an operative term only.[44] I think it is a deeper problem than he conceived.

The elements -*the objects*- of axiom systems are logical "invariants" of their generating relations and internal to the rule itself.[45] Neither "presentation", (nor reference), is implicit in them. They are "entities *whose whole being* is to be bearers of the relations laid down by the system."

I urge that this -the Concept of Implicit Definition- is the *ultimate* logical rule, and the ultimate "ordering". It captures the ultimate functionality, (in Cassirer's sense), of a logical system and generates its extension, (its abstract "domain"), as a virtual embodiment of its *own* (logical) "ordering" -its rule. An axiom system, (conceived mathematically), is a rule which *wholly specifies* its "elements" -by definition.[46] [Note 2010 –in Hilbert's sense.]

I propose, therefore, a new and largest formal "Concept": the Concept of Implicit Definition. I propose it in strict analogy to the case of the mathematical axiom system and in strict extension of Cassirer's Concept. It is the natural extension of Cassirer's Functional Concept of Mathematics, and embodies, I propose, the ultimate rule of order taken in Cassirer's sense of that word.

But it is a *generalization* of Cassirer's formal concept, not an instance of it. Conceptual "dimensionality", (a "series of

---

[44] Which fits perfectly with my earlier conception of a *strategic* rather than a representative brain!

[45] Contrary to this view, Resnik,(Resnik, 1992), criticized an example of such a "structuralist" conception of mathematics in terms of the theory of reference. Under my hypotheses, however, the theory of reference *itself* becomes highly problematic. (cf Quine, 1953, pps.139-159, "Reference and Modality") Also see Chapter 5.

[46] See prior "Elaboration" discussion

series"), implicit in Cassirer's linear function of functions: F(x,y,z..), is a *special case* of the "rule" -and of the formal Concept.

The concept of an axiom system, its "rule" of implicit definition, embodies something absolutely new and unique amongst concepts however. Its "extension" is *precisely* its own analycity! The "being", (and the "meaning"[47]), of its elements are, by definition, *identical* with the purely logical "singularities" of the (complex) rule -and the concept- itself. They "are ... *defined* just by the fact that they satisfy the axioms." [IX] [48]

Implicit Definition vis a vis Presentation:

Like Cassirer's Concept, (its conceptual progenitor), the Concept of Implicit Definition is not *oppositional*: i.e. it does not (logically) presuppose "abstraction" or "attention" either. It too is a "peculiar form of consciousness", an "act of unification ... not reducible to the consciousness of sensation or perception". But *this particular* "act", (unlike Cassirer's), does not presuppose "presentation" either!

It does not just logically specify its extension; it logically encompasses it! The rule of "implicit definition" itself then, following Cassirer, is logical exhaustion and its "objects" are *purely logical* objects. They are "crystallizations" - i.e. logical "invariants"[49] of and internal to the rule itself.[50] This Concept, I

---

[47] see above --Schlick

[48] Wilder quotes Nagel: "Indeed, if geometry is to be deductive ... only the *relations* specified in the propositions and definitions employed may legitimately be taken into account." (Wilder, 1967, p.7)

[49] cf Cassirer, 1923 pps.36-41

[50] Implicit definition is important when something significant is *actually* defined. The "objects" of abstract mathematics, (integers, for instance), are, (in opposition to Mill),"concrete", viable and fruitful. Its element specifies a particular kind of object, and that object is specifically a "crystallization" of a peculiar kind of "*ordering*"! It *embodies* the logical and relational *essence* of that ordering -and that's all! Its "objects" are "crystallizations" of its rule -just like the objects of the

suggest, does not entail "extension" at all –it is a (complex) unity of meaning.

Cassirer's Concept, (the Functional Concept of Mathematics), is unique in that its arguments show that the fundamental logical Concept is not derived from presentation or perception but is a free and independent act of unification. It is a "new form of consciousness" *not* dependent on them. The Concept of Implicit Definition, (an extension of Cassirer's thesis), opens a further possibility, however.

It potentiates the possibility *that objects as well* can be free creations, acts of unification of that same new consciousness, (and of biological organisms I argue), and not derived from presentation or perception either. This is a radical idea admittedly. Though somewhat repugnant and somewhat astounding to our preconceptions, it is certainly consistent with the biological conclusions of Maturana, Edelman, Freeman, and myself wherein perception and consciousness, (whatever those may or may not be for these authors –more generally, the internal biological function), of an organism does *not* derive *information* from the world. But that is just what perceptual presentation would imply.

The positive and the immediate consequence of this new rendering of the Concept, (C.I.D.[51]), is that we now have the tools to understand –completely resolve in fact- the problems of the "homunculus" and the Cartesian theatre.

The virtual objects of implicit definition are *known* to the system as a whole. For it is only *as implicitly defined resolutions of the system as a whole* that they exist at all! [52]

---

training seminar. The rules here, (and there), I argue, define the object, not the converse. But here the actual mechanism of that "crystallization" is transparent. The "calculus" defines the object, and the definitional mechanism is implicit definition.

[51] my "Concept of Implicit Definition"

[52] See the earlier references by Hilbert *to the axiom system as a whole*!

This is a major advance on the problem and enables the only realist solution[53] of the problem yet proposed other than a denial of the problem itself. It was in "presentation" *itself* that the irresolvable paradoxes arose after all. To repeat myself however, the denial of (metaphysical) "presentation" does not result in solipsism, but in *realism* sans information and presentation.[54]

Why is this relevant to mind?

4. Why is this significant to the problem at hand? It is because *this* Concept, (C.I.D.), seems "tailor-made" to the logical problem of mind: It is capable of solving the homunculus problem and that of the Cartesian theatre. It can resolve objects without presentation, (without "the homunculus"), and in itself supplies the "theatre"! It supplies an autonomous theory of meaning as well!.

Cassirer has established the equivalence of "concept" and "rule".

If, (1) following the arguments of chapter 4,[55] we are no longer concerned with representation, (nor, with it, of "presentation"),[56]

and (2) if, tentatively, mind were taken as the unified rule, (the "act of unification"), of brain response,[57] –if it were taken as the *unified rule* of the "structural coupling"[58] of the brain

–then (3), (following Cassirer), "mind" might reasonably be identified with the rule of the "Concept" of the brain,

---

[53] See my heading: "Turning our Perspective Around –a Model of Process!" in Chapter 4
[54] which I call "ontic indeterminism"
[55] and of Maturana and Varela, Edelman and Freeman
[56] See the Raichle citations in Chapter 3, for instance.
[57] i.e. as an organizational rather than a representative model as I argued in chapter 4
[58] See Chapter 6: Maturana and Varela on "structural coupling"

("concept" being taken in the larger constitutive sense of my Concept of Implicit Definition, C.I.D.).

If that particular concept were analogous to the "Concept of Implicit Definition" in mathematical axiom systems furthermore,[59] then it would not just "take account" of the elements of its "extension", it would *know* them![60] Their "meaning" and their "being" would be logically manifest *internal* to that concept, (and rule), itself.

They would be resolved as *virtual* expressions of that very rule. They would "acquire meaning ... and possess only the content that it bestow[ed] upon them." They would be *logical* entities "whose whole being [was] to be bearers of the relations laid down by the system."

I argue that the "logic" just mentioned is a constitutive logic[61]. I will argue presently that it is the schematic calculus of Chapter 4!

But these *particular* entities -as cognitive and perceptual entities- *no longer* (metaphysically) presuppose attention or abstraction -*nor do they presuppose presentation*. Therefore, they do not presuppose that *to which* it would be presented -i.e. a "seer"! The logical problems of "the object" -the problem of the homunculus, the problem of "the mind's eye", the "Cartesian theatre", (which are the principal enigmas of consciousness) -are thereby solved in principle.

The fundamental duality, implicit in classical logic, between "seer" and "seen", "thinker" and "object of thought", "perceiver" and "perceived", or, more fundamentally, *between cognition and presentation*, is bridged. The unity, and the very possibility of cognition of "the object" -the global perspective of

---

[59] This is consistent, certainly, with the "schematic object" presented earlier. How *could* evolution crystallize its (schematic) objects? The implicit definition of process -of "rule"- provides an explicit mechanism and rationale!

[60] If there is a tendency to characterize my thesis as a variation of functionalism, then it should be noted that it involves a totally different notion of "function", (and "relation"). You might want to look at Appendix E.

[61] after Kant's usage

the many in the one- is explained in the unity of its existence as a virtual object of implicit definition. For it is only globally[62] that such a virtual object even *exists* as an object.

In our rational universe, then, the Concept of Implicit Definition seems the most appropriate,[63] as a model, to the logical problem of "consciousness". There *is* no categorical disjunction between the "form of the series", (i.e. the "rule" of implicit definition), and its "elements". They are unified in the concept itself. But the Concept of Implicit Definition, (C.I.D.), is just Hilbert's early concept of "implicit definition" *interpreted operatively*. Its terms and concepts derive *from the whole of the axioms*! In Schlick's words, "they stand for entities *whose whole being* is to be bearers of the relations laid down by the system."

Contra Cassirer:

Cassirer "bent" the focus, however:

> "... there is no danger of hypostasizing the pure concept, of giving it an independent reality along with the particular things. ... Its 'being' consists exclusively in the logical determination by which it is clearly differentiated from other possible serial forms ... and this determination can only be expressed by a synthetic act of definition, and not by a simple sensuous intuition."[x]

I argue that there are two crucial flaws in his argument, however:

(1): In the axiom systems of pure mathematics, the *elements* are also expressed by an "act of definition", (albeit an analytical one) -i.e. that of "implicit definition". They are themselves manifestations of that "peculiar form of consciousness, such as cannot be reduced to the consciousness of sensation or perception."

---

[62] re: Hilbert's "only the whole of the axiom system constitutes a definition of its 'elements' " —see chapter 2

[63] the *only* appropriate yet suggested!

(2): While he states that the application of the Functional Concept is embodied in the concept itself,[64] he argues that concepts are different *in kind* from their extension. These are "objects" of a different world from that of the "particular things" - the objects of "simple sensuous intuition".

I argue, (in concert with my first thesis), that the "objects" of "simple sensuous intuition" are *themselves* ultimately objects of "implicit definition", (within the new context of the yet again enlarged redefintion of the Concept itself, C.I.D.), and part of that *same* "peculiar form of consciousness". They are "positions in a structure".

It follows, then, (given my hypothesis), that there is, (in concert with W.J. Freeman, for instance), no *simple* sensuous intuition at all –it does not exist. It is the *perceptual object* itself instead which has been hypostasized! His dichotomy of the "being" of the pure concept and the "being" of the "particular things" need not stand on either leg.

Cassirer did not generalize the "Functional Concept of Mathematics" into "the Concept of Implicit Definition". The "new consciousness", furthermore, stopped short of "sensuous impressions" themselves. For him, the latter were absolute and unknowable. They were, in effect, the focal point upon which the various forms of knowledge, his "Symbolic Forms",[65] were oriented, but could never reach. They were the rock upon which he erected, in Swabey's characterization, his "epistemological theory of relativity".[66]

His "object of knowledge" was a purely conceptual object, implicitly defined by the fundamental laws of the sciences, -their "generating relations". The "objects of perception", the "particular things", were of a different and untouchable world, the rock splitting the intellect in two.

---

[64] "if I know the relation according to which a b c ... are ordered, I can deduce them by reflection and isolate them as objects of thought" ibid p.26
[65] cf Cassirer 1923, and Chapters 7 through 9 of the current writing.
[66] Op. cit P.v. I will have much more to say about "Symbolic Forms" in Chapter 8. It is a conception of great beauty, but, I think, unfinished.

The Crux of the Issue: Presentation

Cassirer did Promethean work, however. He demonstrated the fundamental inadequacies of the classical Concept, both in its scope and specifically as regards "perception". He illuminated the profound and expressly logical chasm between the Concept and the perceptual realm, (the "material" with which it purportedly deals!), and hence the pervasive *duality* which "perception", i.e. "sensuous impressions", necessitates for mind and logic.

Even Cassirer's "Functional Concept of Mathematics" was insufficient to the fundamental problem however, and he remained inside the "magic circle" of perception. The opposition of "Concept" and "percept", (e.g. "attention/abstraction" and "presentation" or still even the opposition of Cassirer's "Functional Concept" and presentation -"sensuous intuition"), and the dualism which is still implicit in it, is the essence of the issue. It is a genuine antinomy and the actual genesis of the problem.

Already contained in "abstraction", already implicit in "attention", already embodied in "presentation" is the dualistic homunculus: i.e. that *to which* "presentation" is offered. There was no way heretofore that we could even *conceive* of an answer to this problem because it was the formal Concept *itself* which generated it. *This* was the retort in which the "homunculus" was conjured!

"Implicit definition", however, belongs *totally* to the "new form of consciousness" -as do the "objects" which it "orders". But here, (beyond Cassirer), there is no longer the assumption of a *presentation* of "elements", (psychological impressions or otherwise), from one world to an intellectualizing, (cognitive), faculty in another. There remains, therefore, *no implicit need* for the dualistic homunculus in cognition. This explains why the two worlds are compatible. There are not two worlds, but one! This "peculiar form of consciousness", this "new consciousness" I maintain, is the *only* form of consciousness!"

[End snippet –with some current editing]

The whole of that chapter in my earlier book is reasonably cohesive, but now you have seen most of it –and more besides. I began it with a presentation of Hilbert's thesis, (not incorporated here[67]), but I think I have done that better early in this present book. There was always an indecision in me as to whether to start from Cassirer or from Hilbert, they are linked so tightly.

---

[67] See Chapters 1 & 2 of the current writing

Chapter 6: Maturana & Varela & Kant. Biology-Part II: Towards the Where and the What?

Biology & Epistemology, (Maturana and Varela and Kant)[1]

"If in a new science which is wholly isolated and unique in its kind, we started with the prejudice that we can judge of things by means of alleged knowledge previously acquired -though this is precisely what has first to be called in question -we should only fancy we saw everywhere what we had already known, because the expressions have a similar sound.

But everything would appear utterly metamorphosed, senseless, and unintelligible, because we should have as a foundation our own thoughts, made by long habit a second nature, instead of the author's."[1]

From our ordinary way of looking at things, my third and final thesis, (which will be formally stated in Chapter 11), will appear convoluted, esoteric and disturbing. When the inverting glasses of habit are removed and a proper perspective is attained, however, it will become considerably simpler[2], more plausible and profoundly more compatible with modern science than any proposed alternative. To reach that perspective and before I can even begin to properly state this thesis however, I must deal with several seemingly divergent, (but actually closely related), issues. This chapter will discuss the first of them. I must begin to address the epistemological dilemma created by the conclusions of the first two theses.

Nobody writing meaningfully about the mind-body problem today appears to take Immanuel Kant as seriously and as

---

[1] I will begin this Chapter with my original version of it and then come back to make more recent comments. Original MS Numbered Chapter 3 in Iglowitz, 1995.
[2] in a mathematical and organizational sense of the term

literally as I do, and yet he seems to be the thinker most pertinent to it.[3] (I think he must be informed and corrected by Maturana however.) The problem of mind-body is, in one profound respect, the problem of knowing, (epistemology), itself. The questions of what we, as organisms, do know, or even *can* know - and how!- reflect back on the very knowledge by which we judge the problem itself.

In an ancillary and important respect, moreover, the problem Kant faced in attempting to communicate his ideas is very similar to the one Maturana and I face as well. (I referred to this in the introduction.) Both theses totally contravene the common wisdom, and (therefore) make sense only as a whole and not in their parts.

Like Kant's problem "of pure reason", (which is clearly a part of my own problem), my problem:

> "is a sphere so separate and self-contained that we cannot touch a part without affecting all the rest. We can do nothing without first determining the position of each part and its relation to the rest; for, as our judgment within this sphere cannot be corrected by anything without, the validity and use of every part depends upon the relation in which it stands to all the rest within the domain [of reason].
>
> As in the structure of an organized body, the end of each member can only be deduced from the full conception of the whole. It may, then, be said of such [a critique] that it is never trustworthy except it be perfectly complete, down to the minute elements [of pure reason]. In the sphere of this faculty you can determine and define either everything or nothing."[II]

---

[3] "This is an advantage no other science", [than epistemology/metaphysics], "has or can have, because there is none so fully isolated and independent of others *and so exclusively concerned with the faculty of cognition pure and simple*". Kant, "Prolegomena", Lewis Beck translation, Bobs-Merrill, 1950, p.131, my emphasis

The combination of my first two theses provides radical and powerful simplifications to the mind-body problem. But it raises a new and seemingly overwhelming difficulty however. If it is true, then what do we know, and what *can* we know of the reality in which we exist? Since my very arguments depend, moreover, on *accepted knowledge*[4] of that world, have I not reduced my own case to absurdity? The path to my third thesis will answer these questions and supply, (at its conclusion), the single remaining part of my promised *complete* solution to the mind-body problem. The latter is the answer to the problem of the "substance" of the mind. *What* is "mind" and *where* is it? How could it even *be*?

Before I can formally state my third thesis which will answer these questions, (in Chapters 10 and 11), however, we must look at the problem of knowing, (epistemology), and at the broader problem of cognition generally, to include perception. It *demarcates* the problem of "substance". It sets the bounds and defines the very context within which we must consider it. The pivotal issue will be "closure"![5]

Closure:

A mathematical domain D is called "closed" under the arbitrary operations "$\star$" and "#", (let us say), if for every x and y in D, "$x \star y$" and "$x \# y$" are necessarily in D as well. The result of all such operations on the domain, *no matter how far concatenated*, will always remain again within the domain. It never "escapes" itself!

---

[4] e.g. Darwinian evolution

[5] This is, as an emotional issue, the most difficult of my theses and I must expect to lose my credibility with many of you here. It is a strange and esoteric idea, but, I believe, true. It must, on my part, be presented with the utmost delicacy. On your part, I must ask for a very careful reading as it may not be as it seems at first.

I will argue that our human cognitive domain is itself likewise closed, (though bounded),[6] under its operations. This was Kant's, (and Maturana's), essential conclusion as well.

Surprisingly it will *simplify* the problem of "substance" and resolve the intolerable dilemma I (so innocently) raised as well. It is not that the problem of substance is itself so difficult; it is the demands that we make on the answer.

Kant was the earliest scientific, (I might equally say "mathematical" –in the sense of modern mathematics), thinker on this problem, and he is confirmed more recently, from the logical side by Quine,[III] and, from the side of biology, by Maturana and Varela. Though Kant's arguments belong to another era, his *fundamental* conclusions and his rigorous identification of the basic and necessary assumptions remains intact.

Sanity and plausibility depend on just two, (*by definition* "metaphysical"), postulates of absolute existence: "externality" and "experience", ("intuition"). Without them, there is no reason for reason! But those postulates operate solely within the closed domain of reason: "our judgment within this sphere cannot be corrected by anything without."[IV]

While fully affirming the *existence* of our external world, ("substantia phenomena"), as a necessary prerequisite to reason, Kant concluded that we are inherently incapable of knowing any of its independent properties, (to include time, space, extension, tactility, impenetrability …), that is, we are incapable of knowing them *independently of* their revelation in, and in combination with, human cognitive forms.

Kant argued, (in quite a modern vein), that it is impossible to separate our "instrument", (the peculiarities of biological human cognition), from what it "measures", i.e. the

---

[6] A simple mathematical example of a closed and bounded domain would be the domain of the open interval $-1 < x,y < 1$ under the operation of multiplication. Another would be the open domain bounded by unit circle: for all (x,y): $-1 < x,y < 1$ with the operation #: $(x,y)\#(u,v) = (x\star u, y\star v)$. The integers are, of course, closed under addition and multiplication, the rationals under addition, multiplication, and division, …

world it cognates. His genuinely relativistic conclusion gains modern physical credence from the theories of relativity and quantum mechanics, and logical credence, (though it contravenes certain of his own, dated, arguments), from the axiomatic foundations of mathematics. He arrived at a position which I will rename as "ontic indeterminism"[7], (i.e. an indeterminism as to properties, but not as to the *existence* of the "something" –or rather of the "somewhat"- we call "external reality".[8]

More recently, Quine[9] has argued that our "system of knowledge and beliefs" is *logically* closed, and Maturana and Varela[V] have argued that biological organisms are, (by definition!), *operationally and cognitively* closed.

I will argue that our knowledge and, even more broadly, *cognition generally*[10], (to include perception!), is a closed, (i.e. self-referential), domain whose "boundary conditions"[VI] are:

1. the most general, (i.e. the weakest and most abstract), possible assumption of "externality" itself, and

2. "experience" as an *uninterpreted primitive*, i.e. not the interpretation or organization of that "experience" –not, for example, its interpretation as "sense impressions"[11].

---

[7] Kant himself was never satisfied with "critical idealism" but was forced to retain it for historical reasons. "This being the state of the case, I could wish, in order to avoid all misunderstanding, to have named this conception of mine otherwise, but to alter it altogether is probably impossible. It may be permitted me however, in future, as has been above intimated, to term it 'formal' or, better still, 'critical' idealism, to distinguish it from the dogmatic idealism of Berkeley and from the skeptical idealism of Descartes." -"Prolegomena", Pps.124-125

[8] See a more thorough analysis of Kant's "Critical Idealism "later in this chapter.

[9] W.V.O. Quine, 1960. I will elaborate Quine's position in Chapter 7.

[10] Cognition has two aspects. Repeating the definition cited earlier, (Webster's: "cognition: the act or process of knowing, including both awareness and judgment". Also, "Perception: (4a) direct or intuitive cognition.")

[11] But if our perceptual objects *are* cognitions, then how can they be a *boundary condition* of cognition as well? How can our perceptual objects and the things they do be "experience" themselves? I will argue that they are not! "Experience"

The connection between these two assumptions is not necessarily simplistic. This chapter elaborates the first of them. In this chapter, I will examine Maturana and Varela's arguments as set forth in "*The Tree of Knowledge*".[VII]

They consummate the viewpoint of modern biology on the issue of closure. This penetrating work, very much the biological complement of Kant's "Prolegomena" I feel, defines the secure biological context in which they develop a single heuristic principle, (i.e. "structural coupling"), crucial to the mind-body problem.

I will differ strongly with the conclusions they draw from it, however, as they were unwilling to accept the devastating consequences of their own arguments. I do.

Maturana and Varela characterize their book as an argument against a *representative model* of environment in the brain, -against the existence of a current "map" which we use to compute behavior appropriate for survival in our contemporaneous world.

Their argument propounds, instead, a closed, (and evolutionarily determined), reactive parallelism to environment - i.e. "congruent structural coupling". They argue that organisms do not behave as they do *because* of the nature of their current surroundings; they behave *alongside of* it![12]

Organisms, as reactive physical systems, are "operationally closed". The closed ontogenic state of these systems is only "triggered" by their environment. Environment is a "boundary condition" of survival, not a *motivation* for action. (See my illustration "Bounds and Limits" in Chapter 4 which illustrates the lack of need for a parallelism between environment and the organism.) Maturana and Varela conclude there is no *current model* because there is no flow of current "information".

---

is their *invariant relationality* across all orientations including even those which might distribute the "objects" themselves! Does *perceptual cognition* equate with "experience"? No, it is a particular (evolutionarily derived and "pictorial") *orientation* of that relationality! See Chap.7 and the "King of Petrolia".

[12] Their argument is considerably subtler than this as I will detail below.

They develop their fundamental thesis, "structural coupling", at the ground level of primitive evolution. It is a principle of purely mechanistic coexistence between "organism" and "environment" which preserves "autopoiesis", (reproduction). It is, I will argue however, weaker than the strict parallelism, ("congruence"), they demand of it.

Their argument, examined more deeply, is against "information" between an organism and its environment *at any stage* -to include that of natural selection! "Congruence" [13] [14], however, would clearly be *evolutionary* information![15] "Structural coupling" and the "conservation of autopoiesis", (and Darwin's principle of "natural selection" itself), are all quintessentially principles of raw *appropriateness* alone however.[16] They are not informational.

These principles say: "This works!" They do not say: "This is what is!" (They do not exhaust or mirror the whole of possibility). Neither parallelism, ("congruence"), nor embodiment are legitimate consequences of these principles, I will argue, even at the evolutionary level.

There are correlations between domains other than "isomorphism" or "congruence" which preserve pertinency. The mappings and transformations of abstract algebra are obvious counterexamples disproving the inference.[17]

It is only necessary that (some) feature(s) *compatible* with the milieu of the domain be preserved. I will argue that the presumed necessity of "evolutionary congruence" is a *human*

---

[13] Thinking it over, February, 2010, there is a way that their usage of "congruence" could be re-interpreted so as to correspond with my later criticism of their employment of it. It could be re-interpreted as "simple, non-destructive co-existence"! But this is an interpretation of "congruence" that does not involve "parallelism"!

[14] as in "*congruent* structural coupling"

[15] cf Edelman, 1992. He argues that the human genome is simply too small for the purposes of information

[16] i.e. they are boundary conditions, not limits!

[17] Think about Hilbert's "beer mugs" in Chapter 2, for instance.

precept and part of the closed and specifically human cognitive model.

I will now attempt to summarize Maturana and Varela's thesis. Their arguments are profound, subtle, and more concise than any paraphrase. I believe they are, up to a certain point, conclusive.

Maturana and Varela:

Maturana and Varela,[VIII] make a profound and phenomenologically pure[18] argument proceeding from first principles. It leads to severe epistemological consequences. They begin by outlining *minimal and necessary* biological specifications for "living organisms". Those then become an entirely *sufficient* rationale for the *whole* of metacellular organisms and their (nervous) behavior.[19] The argument is wholly operational and constructive.[20] Please forgive the length of the following quotes, but they make their case better than I could. Echoing my comment about one of Cassirer's arguments, I believe that it, too, is Mirabile dictu. It is not my purpose to make their case here, but rather to build upon it!

> "Our intention, therefore, is to proceed scientifically: if we cannot provide a list that characterizes a living being, why

---

[18] i.e. they do not mix their contexts or the origins of their presumptions

[19] "And how can we tell when we have reached a satisfactory explanation of the phenomenon of knowing? ...when we have set forth a conceptual system that can *generate* the cognitive phenomenon as a result of the action of a living being, and when we have shown that this process can produce living beings like ourselves, able to generate descriptions and reflect on them as a result of their fulfillment as living beings operating effectively in their fields of existence." Please note their use of the operative word "conceptual" in "conceptual system" –theirs indeed is a conceptual and a Hertzian "axiomatic" foundation. (op.cit P.30)

[20] Please come back and review Maturana's preamble when you have gotten through Chapter 7, particularly Hertz's reflections on the nature of science. I think the connection is important.

not propose a system that generates all the phenomena proper to a living being?

The evidence that an *autopoietic unity* has exactly all these features becomes evident in the light of what we know about the interdependence between metabolism and cellular structure."[IX]

Plausibly, they characterize a "living organism" as an "autopoietic unity", i.e. a replicating (cellular) physical entity. In so doing, they clarify the inherent nature of biological *phenomenology* itself, (i.e. its innate categories and operative principles)

> "the potential diversification and plasticity in the family of organic molecules has made possible the formation of networks of molecular reactions that produce the same types of molecules that they embody, while at the same time they set the boundaries of the space in which they are formed. These molecular networks and interactions that produce themselves and specify their own limits are ... living beings."[X]

> "*Autopoietic unities specify biological phenomenology as the phenomenology proper of those unities*", (my emphasis), "with features distinct from physical phenomenology... because the phenomena they generate in functioning as autopoietic unities depend on their organization and the way this organization comes about, and *not* on the physical nature of their components."[XI]

The legitimate and minimal principles appropriate to biological process are operational closure and operational independence.

> "Ontogeny is the history of structural changes in a particular living being. In this history each living being begins with an initial structure. This structure conditions the course of its interactions *and restricts the structural changes* that the interactions may trigger in it", (my emphasis). "At the same time, it is born in a particular

place, in a medium that constitutes the ambience in which it emerges and in which it interacts.

This ambience appears to have a structural dynamics of its own, operationally distinct from the living being.

This is a crucial point. As observers, we have distinguished the living system as a unity from its background and have characterized it as a definite organization.

We have thus distinguished two structures that are going to be considered *operationally independent* of each other, "*living being* and *environment.*" [XII] (my emphasis),

Physical science's primal principle of "mechanism", however, leads to a distinct point of view on the interactions of the "autopoietic unity" with its environment: "triggering", "perturbation", and "structural coupling". Organism and environment are *coincident*, not operationally dependent!

"Every ontogeny occurs within an environment; we, as observers, can describe both as having a particular structure such as diffusion, secretion, temperature. In describing autopoietic unity as having a particular structure, it will become clear to us that the interactions (as long as they are recurrent) between unity and environment will consist of reciprocal perturbations. In these interactions, the structure of the environment only *triggers* structural changes in the autopoietic unities (*it does not specify or direct them*)", (my emphasis), "and vice versa for the environment.[21] The result will be a history of mutual congruent[22] structural changes as long as

---

[21] "mutual congruent structural changes" hinges on the meaning of "congruent"! Interpret it as "reciprocal" and I think it is viable, but to interpret it in a geometrical sense of parallelism is totally invalid! -see "Bounds & Limits" diagram.

[22] See prior footnote on another possible interpretation of "congruent"

the autopoietic unity and its containing environment do not disintegrate: there will be a *structural coupling*."[XIII]

(I argue that their phenomenology applies to genetic modification as well as ontogenic modification. A genetic change -randomly and not causally obtained- is retained simply if it is a benefit to the functioning of the organism -i.e. solely on the basis of *appropriateness*. It, and the summation of such genetic changes, therefore, do not actually imply "congruence", [in the sense of *parallelism*], but rather *some* pertinent, (beneficial or at least non-destructive), correlation between domains. "Structural coupling" and "conservation of autopoiesis" are not determinate. They are not "specified or directed" by the environment either; they are *bounded* by it. Structural coupling is therefore a weaker and more abstract condition than they presume —at least under my understanding of their usage of "congruent".)[23]

Between the living being and the environment there is a "necessary structural congruence", [but see my comment above], "(or the unity disappears)." But organisms *must*, (in the innate phenomenology of biology), be considered as *independently reactive to*, rather than determinately, (i.e. informationally), *guided by* their environment. The conclusion is grounded in the structure of science itself:

> "In the interactions between the living being and the environment within this structural congruence, the perturbations of the environment do not determine what happens to the living being; *rather, it is the structure of the living being that determines what change occurs in it.* This interaction is not *instructive*",[24] (my emphasis), "for it does not determine what its effects are going to be.

---

[23] Cognition as a coordination of atomic *primitives*, (as argued in Chapter 4), makes a great deal of sense in this context. The organization is not itself correlative to externality, but is an operative device working on ultimately indeterminate primitives.

[24] i.e. informational

"Triggering" vs "Causation":

> Therefore, we have used the expression 'to trigger' an effect. In this way we refer to the fact that the changes that result from the environment are brought about by the disturbing agent but *determined by the structure of the disturbed system*. The same holds true for the environment: the living being is a source of perturbations and not of instructions."[XIV] [25]

> "The key to understanding all this is indeed simple: as scientists, *we can deal only with unities that are structurally determined.* That is, we can deal only with systems in which all their changes are determined by their structure, whatever it may be, and in which those structural changes are a result of their own dynamics or triggered by their interactions."[XV]

Organisms *react!* They react, moreover, in the operational closure of their *current* (physical) *structure*. The latter is determined by their "ontogeny", (i.e. on their summed history of structural change as individuals), which has modified the original phenotypic structure:

> "This ongoing structural change occurs in the unity from moment to moment, either as a change triggered by interactions coming from the environment in which it exists or as a result of its internal dynamics. As regards its continuous interactions with the environment, the cell unity classifies them and sees them in accordance with its structure at every instant.

---

[25] I cited this passage in my opening chapter at the beginning of my "Argument from Fundamentals".

That structure, in turn continuously changes because of its internal dynamics. The overall result is that the ontogenic transformation of a unity ceases only with its disintegration."[XVI]

Maturana goes on to define "second order" and "third order structural coupling" as the structural coupling of the multicellular organism with its environment, and the coupling of intraspecies' behavioral interaction, (e.g. linguistic behavior), with environment respectively. But these are always dependent upon the necessary conservation of the autopoiesis of the *germ cell*.

The scope of the subsequent development, (the operational range), of the *metacellular* organism[26] is determinate from its unicellular stage, and subject to its phenomenology.

> "The life of a multicellular individual as a unity goes on through the operation of its components, *but it is not determined by their properties.* Each one of these pluricellular individuals...results from the division and segregation of a lineage of cells that originate ... (from) a single cell or zygote. ...It is as simple as this: the logic of the constitution of each metacellular organism demands that it be part of a cycle in which there is a necessary unicellular stage."[XVII]

The Conservation of Autopoiesis:

The conservation of the autopoiesis of that unicellular stage is the necessary boundary condition of the (independent and coincident) function of any organism, unicellular or multicellular.

"Living beings are not unique in their determination nor

---

[26] i.e. the phenotype

in their structural coupling. What is proper to them, however, is that structural determination and coupling in them take place within the framework of ongoing conservation of the autopoiesis that defines them, whether of the first or second order, *and that everything in them is subordinate to that conservation.*

Thus, even the autopoiesis of the cells that make up a metacellular system is subordinate to its autopoiesis as a second-order autopoietic system. *Therefore, every structural change occurs in a living being necessarily limited by the conservation of its autopoiesis; and those interactions that trigger in it structural changes compatible with that conservation are perturbations, whereas those that do not are destructive interactions.*

Ongoing structural change of living beings with conservation of their autopoiesis is occurring at every moment, continuously, in many ways at the same time. It is the throbbing of all life."[XVIII]

Behavior as an Aspect of Structural Coupling:

Behavior, from the biochemical behavior of the amoeba to the nervous behavior of man, is simply an aspect of primary structural coupling. It is the correlation of sensory surfaces with motor surfaces:

> "...the sequence of movements of the amoeba is therefore produced through the maintenance of an *internal correlation* between the degree of change of its membrane and those protoplasmic changes we see as pseudopods.
>
> That is, a recurrent or invariable correlation is established between a perturbed or sensory surface of the organism and an area capable of producing movement (motor surface), which maintains unchanged a set of internal relations in the amoeba."[XIX]

"This basic architecture of the nervous system is universal and valid not only for the hydra, but also for higher vertebrates, including human beings. ... the basic organization of this immensely complicated human nervous system follows essentially the same logic as in the humble hydra ...the nervous tissue understood as a network of neurons has been separated like a compartment inside the animal, with nerves along which pass connections that come and go from the sensory surfaces and motor surfaces.

The sole difference lies not in the fundamental organization of the network that generates sensorimotor correlations, but in the form in which this network is embodied through neurons and connections that vary from one animal species to the other. ...

But we emphasize: ... this is the key mechanism whereby the nervous system expands the realm of interactions of an organism: *it couples the sensory and motor surfaces through a network of neurons whose pattern can be quite varied.* Once established, however, it permits many different realms of behavior in the phylogeny of metazoa. In fact, the nervous systems of varied species essentially differ only in the specific patterns of their interneuronal networks."[xx]

Brain cells do not connect only to motor and receptor cells, however, most of them connect to *other brain cells*:

"in humans, some 10^{11} (one hundred billion) interneurons interconnect some 10^6 (one million) motoneurons that activate a few thousand muscles, with some 10^7 (ten million) sensory cells distributed as receptor surfaces throughout the body. Between motor and sensory neurons lies the brain, like a gigantic mass of interneurons that interconnects them (at a ratio 10:100,000:1) in an ever changing dynamic."[xxi]

The sensory surface includes, however, not only those cells that we see externally as receptors capable of being perturbed by the environment, "but also those cells capable of being perturbed by the organism itself, including the neuronal network."

> "Thus the nervous system participates in the operation of a metacellular as a mechanism that maintains within certain limits the structural changes of the organism. This occurs through multiple circuits of neuronal activity structurally coupled to the medium.

Operational Closure:

> In this sense, the nervous system can be characterized as having *operational closure*", (my emphasis). "In other words, the nervous system's organization is a network of active components in which every change of relations of activity leads to further changes of relations of activity. Some of these relationships remain invariant through continuous perturbation both due to the nervous system's own dynamics and due to the interactions of the organism it integrates. In other words, the nervous system functions as a closed network of changes in relations of activity between its components."[XXII]

External perturbations only *modulate* the constant interplay of internal balances of sensorimotor correlations:

> "It is enough to contemplate this structure of the nervous system... to be convinced that the effect of projecting an image on the retina is not like an incoming telephone line. Rather, it is like a voice (perturbation) added to many voices during a hectic family discussion (relations of activity among all incoming convergent connections) in which the consensus of actions reached will not depend on what any particular member of the family says."[27]

---

[27] ibid Pps. 161-163. Also consider Edelman's comment on this same issue: "... To make matters even more complicated, neurons generally send branches of

"a nervous system...as part of an organism, will have to function in it by contributing to its structural determination from moment to moment. This contribution will be due both to its very structure and to the fact that the result of its operation (e.g., language) forms part of the environment which, from instant to instant, will operate as a selector in the structural drift of the organism with conservation of adaptation."

The Structural Present:

Living beings (with or without a nervous system), therefore, function always in their structural present. The past as a reference to interactions gone by and the future as a reference to interactions yet to come are valuable dimensions for us to communicate...however, they do not operate in the structural determinism of the organism at every moment. *With or without a nervous system, all organisms (ourselves included) function as they function and are where they are at each instant, because of their structural coupling.*"[XXIII]

Maturana presents a sufficient and scientifically necessary rationale for the whole of "living organisms" –to include their "behavior". It is convincing because of the *purity* and the *correctness* of his phenomenology *as biology*. At each step of evolution, on each fundamental aspect of the functioning of an "organism", on the reconciliation of the metacellular, (in all its functions), with the germ cell, these are the *biologically definitive* categories and principles proper to a "living being".

---

their axons out in diverging arbors that *overlap* with those of other neurons, and the same is true of processes called dendrites on recipient neurons .... To put it figuratively, if we 'asked' a neuron which input came from which other neuron contributing to the overlapping set of its dendritic connections, it could not 'know'." Edelman, 1992, p.27

Its "purity" lies in the fact that he never, (and never *has* to), step *outside* this phenomenology -this particular context- to complete his thesis. It is necessary and sufficient, -and *legitimate*, (in the legal sense),- to the whole of "living beings". It is, therefore, completely plausible.[28]

Nowhere does his mechanics involve "representation", however! Indeed, "representation" is inconsistent with the mechanics itself. He concludes as a necessary consequence of scientific principle that neither organisms, nor their brains, operate with *representations* of their surroundings.

"Representation" is inconsistent with the necessary phenomenology of organisms -and extrinsic, (and inessential), to the "mechanism" of science. The principle of parsimony, (i.e. least cause), dictates his conclusion. *Organisms are structurally closed systems*, only "perturbed" by their environment, *never* "in knowledge" of it!

> "The most popular and current view of the nervous system considers it an instrument whereby the organism gets information from the environment which it then uses to build a *representation* of the world that it uses to compute behavior adequate for its survival in the world.
>
> This view requires that the environment imprint in the nervous system the characteristics proper to it and that the nervous system use them to generate behavior, much the same as we use a map to plot a route. We know, however, that the nervous system as part of an organism operates with structural determination. Therefore, the structure of the environment cannot specify its changes, but can only trigger them. ...
>
> Our first tendency to describe what happens .." (is in) "... some form of the metaphor of 'getting information' from the environment represented 'within'. *Our course of reasoning, however, has made it clear that to use this type*

---

[28] Compare this to Hertz's axiomatic characterization of "the object" of science.

> *of metaphor contradicts everything we know about living beings.*"[XXIV]

His argument is not against models in general, however, but, rather, against *representative* models, and in this I think it is conclusive.[29] It leaves very little room for objection. It is consistent, convincing and in the mainstream of science. It leads, perplexingly, to a disastrous paradox:

Maturana's Paradox

> "We are faced with a formidable snag because it seems that the only alternative to a view of the nervous system as operating with representations is to deny the surrounding reality"!
>
> "Indeed, if the nervous system does not operate –and cannot operate –with a representation of the surrounding world, what brings about the extraordinary functional effectiveness of man and animal and their enormous capacity to learn and manipulate the world?[30] If we deny the objectivity of a knowable world, are we not in the chaos of total arbitrariness because everything is possible?
>
> This is like walking on the razor's edge. On one side there is a trap: the impossibility of understanding cognitive

---

[29] I have proposed a very different, and plausible, alternative model in Chapter 4. I proposed that organisms *do* use models, but that those models are schematic; their "objects" schematic objects only, aspects of operationally closed process. The "objects" of that model are not "entities" in reality; they are optimizing loci of process itself.
I propose that models do, in fact, exist in the human brain, but they are *schematic* models. Their *virtual* "objects", (in no necessarily simple correlation with externality), are evolutionarily derived *schematic artifacts* of process like the "objects" of the training seminar of chapter 4. They effectively coordinate the sensory and motor faculties of the brain!

[30] This was the same dilemma that Kant was faced with –it led to his "categories". This, I believe, was where they both went wrong!

phenomena if we assume a world of objects that informs us *because there is no mechanism that makes that 'information' possible*", (my emphasis). On the other side, there is another trap: the chaos and arbitrariness of nonobjectivity, where everything seems possible."[xxv]

"In fact, on the one hand there is the trap of assuming that the nervous system operates with representations of the world. And it *is* a trap, because it blinds us to the possibility of realizing how the nervous system functions from moment to moment as a definite system with operational closure. ... On the other hand, there is the other trap: denying the surrounding environment on the assumption that the nervous system functions completely in a vacuum, where everything is valid and everything is possible. This is the other extreme: absolute cognitive solitude or solipsism. ... And it is a trap because it does not allow us to explain how there is a due proportion or commensurability between the operation of the organism and its world."[xxvi]

Maturana and Varela have honed their "razor's edge" with the same care and meticulous skill with which, as biologists, they would undoubtedly hone a microtome. I suggest they are proposing that we stand, therefore, not on a razor's edge, but on a *microtome's*! That, as any biologist should surely know, is an invitation to suicide. It is likely to result, depending on the angle of fall, in decapitation or, as seems to have happened here, in a severing of the corpus callosum. [;-)]

They have created a full-blown antinomy. The usual method of dealing with antinomies is to examine the presuppositions.

Wait though, you must surely be thinking! Couldn't we just *deny* "mind" in its ordinary sense, then? Isn't that the simplest solution to the difficulty? Why not just abandon (organic) "cognition" entirely, and "experience" and "externality", (in our normal meanings of them), right along with it- and go back solely to parallel and congruent *behavior* itself -i.e. to parallel *reactivity*, predetermined by evolution? Why not just deal with the reactivity and the (reductionist) process of the brain *as part of* the world, (as most current Naturalists, in fact, actually do), accepting the arguments for the inadequacy and the inconsistency of organic cognition as a final reductio ad absurdum of "mental states" and deal only with organisms' (behavioral) function?

Maturana and Varela have, you might correctly continue, specified a phenomenology specific to organisms, but they have specified it *within* the context of an actual physical world. Couldn't we, therefore, just deny the "figment"[XXVII] of the mind, (the "consciousness", the "awareness" of the brain -or organism), as "folk psychology" and myth?[XXVIII] Couldn't we consider "mind" as just a linguistic and behavioral phenomenon? Sure we could, and it is a necessary consequence of ordinary Naturalism. But then we are right back, (necessarily), in Maturana's dilemma, (and Quine's and Kant's which are themselves the children of an ancient line of legitimate skepticism), but invoked at a deeper level!

For how then does even the behavioral, and especially the linguistic[31] function, (our descriptive language), of (human) organisms, *as behavior*, come to be *specifically*, (i.e. informationally), relevant to the world? Is this not *linguistic idealism*?[32] Maturana's whole primary argument -and Darwin's as well - is instead one of simple appropriateness. It is "survival" and "structural coupling", not "information".

This Naturalist argument presumes that organisms' reactivity -third order coupling, (language), and behavior- determined from the beginning by evolution *for the phenotype* and operationally closed thereafter, *is categorical*[33]! This is an astounding conclusion and more than the principles, (and Occam's razor), will bear! At best it is petitio principii, (assuming what you have to prove); at worst it is magic!

---

[31] for behavioral "knowledge"
[32] As I will suggest in Appendix A later it is also the case with Dennett's thesis
[33] i.e. any two models are isomorphic

This, however, is the *only* plausible course left to ordinary[34] Naturalism after Maturana, but it is a difficult one. It assumes that *whatever* evolution determines, (whatever "parallelism" or "congruency" or "adaptability" that evolution gets for an organism), *is embodied in the genotype* -and subsequently in the phenotype. From that point on, the argument is necessarily entrapped in the operational closure of the organism. That closed system must determine its reactivity, (its supposed "parallel reactivity"), *forever after* throughout its subsequent ontogenic history.[35]

But if even the weather is not determinate from a fixed set of principles and starting point, then how are we to believe that evolution has embodied the complexity of day to day, week to week, or year to year physical reality in such a fixed beginning?

What model does evolution, (as embodied in the genotype), *itself* have that it is trying to parallel? If a butterfly in Australia can cause a hurricane in Florida then how are we to believe that evolution has a model *at all*, much less that it can embody such in closed (behavioral or linguistic) principles and laws of reactivity *for the phenotype.*

The argument assumes that evolution launched a closed operational system, (the phenotype), out into the world. But evolution could not know what that phenotype must be functional *with* -i.e. evolution has no model *itself.*[36] Evolution cannot predict the world -especially in its human-scale features. It cannot predict the weather, the pattern of rocks, foliage, water and heat -i.e. "*the facts*"- in an ecosystem, and, if not them, then

---

[34] cf Chapter 7 for my distinction of "ordinary Naturalism" from "relativized Naturalism".

[35] February, 2010. Another possibility occurs to me at this time. It is that genes for a communicating entity might serve. I.e. a "linguistic" entity in the broad sense of any passage of "memes" for instance —by whatever route. This does not invalidate my central thesis in any sense however.

[36] February, 2010. Note: See "other minds" discussion and graphic in Chapter 1. It gives a clue to this problem, consistent with my just prior footnote.

it surely cannot predict the more complex reactivity of the organism's fellow biological creatures -pinching claws, a stalking tiger, or an infection by vibrio comma, (cholera).

"Chaos theory", (for instance), argues that while cyclical processes, (e.g. the large-scale motions of the planets and stars), produce regular and predictable results, non-linear processes do not. But physical process, (the ongoing world), especially at the human scale, *is*, in fact, dynamic and non-linear. Moreover it is, by and large, *not* cyclical. It is, therefore, not predictable in a determinate model.

To assume that such a correspondence to the physical world can be implemented throughout the lifespan of an organism in a fixed and determinate, and specifically *a parallel* operative model, (an informational model), is a difficult premise. For the specifically biological world, the biological ecosystem, it is *more* than difficult. More plausible is that evolution works by the creation of dynamic and operative local, (primitive) -and not *informational* -functions that are intimately and locally connected to changing process –[that affect it "at the system level"]![37]

The creation of a multitude of these "atomic" functions that track, (i.e. trigger from), incremental change in the physical world is a more plausible evolutionary scenario than the representationist one. But this is exactly my first hypothesis: that evolution created local functions like this at the cellular level. The *organization* of these atomic processes then becomes the real problem for the "evolutionary engineer", and it is this organization which, I propose, was accomplished incrementally by the schematic model.

Our primitive (biological) "objects" are organizers, I argue, *organizing loci* of these atomic processes and not informational

---

[37] February 2010. It gives rise not to an "informational" model, but rather to an ongoing refinement of a strategic model which is perfectly consistent with my thesis.

representations. The schematic object is an organization of atomic processes, which latter track we-know-not-what.

For how could even evolution know what that "what" might be? Evolution produces the operationally closed structural coupling of the phenotype, but that structural coupling must be specifically dynamic rather than informational. What evolution *can* deal with are such processes, not information. It can deal with processes that *work* on the local, tactical level.

The representationalist schema, (of ordinary Naturalism), is not plausible. No, that is not quite true; it is plausible *inside* of our own human cognitive model. It is plausible *because it happens that way*! My argument is that it happens *that* way because it *is* inside of our model!

To quote Dennett, (a surprising passage for me):

> "it is not the point of our sensory systems that they should detect 'basic' or 'natural' properties of the environment, but just that they should serve our 'narcissistic' purposes in staying alive; *nature doesn't build epistemic engines.*" [xxix] I find this a very curious statement —coming from Dennett.

This is an antinomy. No, more accurately, it is a specific and pointed reductio ad absurdum of the (ordinary) Naturalist premise![38] What Bertrand Russell says of naïve realism applies to ordinary Naturalism, its (natural) child:

> "We all start from 'naïve realism'. We think that grass is green, that stones are hard, and that snow is cold. But physics assures us that the greenness of grass, the hardness of stones, and the coldness of snow are not the greenness, hardness, and coldness that we know in our own experience, but something very different.
>
> The observer, when he seems to himself to be observing a stone, is really, if physics is to be believed, observing the

---

[38] but not of relativized Naturalism! cf Chapter 7

effects of the stone upon himself. Thus science seems to be at war with itself: when it most means to be objective, it finds itself plunged into subjectivity against its will. Naïve realism leads to physics, and physics, if true, shows that naïve realism is false. Therefore naïve realism, if true, is false; therefore it is false."[xxx]

To paraphrase Russell, if we *know*, then we *can't* know. Therefore we *do not* know.

Maturana and Varela characterized the dilemma incorrectly, however. They specified a necessary choice between solipsism on the one hand, and representationalism/realism on the other, and this is not the case.

The Axiom of Externality

As realists, we needn't deny *reality* based on their arguments, just our specific *knowledge* of it! Nor need we deny "mind". It is the acceptance of an *intentional* "Axiom of Externality", in its *most abstract* form, *taken axiomatically*, that is demanded here,[39] and *that* is not denied by their arguments. It is the improper extension of that demand, and its confusion with the particulars of our specifically human organic process, (to include cognition), that generates the difficulty.

As realists we *must* grant the (intentional) presumption of "externality": i.e. we must grant the simple *posit* of an ontic existence. It is fundamental to sanity and to plausibility. The posit of *our world*: men and baseballs and trees and planets as *necessary ontic entities*, however, is *not!*[40] Even our perceptual world is a part of our closed cognitive process. I have argued, (in Chapter 4), that it is an operative, (and dynamic), artifact.

---

[39] both here and in the foundations of physics
[40] See d'Espagnat in Chapter 13 on "multitudinism".

But, you surely object once again, we *cannot* deny the "objects of our experience" and their apparent relationality![41] I agree, it is these objects which provide the stability of our life experience and ground the very essence of sanity, (my thesis is *not* solipsism). In the next chapter, I will show why we need not.

We all want our naïve world to be real: trucks, men, planets and baseballs, and all our normal relations between them - i.e. all the things they do. It is a necessary component of "sanity", and distinguishes it from dreams, fantasies, and, baldly, insanity. If a rock hits me on the head, it *will* hurt! But, contrarily, our *best science* says that our naïve world is *not* real! What is real for science are atoms, forces, photons, quarks,... all embedded in some mathematically esoteric spatial context.

For it, myself and the man in front of me are, in fact, biological pluralities, or, deeper still, atomic amalgams... down to the deepest levels of physical conception. Naturalism, (the scientifically extended[42] form of our naïve conception and the verity Maturana is loathe to lose), allows this heresy only because it says that our natural world is hierarchically,[43] (and

---

[41] You might want to look at the section of Chapter 4:: "The Realism of the Accidents". Also d'Espagnat 2006 covers it pretty well.

[42] to *whatever* level of sophistication!

[43] See the discussion in the Preamble to this work for a detailed discussion of hierarchy. The reduction of scientific theories, (and theoretic reduction in general), is subject to a fundamental logical limitation under the classical, (pre-Cassirerian), concept. In Chapters 3 and 5, I exhibited Cassirer's arguments that the whole root of the classical formal concept is set-theoretical. Concepts, or concepts of "things", (to include, for instance, our ordinary objects), were reducible only in a set-theoretic sense, i.e. by abstraction, (intersection), of common properties. They are, therefore, subject to Russell's "theory of types". At the bottom level, and there *must* be a bottom level according to the theory of types, there are atomic primitives. Each of the levels above that must be hierarchically oriented, each containing the one above it, (i.e. the "things" of the next higher level are abstractions -intersections- of the ones below). This theory of types was the logically necessary result of the antinomies discovered in the roots of set theory. The most famous is, of course, Russell's paradox. Cassirer's fundamental advance on the classical formal concept, "the mathematical concept of function" however, provides an escape. There is no "Cassirer's

isomorphically), *embedded* in that primitive existence which science posits, and that those hierarchical entities, (our normal "objects"), act *as units.*[44] It maintains that this reduction is specifically a *hierarchical*[45] one which maintains all the spatial and material relationships down through each and all of the depths of scale -*hence their reality*!

Modern science has not confirmed, but rather has seriously questioned, that assertion. What are we to embed them *in*? At the bottom level of physics, "matter", "space", even "existence", in the sense in which naïve realism uses them, are anomalous terms. Even "cardinality" *as such* -the "how many of it"- is dubious![46]

Even *ordinary* Naturalism[47] does not, therefore, maintain the integrity of our naïve objects! But is its insistence on the maintenance of the *hierarchical* integrity of those objects a

---

paradox" in the universal formation of concepts. There is no "concept of all concepts", because concepts are now constituted as an assemblage of (consistent) generative rules, not as a (set-theoretic) abstraction (intersection) of properties - which currently stands for the process of scientific reduction. There is clearly no "rule of all rules" as some rules obviously contravene others. At the level of my "concept of implicit definition", concepts are assemblages of rules, (actually of the *meaning* of those rules), of "axioms", (i.e. fundamental and consistent generative rules), and the same situation obtains. But, just as is the well demonstrated case for mathematical axiom systems, it is possible to exchange an appropriate subset of theorems for the pre-existing axioms, (while still absolutely preserving the integrity -the interior relationality- of the mathematical subject), so is it possible to "cross-reduce" theories. We do not have one single preferred perspective.

This is the relativism of Cassirer's "symbolic forms". What remains is the "web" of relationality, the "invariants" of experience that must be preserved under *all* comprehensive perspectives. But that web, those invariants must be viewed, in Van Fraassen's term, in a "coordinate-free" sense, i.e. they must be viewed in their abstract relationality, not from any *particular* orientation. cf. Chapters 7 & 8 and Appendix B: Lakoff / Edelman. See also the "mathematical ideals" discussion in Chapter 9.

[44] See Chapter 13 (d'Espagnat) on "multitudinism" which he argues is disproved by modern physics
[45] Please consider Bell's comments in Chapter 2 on hierarchy
[46] Cf Penrose on the twin-slit experiment, for instance
[47] i.e. scientific naturalism = "scientific realism"

necessary, or even a plausible presupposition at this juncture in our intellectual history?

My hypothesis of the schematic object, contrarily, says that our naïve world -to include its relationality, (its laws and happenings),-is more probably *unhierarchically*, (but rather transformationally), correspondent with absolute externality, *whatever* and *however* the latter may be. That is, using Maturana's term, it is merely "structurally coupled" with the latter.

Ultimately it says that our naïve world is in correspondence to "points" of *atomic biological process*,[48] and not to "points" of ontology. It is a *metaphor of response*. It says that the further correspondence between those atomic processes themselves and ontology is *completely indeterminate* to us as biological and cognitive entities!

The acceptance of this, the acceptance of the bare, raw *existence*[49] of such a correlation, however, constitutes a necessary requirement for any sane or plausible argument -to include my own. This then is my assertion, the "Axiom of Externality" in its most abstract and precise form, and constitutes the first of the two necessary, (apodictic, intentional), premises for realist reason.[50] (The other is the "Axiom of Experience" which I will treat in the following chapter.)

The "realism" Maturana impeaches is, in fact, (ordinary) "Naturalism". Nor has he really made a case that solipsism is the only other alternative.[51] While his case against representationalism *does* destroy the claims of ordinary Naturalism,[52] a *realistic* case is

---

[48] It is an optimizing organization of primitive, organic *process* -i.e. of primitive *operational* process.

[49] which assumes, therefore, both the axiom of existence and the reality of experience

[50] Is the "axiom of externality" the same as the "realistic imperative" of Hume? Is it an *emotional* imperative? It *orients* world-views.

[51] Theirs is a *structured* isolation. It does not support the implication that "everything is valid and everything is possible"!

[52] Since it assumes the premise of naturalism and ends in a contradiction, it is, in fact, a reductio ad absurdum of that premise!

still possible –but it must be a theoretically mature one. Einstein's realism[53] is more plausible.

That brand of realism involves simply that "theory be organized around a [some] conceptual model of an observer-independent realm".[XXXI] My thesis takes this "some" in its most abstract form, as the (pure) limit of reason. *This* "realism" is certainly more credible in light of today's physics. Realism is more robust than Maturana assumes, and is capable of greater sophistication than a mere linear extension of the naïve world-view. In Fine's words, it is an "attitude". In disagreement with Fine however, I believe it is a *robust* attitude.

Maturana came very close to the answer I propose however. His "object" of cognition[54] is an object of *process*: "cognition does not concern" [external] "objects, for cognition is effective action." (Please savor these words!) He relapses, however, into [the language of] the "objects" of the Naturalistic context in which he framed the problem:

> "Thus, *human cognition as effective action* pertains to the biological domain, but it is always lived in a cultural tradition. The explanation of cognitive phenomena that we have presented in this book is based on the tradition of science and is valid insofar as it satisfies scientific criteria. It is singular within that tradition, however, in that it brings forth a basic conceptual change: *cognition does not concern objects, for cognition is effective action...*'[55]

---

[53] "It is existence and reality that one wishes to comprehend. ... When we strip the (this) statement of its mystical elements we mean that we are seeking for the simplest possible system of thought which will bind together the observed facts." (Einstein 1934, Pps. 112-113) I believe mine is such a theory.

[54] In fact, they do not actually allow an "object" of cognition, as the following citation shows. I am referring here to that aspect of brain process –the effective action– which corresponds to their object of linguistic coupling –which latter is the only "object" they will explicitly allow.

[55] How close this is to my suggestion that "objects" are the a/d converters of the brain.

> "At the same time, as a phenomenon of languaging in the network of social and linguistic coupling, *the mind is not something that is within my brain.*[56]
>
> *Consciousness and mind belong to the realm of social coupling.*[57] That is the locus of their dynamics....Language was never invented by anyone only to take in an outside world. *Therefore, it cannot be used as a tool to reveal that world.* Rather, it is by languaging that the act of knowing, in the behavioral coordination which is *language, brings forth a world.*"

No, I think it brings forth a common intentional strategy *towards* "a world"!

> "...We find ourselves in this co-ontogenic coupling, not as a preexisting reference nor in reference to an origin, but as an ongoing transformation in the becoming of the linguistic world that we build with other human beings", (metacellular organisms).[xxxii]

But "*language ... cannot be used as a tool to reveal [the] world.*" Hence, (accepting his own conclusion), *all* his primitives at the final telling are "entities" *solely* of linguistic (and ontogenic) coupling, and, *as such*, have no absolute referent! He maintains that we are *wrong* in characterizing the actual world "in reference to an origin".

Yet he does exactly that himself. He frames his primitives —*all of them*: structural coupling, metacellular coupling, intraspecies' coupling, ("third order coupling"), and linguistic coupling as interactions of "*autopoietic [biological] unities*"!

What "autopoietic unities"? And *where*? Where do these linguistic domains exist -and between what and whom? Where

---

[56] See prior reference to "other minds"

[57] To repeat a prior reference, they display here a problem that is ubiquitous amongst epistemologists, (to include even Kant himself), who *always* posit "a God's Eye View".

does *his book* exist? *Does* it, and, if so, how is it relevant to anything at all? *What "history of evolution"?*

These linguistic terms supposedly do not "reveal the world"!

He *is*, in fact, committed to a Naturalist ground, and it contains real organisms, i.e. "objects". His "object" is ambiguous however. On the one hand it is solely a product of linguistic coupling, (the object of language), but, on the other hand, (in his presupposition of objects/biological unities which are coupled), it is also the basis of his ontology. This is an explicit and fatal self-contradiction.

Either the object, i.e. the organism, actually *exists* - providing the *ground* of this linguistic coupling, -or it *does not* - in which case "linguistic coupling" is vacuous!

Does my *own* thesis make our objects *not real*, then?[58] Does it mean that there is no connection between them and the "externality" we must assume? The answer is emphatically "No!" The connection is in *the interface itself*, ("structural coupling") and "experience". But the latter must be understood in terms of the former. We are not justified in assigning a *particular* ontic interpretation to "experience".[59]

In my next chapter I will "slice" this problem from another side, (citing Quine and Cassirer), and argue that "experience", as an ontic posit -and a cognitive primitive -while absolutely justified as such, can be legitimately described only as that which *remains invariant* under all possible (viable) interpretations, (and I will argue there is always more than one interpretation). But "invariants" are in themselves a very concrete

---

[58] I will make this case in greater detail in the next chapter.

[59] Naturalism's mistake is in trying to assign an ontic reference to our *whole* cognitive domain. As I have argued, we are justified in making only two primitive ontic, (metaphysical), assertions: "externality" and "experience". These are the minimal and the maximal legitimate ontic posits. Maturana will contribute a third: i.e. "structural coupling" which I will identify with "interface"! See Chapters 9 and 10.

form: they stand, for instance, as the foundation of the Theory of Relativity.

Our human cognitive world, and specifically our perceptual world: people and baseballs and the things they do, *are real*, but they are real in the most general interpretation of their relationality, (them and the things they do). This is not so strange a conception –it is implicit in the reductions of science already. But the latter's requirements of hierarchy and isomorphism are *not* inherent; they constitute the *crux* of the problem. It is those requirements which lead to the disastrous end of Maturana's noble and profound enterprise.

Beneficial connection, *pertinent* connection between domains, (i.e. "structural coupling"), does *not* require "parallelism", it does not imply "congruence", it does not require "hierarchy".[60] Virtual embodiment demonstrates another, non-hierarchical yet exhaustive possibility of compatibility, and it is this that I have argued in my first thesis.

Maturana's thesis of "structural coupling" is of profound importance. It is an epistemological principle of the highest significance.[61] It is a necessary consequence of his Naturalist beginnings –and impeaches them! It precedes and supersedes even its biological origin in its relation to the fundamental problem of knowledge.

Biology, therefore, must integrate into a new and *larger* frame, a new orientation of the whole context of our world and our reality. But the Copernican center of that frame must be structural coupling *itself.* (Think of the connection between "structural coupling" and Kant's brilliant vision!) It is "structural coupling" which must ground biology; it is not biology which

---

[60] *Could* there be a congruent correspondence, (though admittedly not apodictic), however? Sure, but it would be "magic" of a high order- "and then a miracle occurs"! Churchland, 1986)

[61] It is, in fact, a biological and epistemological principle of relativity. This does not imply that it is a *frivolous* relativity, (i.e. solipsism), however, no more than did Einstein's Relativity imply a lawlessness in physics!

must ground "structural coupling"!⁶² (This possibility will be argued in the next chapter within the context of Cassirer's "Symbolic Forms".)

I propose to accept *absolutely* the consequences of "structural coupling": that the "object" of biological cognition is a function of brain process itself –it is "an object of effective action"! It is not an embodiment of its environment.⁶³ But this must *necessarily* translate into a Copernican revolution in our very *world-view*: if *we* are biological organisms, then the objects of *our own* human world-view are objects of process, of response as well. They are "objects" of "effective action"!

Maturana and Varela's profound heuristic principle reduces their premise to absurdity -i.e. the metaphysical certitude of the ordinary Naturalist world-view from which they started. The naïve-realistic world, (the represented "naturalist" world), can have no internal relevance to the organism, *as* organism. But this does *not* impeach the science, (evolution and biology), which is their ground -no more than did Einstein's Relativity impeach the physics which was *his* ground! The *viable relationality*, (the viable system of predictivity), of biology and evolution, (and of science generally), can be, (*must be!*), preserved, (as was the observed relationality of Ptolemean astronomy -times and angles and relative positions- within the Copernican system which replaced it), but they must be interpreted as transformations rather than as reductions.⁶⁴

Are we to throw away the whole of our human enterprise then -to include its science? Of course not -that would be preposterous! But the most profound and most radical advances

---

⁶² It is not an unusual, (nor inconsistent), practice in mathematics to begin by constructing a new mathematical discipline from one set of premises, and then to start all over with what were originally derivative consequences as the new, (and more appropriate), primitives.

⁶³ Though this might still seem self-contradictory, please bear with me. I will explain myself fully in the next chapter.

⁶⁴ You might want to refer back to the section of Chapter 3: "Modern Ptolomean Physics" which discusses this problem.

in human thought, its "Copernican revolutions" and "SUPERB^{XXXIII} theories", have always, (by necessity), subsumed the viable parts of pre-existing knowledge. In the present case, the subsumption of the *preponderance* of naïve realism and the *preponderance* of naturalist science stand as necessities. They *work*, after all, with a power and effectiveness which is awe inspiring.

My proposal does not suggest or imply that they be considered any less important. It subsumes the whole of those vistas, but it subsumes them in their viable *relationality*,[65] and not in their specific ontic (*metaphysical*) reference! Their connection to externality is operational, and not referential. In their whole, they constitute a profoundly effective and complex *algorithm* of unparalleled significance whose link to externality is "structural coupling".

Relativized Materialism

The latter, however, is *referentially* indeterminate, (i.e. metaphysically so). Science turns recursively back on itself in biology and finds that there is a limitation to knowledge itself. Structural coupling is the antinomy which forces the absolute relativization of all knowing -to include "biology" and "evolution" -and even "perception" - themselves. These are "creatures" of *human* knowledge, of cognition. They are *organizers*, not primitives.[66]

Our true primitive is "*experience*", (under the necessary prior premise of "externality"), not any particular interpretation - or organization of it. My hypothesis implies, then, a relativization

---

[65] i.e. their *predictivity*! I will clarify this point in my next chapter.

[66] It is explicit in Maturana's argument, (as we have seen), that "structural coupling" and "the conservation of autopoiesis", (and "congruence" itself), are specifically part of the closed, human (biological) cognitive process.

of epistemology precisely equivalent to Einstein's relativization of physics. This is what Cassirer concluded as well.[xxxiv]

An Answer to the New Dilemma:

At last I can give a preliminary answer, (which I will complete in the next chapter), to the disturbing question raised at the beginning of the chapter. How can I presume the naturalistic world -with its "evolution"- to prove a hypothesis which severely questions them?[xxxv] How can I use a (Darwinian) biological argument, (which *presumes* a simple correspondence between our cognitions and the real physical world), *against* that very simplicity -and embodiment- itself?

If my thesis is true, then our ultimate external reality, (ontology), is *not* necessarily, (nor even probably), like the reality of our cognitive model! The answer is that "evolution" is as much an organizing principle as is "causation". It, (and the objects it treats), is part of the (closed) model itself. It is not a necessary, (or proper!), *metaphysical* presumption, but is, in Kant's words, a "synthetic a priori" proposition. It is not a necessary part of reality; it is instead a necessary (plausible), part of our *cognition* of reality. As such, I can use it with perfect legitimacy within that closed domain. But I use it, (modifying but keeping the sense of Dennett's word), "*heterophenomenologically*", i.e. with a neutral ontic reference!

My epistemological and metaphysical position, therefore, corresponds very much to Kant's, and ultimately, to Cassirer's. It is neither idealism nor solipsism, but a genuine, (*and realistic*), ontic indeterminism.

Kant's "Critical Idealism": What it *really* means!

"Idealism consists in the assertion that there are none but thinking beings, all other things which we think are perceived in intuition, being nothing but representations in the thinking beings, to which no object external to them in fact corresponds. I, on the contrary, say that things as objects of our senses existing outside us *are*", (my emphasis), "given, *but we know nothing of what they may be in themselves, knowing only their appearances, that is, the representations which they cause in us by affecting our senses.*

Consequently *I grant by all means that there are bodies[67] without us, that is, things which, though quite unknown to us as to what they are in themselves, we yet know by the representations which their influence on our sensibility procures us. These representations we call 'bodies', a term signifying merely the appearance of the thing which is unknown to us, but not therefore less actual. Can this be termed idealism?*' (my emphases)

Is he an idealist, then? I think his recharacterization of himself as a "critical idealist" was a profound and misleading mistake —probably his greatest!

"Long before Locke's time, but assuredly since him, it has generally assumed and granted without detriment to the actual existence of external things that many of their predicates may be said to belong, not to the things in themselves, but to their appearances, and to have no proper existence outside of our representation.

Heat, color and taste, for instance, are of this kind. *Now, if I go farther and, for weighty reasons, rank as mere appearances the remaining qualities of bodies also, which are called primary -such as extension, place, and, in general, space... with all that which belongs to it (impenetrability or materiality, shape, etc.)*", (my emphasis), "-no one in the least can adduce the reason of its being inadmissible.

As little as the man who admits colors not to be properties of the object in itself, but only as modifications of the

---

[67] I differ with this obviously. I have concluded that externality is a "somewhat", not a "something" or "somethings". Ding an Sich, (the "thing in itself"), is an unnecessary assumption.

sense of sight, should on that account be called an idealist, so little can my thesis be named idealistic merely because I find that more, nay, *all the properties which constitute the intuition of a body belong merely to its appearance.*" [His emphasis].

It is on such points that I claim that Kant is in no way an "idealist", but was rather, in my own terminology, an "ontic indeterminist" which I think is a more accurate description.

> "The existence of the thing that appears is thereby not destroyed as in genuine idealism, but it is only shown that we cannot possibly know it by the senses as it is in itself."[XXXVI]

The "world" as ontic reality certainly *does* exist for Kant, and he acknowledges it as "substantia phenomenon". Our *knowledge* of the world however is necessarily indeterminate. His is a world of ontic indeterminism, but not a denial of the ontic existence of "the world" itself. The ontic world, for Kant, is most certainly *not* "his idea".

The term "indeterminism" refers to the impossibility of knowing the nature of that ontic reality *independent of* our cognition. It does not, however, assert a doubt as to, but rather affirms, its *existence*.

> "Matter is *substantia phaenomenon*. Whatever is intrinsic to it I seek in all parts of the space that it occupies and in all effects that it exerts, which, after all, can never be anything but phenomena of the outer sense.

> Thus I have nothing absolute but merely something comparatively internal which, in its turn consists only of external relationships. But what appears to the mere understanding as the absolute essence of matter is again simply a fancy, for matter is never an object of pure understanding; but the transcendental object that may be the ground of this appearance called matter is a bare Something," [Note: I would use the term '*somewhat*' instead!], "whose nature we should never be able to understand even though someone could tell us about it. ...

> The observation and analysis of phenomena press toward a knowledge of the secrets of nature and there is no knowing how far they may penetrate in time. But for all that we shall never succeed in answering those transcendental questions that reach out beyond nature, though all nature were to be revealed to our gaze."[xxxvii]

I will, (in chapters 7 through 10), however, make the limiting step that Kant did not. I will (intentionally) posit our cognitive interface, (whatever that may ontically be"![68]), as *itself* a metaphysical entity. It is a part of the minimal (realistic) ontic posit. It is the synthesis of "externality" and "experience".[xxxviii] It is the generalization of "structural coupling"!

Knowledge is cognitively closed. It is an organizational system that *works*. It is Quine's "body of statements and beliefs", (see Preface or Chapter 5), constrained only by its "boundary conditions", ("experience"). But it exists always within the human (biological) cognitive frame. It can never achieve a "God's eye view"!

> "It is by languaging that the act of knowing, in the behavioral coordination which is language, brings forth a world. ...We find ourselves in this co-ontogenic coupling, not as a preexisting reference nor in reference to an origin, but as an ongoing transformation in the becoming of the linguistic world that we build with other human beings."[xxxix]

A New and More Recent Perspective on Maturana:

I said at the beginning of this chapter that though I have not changed my conclusions on its original essence, I had a significant and clarifying insight on it as seen within the context of this current writing. What is it that is substantially new in

---

[68] i.e. "heterophenomenologically"

Maturana's perspective that is different from Kant's? And what was wrong with Kant's?

To review: I think there was a lot still wrong with Kant's vision. For instance, he still maintained that there was a logical necessity of ontological "things" -of "objects" *per se* "out there", ("ding an sich"). This conception was inherited, though modified by Cassirer -and I think they were *both* wrong. Maturana and Varela[XL] exposed the crucial factor in dealing with this part of the problem, i.e. "structural coupling". My overall conception of mind as an organization of blind process —see Chapter 4, and the heading: "Turning our Perspective Around —a Model of Process!" makes the same contrary argument. D'Espagnat's arguments in chapter 13 against "multitudinism" make the same case based on the results of modern experimental physics.

Hear Kant:

> "...though we cannot know these objects as things in themselves, we must yet be in a position at least to think them as things in themselves", ["ding an sich"], "otherwise we should be landed in the absurd conclusion that there can be appearance without anything that appears."[XLI]

[Note: And why not? My thesis argues that this is precisely the case, -that "appearance" is an organizational property rather than a referential one, which is no way inherently "absurd".]

This passage distinguishes Kant's position from my own, from Maturana's "structural coupling"[69], from my "concept of implicit definition", and from my "schematic GUI". Maturana's "structural coupling" connects two absolutely distinct operative domains in the most abstract conceptual manner. My "concept of implicit definition" as combined with the "schematic GUI" shows how there can be appearance "without any*thing*", (ontologically interpreted, e.g. —"ding an sich"), "that appears", and how that

---

[69] *sans* "congruent"

"appearance" can, in fact, be efficacious and pragmatic without requiring representation in whatever guise.[70]

I don't think Maturana and Varela finished their task however. They made mistakes,[71] but they actually did the essential work, and it is profoundly brilliant and important. They showed that the basic (conceptual) operational domains: "environment" and "organism" are distinct and separate, lacking any possible transfer of "information"!

> "The key to understanding all this is indeed simple: *as scientists, we can deal only with unities that are structurally determined.* That is, we can deal only with systems in which all their changes are determined by their structure, whatever it may be, and in which those structural changes are a result of their own dynamics or triggered by their interactions."[XLII]

But then comes the crucial point:

> "This is a crucial point. As observers, we have distinguished the living system as a unity from its background and have characterized it as a definite organization. *We have thus distinguished two structures that are going to be considered operationally independent of each other,* (my emphasis): "living being and environment...
>
> Therefore, we have used the expression 'to trigger' an effect. In this way we refer to the fact that the changes that result from the environment are brought about by the disturbing agent *but determined by the structure of the disturbed system.* The same holds true for the

---

[70] We might correct Kant's citation above by substituting "anywhat" for "anything".

[71] in their progression to "congruent" structural coupling which I argue is unnecessary to their perspective.

environment: the living being is a source of perturbations and not of instructions."[XLIII]

They describe the structural coupling of two domains, two *absolutely isolated* operative domains, and this allows a total disassociation of the brain's, (organism's), "things", -of its "objects" from what was thought to be the *bare logical necessities* of "externality".

> "In the interactions between the living being and the environment within this structural congruence[72], the perturbations of the environment do not determine what happens to the living being; *rather, it is the structure of the living being that determines what change occurs in it.*

I consider Maturana's writing to be as profound as Kant's. I consider it to be an extension and a logical consequence of Kant's profound biological insight.

Maturana's absolute primitives, "living being", ("autopoietic entity") and "environment" are defined as pure concepts, not as classes, (or "objects"), however. Nowhere in his development has Maturana been forced to specify referents *across* these domains. He deals, at least as far as the interaction goes, always with the pure concepts, *as concepts*, themselves.

But Cassirer has forced us to a new understanding of "concepts"!. Even Cassirer's concept, his "functional concept of mathematics" is defined, at bottom however, referentially like Kant's. It is the "concept of implicit definition" which makes sense of this situation. It allows a non-referential view of the concept itself by incorporating Hilbert's perspective. It allows the notion of a purely *operational* concept!

It is not logically required, (after Maturana), that we have ontological "things", nor do we need to have "something". What we do require, (as realists), is a *somewhat,* some unknown,

---

[72] Remember my comments on "congruence"

input and output domains[73] –i.e. a *concept* of "externality" in Maturana's sense! This is my assertion of an "axiom of externality", which stands as the first of three axioms that I argue constitute the minimal and necessary intentional requirements of realist reason.

We require a domain of "externality" that is somehow related to the domain of the brain. My concept of implicit definition does not require a functional, set-theoretic correlation but instead allows *any* beneficial correlation of domains.[74] Chaos theory, complexity theory, Freeman's dispersive mapping, Bell's "local mathematics" ... suggest just some of the possibilities.

But Maturana did not accept the consequences of his own profound paradigm shift, and he proceeded to develop his conception of "*congruent* structural coupling". I have argued that he went too far. Structural coupling alone, but not "congruent structural coupling" is the actual consequence of his arguments, as "congruence" presumes, but does not justify, an out and out parallelism. I think that "structural coupling" and "triggering" specifically contradict the conception of "*parallel* structural coupling" as the latter assumes that we, as metacellular entities can have knowledge of what is on "the other side"!

The Parallel Postulate

I have called "parallel structural coupling" the "parallel postulate" of biology in analogy to the famous mathematical problem. *Parallel* structural coupling *per se* specifically characterizes a "God's Eye" view of ontology, (it purports to show ontology "from the top down"), but "structural coupling" in itself does not. The reactive and (mathematically/structurally) closed system of the autopoietic entity, (and which were

---

[73] and, I argue from our beginning pages, that it is inherently unknowable at all coming from the materialistic perspective of "brain as machine"!

[74] Consider category theory's "morphisms" for instance –see footnote in Chapter 7.

specifically characterized by Maturana as such), –i.e. structural coupling between the organism and some structurally coupled "outside", (ontology) is a *specifically intentional*, and beginning realist postulate necessary for any adequate (realist) organization of our world.

Evolutionary theory teaches us otherwise than "parallelism" however. What we require is mere *appropriateness* pure and simple –i.e. anything that works pragmatically.[75] It corresponds to Maturana's *primitive* "structural coupling, not data or information! We *too* must "kick away the ladder". (See Figure 22)

---

[75] Repeating and earlier footnote: The existence and the fulfilling of "evolutionary niches" is the obvious pragmatic example of this idea.

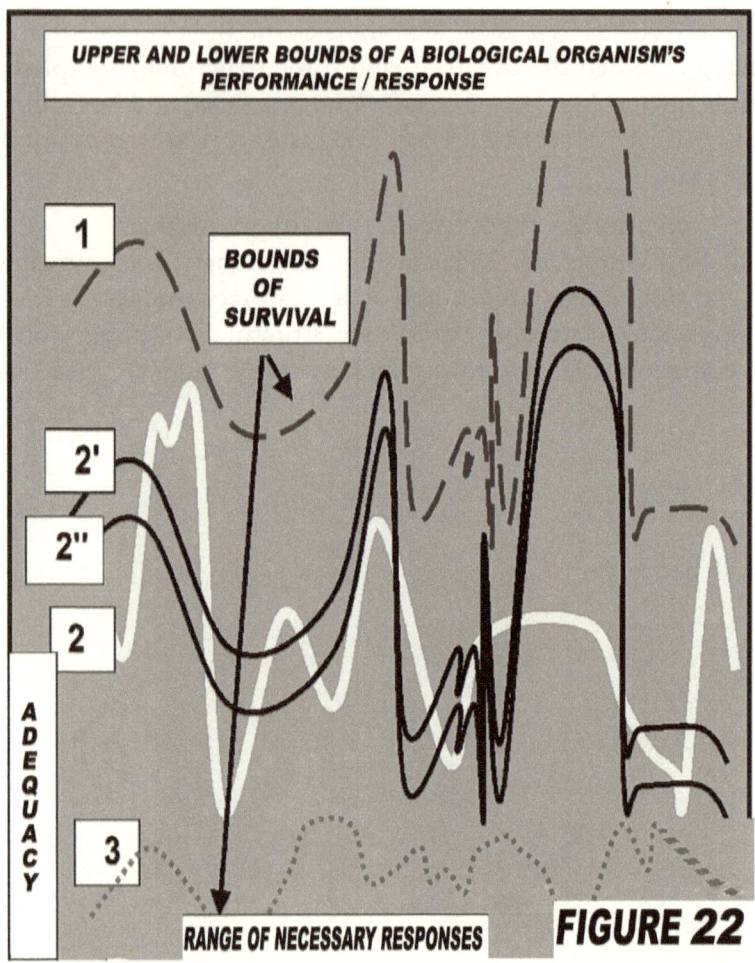

FIGURE 22

(Repeating the graph legend: (1) and (3) represent the best and the least possible performance for an organism over the domain of its behavior in absolute (ontic) reality. Less than (3) results in lessened survivability or death; greater than (1) is impossible as it is envisioned as perfect performance with perfect knowledge in actual reality. Between the two bounds, "adequate performance", (… (2), (2'), (2"), …) need not match, *nor even parallel* these outer bounds. [Note: 2' and 2" parallel 1, but 2 does not!] Any curve within them is consistent with evolution.)

To repeat myself, Maturana's is a fault with just about *all* of the epistemological philosophers, (to include even Kant himself —and Edelman, and W.J. Freeman, the Churchlands, et al…).

They *always* posit "a God's eye view" at some point and then accept their own basic terms *as referential* in some real sense.

The only plausible —and *truly scientific* alternative to it that I can see is a relativism, (albeit a *rigid* relativism), of epistemology itself. Cassirer supplied just such a relativism in his "Theory of Symbolic Forms"[76], and mathematics, in its conception of "mathematical ideals"[77] confirms its essence. But Cassirer's thesis at its bottom is conceptual[78]; it is not based in classes or "objects", (extensionality). It deals with perspectives. It is a conceptual scheme. Reconsider Hertz:

> "The images of which we are speaking are our ideas of things; they have with things the one essential agreement which lies in the fulfillment of the stated requirement, [of successful consequences], but further agreement with things is not necessary to their purpose. *Actually we do not know and have no means of finding out whether our ideas of things accord with them in any other respect than in this one fundamental relation.*"[XLIV]

It is just Cassirer's theme[79] —*as modified with Maturana's*— which I pursued en route to my third hypothesis. It is the only philosophical perspective that allows us to use ordinary descriptive language "heterophenomenologically"[80], (using Dennett's term again), i.e. without an absolute ontic commitment. This is the conception that allowed my idea of a "relativized naturalism", (equivalently "relativized matter"), stated earlier.

---

[76] See Chapters 7 & 8

[77] See Chapter 9

[78] As is Maturana's

[79] The passage above from Hertz is cited in Cassirer's "Symbolic Forms" as part of its rationale.

[80] You might want to look at Appendix D, (Niels Bohr's epistemological foundations), for a confirmation of this perspective from the standpoint of modern physics!

The trick here is to understand Cassirer's "Theory of Symbolic Forms" in terms of Swabey's characterization of him as having created an honest-to-god epistemological theory of relativity. The keys words here are "epistemological" and "relativity".

"Relativity" in a *scientific* sense means a rigid translation of invariants. "Epistemological" means how we describe the world. The key to understanding Cassirer is that he asserts that we have some kind of a constant set of invariants across *all* viable epistemological descriptions[81] of reality. This is where we are coming to and it is the crucial point.

This is the theme that we will revisit in the summation at the end of the chapters 9 and 10: -in the characterization of "the interface" and the "contemplate your navel" sections of it. It leads into Swabey's characterization. Cassirer asserts that there is, in fact, a set of invariants, but those invariants cannot be definitively described from any *particular* perspective. That is, they cannot be exclusively described from any single particular *epistemological* perspective —not even from mathematical physics! This is the "contemplate your navel" part of my answer. (See also the "Where Cassirer and I Fundamentally Differ" heading in Chapter 12).

In the next chapter I will explore the other axiom of realist reason, the Axiom of Experience, and conclude my answer to the epistemological problem I have raised. Quine and Cassirer show the way. This will then allow a brief and succinct statement of my third and final thesis in Chapter 11.

---

[81] At least for those that qualify as "science"!

Chapter 7: Cognition and Experience

Quine and Cassirer
(The Epistemological Problem: *What* do we know?)

Let me begin this crucial chapter by repeating my quote from W.V.O. Quine presented in the Preface in its entirety:

> "The totality of our so-called knowledge or beliefs, from the most casual matters of geography and history to the profoundest laws of atomic physics or even of pure mathematics and logic, is a man-made fabric which impinges on experience only along the edges. Or, to change the figure, total science is like a field of force whose boundary conditions are experience.
>
> A conflict with experience at the periphery occasions readjustments in the interior of the field. Truth values have to be redistributed over some of our statements. Reevaluation of some statements entails reevaluation of others, because of their logical interconnections- the logical laws being in turn simply certain further statements of the system, certain further elements of the field. Having reevaluated one statement we must reevaluate some others, which may be statements logically connected with the first or may be the statements of logical connections themselves.
>
> But the total field is so underdetermined by its boundary conditions, experience, that there is much latitude of choice as to what statements to reevaluate in the light of any single contrary experience. No particular experiences are linked with any particular statements in the interior of the field, except indirectly through considerations of equilibrium affecting the field as a whole.......
>
> Furthermore it becomes folly to see a boundary between synthetic statements.. and analytic statements...Any statement can be held true come what may, if we make drastic enough adjustments elsewhere in the system... Conversely... no statement is immune to revision... even

the logical law of the excluded middle... and what difference is there in principle between such a shift and the shift whereby Kepler superseded Ptolemy, or Einstein Newton, or Darwin Aristotle?"[1]

"Experience"! I have argued it as an axiom of sanity, and a minimal assumption of realist reason. But *what* is it and what does it mean? Is it the same as "sensuous impressions"? Does the posit of absolute experience demand an immediate further commitment to reference?

In this chapter I will examine these questions in the light of Quine's and Cassirer's ideas, (and, of course, of Maturana's), and conclude that the answer to each is "no". I will propose an answer of rigorous and scientific epistemological relativism[1], (a delimitation and extension of Cassirer's), which preserves both the phenomena *and the validity of the whole dialogue of Naturalism*, (including, therefore, that of my first two theses), -but *as organization* only!

It will preserve them without a commitment to metaphysical reference however. "Experience", I will argue, is identifiable with *exactly that* which remains (relativistically) invariant under all consistent and comprehensive worldviews.[2]

Experience is the *phenomena* we must preserve and account for, but it is not the specific *organization* by which we do so. The primitives of a given organization are not legitimized, therefore, on the basis of reference, but on a (relativistic) basis of empirical adequacy.

In the previous chapter, I began a discussion of cognitive closure and asserted an "Axiom of Externality". In this chapter I will continue with the issue of closure and confirm the second necessary, (apodictic), realist prerequisite of cognition, i.e. the

---

[1] This is an *explicit meaning* of the term. I do not mean by it a cultural interpretation, but rather a scientific epistemological relativism totally bound to the mathematical invariants of the phenomena as interpreted under varied scientific perspectives —see later and especially Chapter 9.
[2] Which is essentially a restatement of Quine's position as just re-iterated

Chapter 7 — Cognition & Experience

"Axiom of Experience". Quine's epigram illuminates both. It validates an absolute and ineradicable *multiplicity* of interpretations for both scientific experiment and ordinary experience as well.

To start, let me propose a somewhat whimsical fantasy[3] but which I think nevertheless clarifies the relationship between knowledge, cognition generally, and "experience". It will suggest a viable working definition of the latter.

A Fantasy:

The remote and newly discovered atoll of Petrolia, deep in the south pacific islands and never before touched by modern civilization, was visited by a geological survey party. It was found to lie above enormous undersea oil reserves. Its King and high priest, a primitive but highly intelligent man, asked to see *our* "magic".[4] Seeking to humor him, (and, I am ashamed to tell, selfishly induce him to assign drilling rights to an American company at a ridiculously low price), he was given a "red carpet" tour of the Supercollider Accelerator, our greatest scientific marvel.[II]

The king was mightily impressed. He saw "magical worms", (traces on oscilloscopes), "dancing arrows", (pointers on analog gauges), and tiny "animal tracks", (particle tracks under a microscope), in this "cavern of the gods". He was convinced that the whim of *our* gods provided the "magic", (the "physical laws"), of his experience there, as it, (they), seemed quite different from his own!

He subsequently engaged in a long and heated debate with one of the technicians over the *significance* of it all, ending, sad to say, with his casting a set of boar's knuckles and a shrunken head, (hidden in a bag under his robe), onto the cable-strewn floor with *disastrous* consequences!

---

[3] otherwise termed: a "thought experiment" = gedanken experiment
[4] He was awed when watching reruns of "Gilligan's Island" on the exploratory party's television!

Though whimsical, this little fable helps to clarify the purest, (weakest), and the minimum, (necessary), assumption of "experience". There are clearly aspects of the situation that the king may have considered significant, (i.e. explanatory), that the scientist did not, (and conversely). The color or shape of an instrument, or the particular way the technician cleaned his glasses before initiating the experiment, for instance, are things that the king might have considered as ritual, (or physical), necessities, essential to the result.

Even the number of floors of the facility, the time of day, or the route by which he entered might actually be relevant.[5] The technician, of course, considered the king's multicolored ritual headdress, and his pouch of magic bones totally irrelevant, (the king was doing *his* best to be of help).

What I will call the "abstract frame" of the experiment he witnessed, however, *was absolutely the same* for him as for the scientist conducting it. The abstract frame, (the total data and Quine's "boundary condition"), for both the scientist and for the King of Petrolia was identical with the *abstract*, (from interpretation),[6] of the *whole* of the actual experiment itself –i.e. *the whole* of the experimental situation! ("Experiment" is clearly an extension, albeit a refined and defined one, of "experience" itself.)

The "abstract frame" *must include* the "background situation" however, i.e. *all* the details –to include the observers! We do not know, a priori, *which* of these or *what* of these is relevant. This is one reason why, (other than the issues of personal integrity or error), experiments must the reproducible.

It is to eliminate unique factors deriving from the particular experimental context, (e.g. a magnetic field from the coffee-maker, a power surge from the factory down the block, the crumb from an assistant's lunch contaminating a culture, …), and

---

[5] Consider the fate of a Shaman making a mistake in the rigid ritual of a "Rain dance"!

[6] alternatively, the *experiential or the phenomenological invariant*

to isolate the essentials through a multiplicitous duplication, hopefully random regarding what is (unknowably) extraneous that it must be repeated.

We are never on certain ground in that process however. We are never sure that our historically dictated -and contextually limited- design of an experiment does not implicitly incorporate such factors, or that there are not *broader*, (or different), frames, isolating, (or incorporating), *other* factors as incidental and irrelevant, (or pertinent and important), in which it could be implemented.[7]

Following Quine, we are in a process of dynamic reorientation only *bounded* by the abstract frame! *Any* theoretical description really compatible with the overall experimental situation[8], however, is clearly a *legitimate*, (i.e. logical, "legal"), interpretation in Quine's broad sense!

Consider: was the King of Petrolia's interpretation of the data of the experiment into *his* theoretical scheme, (worldview), *patently* false? Not necessarily, according to Quine. Was the scientist's translation into "laws of physics", "particles of matter" - or as an expression of the "primitive building blocks of reality" *inherently*, (i.e. logically), better? *Also* not necessarily!

Each could use the data to integrate, reinforce or modify his theoretical basis -his world-view. Even the *cumulative* body of scientific experiment can be accounted for by the King. Given an unending stream of counterexamples, he can, via Quine, incrementally account for each. The presumption that this *cumulative* body of counterexamples rules out any other consistent world-view, that eventually he will be backed into a contradiction is not justified.

---

[7] The lack of free ferrous iron in ordinary differential bacteriology plates when looking for Legionnaire's Disease was an example of a too limited context and was the reason for its long mystery. Penrose's "Anthropic Principle" cites the extreme case.

[8] including one which might dissolve -i.e. redistribute- but exhaustively account for- the apparent relationality of our primitives. Virtual systems clearly suggest a new logical possibility.

This is *not* to say that any consistent theory is just as good as any other consistent theory. The King's theory, spirits and witchcraft,[9] let us say, while it may very well be consistent and capable of accounting for any given fact, clearly falls far short in many aspects, perhaps the most important of which is predictability.

The scientist will make strong and definite projections into the future which, by and large, will be clearly and precisely confirmed. He will be able to predict wide ranges of phenomena correctly and efficiently. There are other criteria of good theories as well. Roger Penrose, in his "Emperor's New Mind" has outlined a reasonable standard very concisely.[III]

The issue, which I will postpone for a little, is whether there cannot be, under the thesis of epistemological relativism which I will assert shortly, *multiple* equipotent, comprehensive and *equally* "SUPERB"[10] theories of reality, (using Penrose's classification).[11] The proven equivalence, for example, between Heisenberg's, Schrödinger's and Bohr's (widely divergent) conceptions of quantum mechanics seems to imply that this may be the case.

The fable, (in concert with Quine I maintain), helps us to see that "experience" *as such* is not, (a priori *or* a posteriori), identifiable with any *particular* one of its specific organizations or orientations. (Hilbert claimed as much for pure mathematics![12]) Rather, experience must be identified with the *invariant relationality* -i.e. with that which remains fixed- under *all* global, comprehensive and (even barely) consistent worldviews.

---

[9] Was ancient Greek mythology so very different from this?

[10] Penrose's CAPS

[11] I think Penrose's criteria, and "theoretical beauty" explicitly expose a new dimension on physical science itself. In this "fantasy" the King is continually forced into *ad hoc* positions which considerably lessen his credibility. I think the same may be said of Quantum Physics beyond a certain point —see especially Chapter 14 and Appendix D.

[12] See Chapter 2

The Axiom of Experience:

"Experience", (*tentative working definition*), then is "that", (elephants, atomic bombs, watercress sandwiches,,....), for which both the king and the technician *must account* in *some* manner![13] It is not *itself* an orientation, however. It is, rather, *that* ("thing and its relationality"[14]) *which must remain fixed*, and I argue that it is a *primitive* of reason.[15]

Scientific experiment extends, (generates), experience and thereby bounds (and shapes) the scope of consistent theories. It adds new invariant relationality to the abstract frame, (and the history of abstract frames). Following Quine however, it never *determines* them!

The Epistemological Problem:

At the conclusion of Chapter 4, I asserted the definition: The mind is the "bio-logical", (i.e. as reduced in the context of materialism), "concept" of the brain. (Alternatively, mind is the rule[16] of the brain.) This scientific conclusion, (and the schematic model), of my first chapters, however, raises profound philosophical and epistemological difficulties, seemingly contradicting itself.

It raises questions, moreover, which offend the very foundations of our rational sensibilities. This, however, is not so unusual a circumstance but has *always* been the case, historically, at the major turning points of science. Deep progress has *always* necessitated radical, (and often distasteful), reorientations, (rather

---

[13] This identifies, I propose, a viable and legitimate -and theory independent- working definition of experience.

[14] Alternatively, "the phenomena".

[15] In Chapter 13 I will argue that it is identifiable with our evolutionarily fixed, decoherent artifacts of the schematic model in the brain.

[16] following Cassirer's redefinition of the formal "Concept" as the *rule* of the "functional concept of mathematics" as examined in chapters 3 & 5, coupled with Maturana's profound insight which we examined in the preceding chapter.

than mere polishings), of our fundamental worldview -often with the loss of our cherished convictions.[17]

Most recently, this is seen very clearly at the invocations of Relativity and Quantum Mechanics in modern physics which, incidentally, raise much the same sorts of questions as does my thesis, i.e. "realism vs. empiricism/algorithmic" questions. I urge that the problems raised by my thesis are not inherently more difficult -or of a radically new and different type- than have been raised, (and answered), before in the cause of science.

The real issue is *productivity* -to whose ultimate judgment I hereby submit my thesis. It is to legitimize and justify my conclusions, however, that I am forced to philosophy and a study of the metaphysical and epistemological presumptions of science *itself* -and there *are* such!

---

[17] Here was a negative and sarcastic reaction from a 16th century non-astronomer to Copernicus' revelation as cited by Kuhn:

"Those clerks who think (think how absurd a jest)
That neither heav'ns nor stars do turn at all,
Nor dance about this great round earthly ball;
But th'earth itself, this massy globe of ours,
Turns round-about once every twice-twelve hours;
... So should the fowls that take their nimble flight
From western marches towards morning's light, ...
And bullets thundered from the cannon's throat
(Whose roaring drowns the heav'nly thunder's note)
Should seem recoil; since the quick career,
That our round earth should daily gallop here,
Must needs exceed a hundred-fold, for swift,
Birds, bullets, winds; their wings, their force, their drift,
   Arm'd with these reasons, 'twere superfluous
T'assail the reasons of Copernicus;
Who, to save better of the stars th'appearance,
Unto the earth a three-fold motion warrants"

Kuhn, Thomas "The Copernican Revolution" Harvard Press, 1957
–pps 189-190 Originally from Francis R. Johnson : "Thought in Renaissance England"

Though admittedly painful, how are the epistemological implications of my thesis so much more difficult than those of modern physics, for instance? At the scale of the very small and at the scale of the very large, modern physics says that our physical world is *profoundly* strange and, at the small scale at least, that the picture of science is essentially algorithmic.

My thesis proposes *that our human scale world is very much the same* - that it is *itself* a biological and organic algorithm! It is an *internally* (and *virtually "tactile"* algorithm[18] wherein the "data" we receive and the "instrument" we manipulate to control it are one and the same.[19] Its elements *themselves* however, are purely and abstractly logical, (alternatively "operational"), elements taken in the sense of Hilbert's profound insight!

This is a very different and radical way to look at our "objects", (to include perceptual objects), to be sure.
It is, I believe however, far more compatible with the outlook of modern physics than is ordinary Naturalism. I maintain that our very perceptual "tactile", "spatial", "extensive" et al. "objects" themselves are *logical*, (alternatively, clearly now –since Maturana- "operational"[20] *rather than representative*! But the "logical" here is that of a (Kantian) "constitutive logic" via the "concept of implicit definition" as presented in Chapter 5, rather than one of ordinary classical abstractive logic, (i.e. rather than one of an "associationist logic" –using Dreyfus' term again).

There are really two problems involved with the mind-brain problem. There is a scientific and empirical one, and there is a philosophical and metaphysical one. The combination of my first two theses actually solves the scientific problem I argue, and my third thesis will explicate the metaphysical and philosophical problems. This chapter will resolve just the apparent paradox

---

[18] such things, e.g. tactile VR, are currently being actively explored and developed in the field of virtual reality! See, for instance: http://science.nasa.gov/science-news/science-at-nasa/2004/21jun_vr/
[19] This, of course, is a reference to "The Engineering Argument" of Chapter 4.
[20] See Chapter 4 –"Logic as Biology"

created by the first two hypotheses, i.e. their specifically epistemological problem.

I shall now propose a specific answer to the problems which I have raised. My philosophic answer will lead, (in Chapters 8 through 10), to a plausible and pointed answer to the question of what I propose to be the actual substance of mind itself.

Let me emphasize, however, that my real and central claim remains the scientific one, i.e. the result of the combination of my first two theses; –my philosophic answer is solely its rationale. It is, I think however, a very good one.

Suppose that my scientific conclusion were true, (and I believe the concordance of my first two theses, amongst numerous other reasons, strongly suggests it is), then there seems to be an inherent paradox in knowledge itself, –and in my (tentative) Naturalist premises themselves! If both our perceptual and intellectual objects are solely artifacts of *biological coordination*, then on what ground can knowledge, (and *my own* argument itself), stand? If the *very language*, (to include the very "biological coordination" and "evolution" of my argument),[21] in which I describe the problem, (being *part* of that self-same human reality), is only internally organizational and not referential, –then *what* is it that I am describing? How can I even discuss the problem itself?

Doesn't my theory actually "eat" itself? *How, then, could there be science at all?* Notwithstanding the apparent paradox, (which is not unique to my thesis[22] and to which I will here

---

[21] I will repeat a footnote I inserted much earlier in book: I think it would be wise to explicitly state that in any discussion such as this, where the very meanings of *all* the common terms are questioned, that you must assume just about every term as being in quotes. In Dennett's terminology, every term must be interpreted heterophenomenologically. This is the meaning of the "Kantian Caution" in the "Note for Impatient or Skeptical Minds" of my opening pages!

[22] This problem is inherent in pretty much the same terms in the whole of Kantian and Neo-Kantian philosophy of science, and in the philosophical dilemmas of modern physics as well. I urge that my solution, in a form very close to that offered by Cassirer, fits with the whole of modern science in a way that

propose a solution), I maintain that mine is a very strong and a very pure Naturalist argument and that its conclusion, *as such*, is valid.[23]

Chapters 1 through 5 might be considered as a constructive reductio ad absurdum of the absolutist Naturalist premise, (though not of its relativized equivalent). Chapter 6 is a direct argument to the same effect, building on Kant and Maturana. Less kindly, they might be considered as constituting a "straw man". Combined, however, they are much more powerful than that as they actually do resolve the whole of the Naturalist dilemma, (other than the epistemological one I just raised), and explicate the actual mind-brain problem in absolutely legitimate and empirically promising Naturalist terms. Clearly, there might be something wrong with the Naturalist program, but need it be *fatal?*

My argument turns now then, not to argue against the whole sense of Naturalism, but against the *part* of it I believe is flawed. I base those arguments in an extension of Kant's,[24] and, ultimately, in a delimitation and extension of Cassirer's Neo-Kantian position, i.e. of his "Theory of Symbolic Forms". The thrust is to split Naturalism from its over-strong metaphysical presumptions.

---

none other does. Dogmatic materialism, on the other hand, leads to a linguistic idealism —that somehow our automaton-generated language leads to truth. But then *what* is truth? Is it then transformed to the automaton pragmatic truth of William James? (See Chapter 12) Is it only the"cash value of an idea"?

[23] See the Anthropic Principle discussion in Chapter 12!

[24] Kant's work was concerned *primarily* with the problem of cognition and therefore has a special relevance here.

"This is an advantage no other science", [than epistemology], "has or can have, because there is none so fully isolated and independent of others *and so exclusively concerned with the faculty of cognition pure and simple.*" Prolegomena, P.131

Cassirer Revisited:

My prior arguments do not, however, reduce the system of Naturalist *organization*, (i.e. its predictive schema), to absurdity, (nor, therefore, the corresponding organizational, i.e. Naturalist, validity of my own first two theses which are framed within it), but only its claim of absolute, (i.e. metaphysical), *reference*.[25] Nor do they question the profound *effectiveness* of Naturalist science.[26]

Cassirer suggests a way to preserve that overwhelmingly successful relationality, (i.e. the predictive efficacy), of Naturalism in a *relativized* sense, not as reference, but *as organization*, i.e. in his thesis of rigorous and scientific epistemological relativism.[27] He proposes Naturalism, (and materialism[28]), as just one (among several) of the possible *-and equipotent-* "Symbolic Forms" comprehensively *organizing* the whole of experience.

It is only *experience itself*,[29] (the phenomena), that is preserved as a known metaphysical absolute and to which (relativized) reference can be made. Under Cassirer's "Symbolic Forms", "experience", (Naturalist connotations notwithstanding),

---

[25] again, at *whatever* level of "sophistication" the latter is postulated

[26] The Naturalist *organization* can be taken *within* contemporary anti-realism, (i.e. anti "scientific-realism" -the position that scientific theories do not directly describe ultimate, metaphysical reality). I am making a distinction between *naturalist organization* and *naturalist metaphysics*. Cassirer I believe, like Van Fraassen, is essentially an antirealist. This is not so surprising, given the fact that they both have Kantian roots, (cf., for instance, Van Fraassen's "Laws and Symmetry".) I will most definitely *not* argue in favor of Naturalism, (i.e. metaphysical naturalism ==scientific realism), but will argue for the (relativized and equipotent) naturalist organization. I will argue, therefore, for the structure, but not the reference of that organization. I call it my conclusion a *"relativized naturalism"*. But isn't this just "structuralism" again?

[27] Cassirer's is clearly a mathematical perspective, with its roots in modern algebra. See Appendix E: "Cassirer speaks to Shapiro & Mac Lane".

[28] as embodied in mathematical physics

[29] Experience is not necessarily, therefore, the same as its ordinary organizational Naturalist interpretation, as "sense impressions". Nor, under my thesis, does experience *refer to* externality. It is an expression of process.

must not be confused and identified with its characterization under any *particular* one of the possible symbolic forms however.

It is the confusion of a *particular* "frame of reference", (i.e. symbolic form -and the assumption that there is only one comprehensive frame possible[30]), with the *invariant relationality of experience* in the abstract, -i.e. *under all consistent "Quinean" frames* that is the heart of the issue. It results in a confusion of a *specific organization* of experience with the experience itself[31] which is organized.

It results in an improper assignment of unique metaphysical reference rather than a legitimate judgment of empirical, (i.e. experiential), adequacy for the primitives of the theory. Cassirer's reformulation of the formal logical concept, (and his "new form of consciousness" based in "ordering"[32] within it), allows a new logical possibility and an escape from the dilemma.

Just as Einstein relativized measurement and disembodied the ether, so did Cassirer argue in his "Theory of Symbolic Forms" for a relativization of knowledge itself, and a disembodiment of direct reference.

But Cassirer's is not a frivolous, laissez-faire relativism, (nor is it solipsism); it is an explicit and technical *epistemological* relativism rigorously grounded in the phenomenology of science.[33] What, exactly, is the length of a rod to a physicist? It

---

[30] i.e. Naturalism = scientific realism

[31] to include scientific experiment as an extension of ordinary experience

[32] I have broadened the concept of "ordering" itself in view of Hilbert's profound idea and within a structuralist perspective as I will discuss later.

[33] Why is Einstein not saying that *any* measurements, (whatsoever), would be valid? Why is Einstein's *itself* not a laissez-faire *physical* relativism? It is because there is a *rigid* structure at the core of his assertion -i.e. the specific, (and precise), invariant equations of relativity. It is the rigid and invariant "equations", (alternatively "the topology", so to speak), of experience that structure valid theories. These "equations", this "topology", must be retained as invariant(s) under all viable theories. This is why neither mine, nor Cassirer's, is an irenic relativism.

depends on the measurements, the frames of reference and the *absolute equations* of the theory of relativity relating them!

What is the relevance of a theory, (including a scientific one)? It depends on the experience, the particular "symbolic form", (e.g. physics/biology/chemistry/... Naturalist science), and the (absolute/invariant) relations, ("equations" -i.e. the web of implication), which must be preserved in it. (See the Rosen letter of Chapter 3, or the notion of "mathematical ideals" elaborated in Chapter 9 for examples.)

The Axiom of Experience

What is constant, under all frames, are the invariants, (in a mathematical sense), which must be preserved in them, i.e. "experience" —which I have identified with that *"somewhat" which must remain fixed*. I argue that it is a *primitive* of (intentional) realist reasoning itself.

"Experience", I claim again, is *"that"* for which *both* the King and the technician *must account* in some manner! It is not *itself* an orientation per se, however. Scientific experiment extends, (generates), experience and thereby bounds and shapes the scope of subsequent consistent theories. It adds new invariant relationality to the abstract frame, (and the history of abstract frames). Following Quine however, it never *determines* them!

I argue therefore for a working (and non-referential) definition of "experience": as *that* which must be maintained under *all* comprehensive worldviews.[34]

But what exactly could a materialist's relativized *"substance"* be then? What could Naturalism's *"material"* be under such a conception?

It would be an implicitly defined *term*, (alternatively a "symbol"), under a particular interpretation -i.e. it would itself be

---

[34] Though this is clearly somewhat circular, it is perfectly consistent with my assertion that "experience" is, in fact, an epistemic primitive. Afterthought: look again at Bell's "local mathematics" and "invariants".

an "object" implicitly defined by the "generating relations" of the science which specifies it.

Even materialism need not, therefore, *necessarily* carry a metaphysical commitment. It is, rather, an organization of experience using the (implicitly defined) terms/symbols of "substance". It was the classical, "associationist"[IV] logic of reference unquestioningly incorporated therein which made it seem otherwise.

Cassirer's Theory of Symbolic Forms, an Analysis:

Cassirer suggested a new way to look at the relation between theory and experience. He proposed a rigorous *epistemological* relativism innate in the phenomenology of modern science. [35]

> "Mathematicians and physicists were first to gain a clear awareness of this [the] symbolic character of their basic implements. The new ideal of knowledge, to which this whole development points, was brilliantly formulated by Heinrich Hertz in the introduction to his 'Principles of Mechanics'. He declares that the most pressing and important function of our natural science is" [simply] "to enable us to foresee future experience"[V]

It is the *method* by which it derives the future from the past which is significant, however. We make "inner fictions or symbols" of outward objects, and these symbols are "so constituted that the necessary *specifically logical* consequences[VI] of the images are always images of the necessary natural

---

[35] An absolute epistemological relativism is a very difficult conception to understand and accept. I spent about 6 complete months while I was in my early 20's trying to understand it and explore the possibilities it exposed. But please note how well it fits with the perspective I propounded in Chapter One of the brain as machine. But also remember that it is grounded in the preservation of invariants *across* the various forms!

consequences of the imaged objects".^{VII} But this analysis –and "image"– must be interpreted carefully:

> "… [though] still couched in the language of the *copy theory of knowledge* –… the concept of the 'image' [itself] had undergone an inner change. In place of the *vague demand* for *a similarity of content* between image and thing, we now find expressed a highly complex *logical* relation, a general intellectual *condition*, which the basic concepts of physical knowledge must satisfy.^{VIII} "^{IX}

Its value lies "not in the *reflection* of a given existence, *but in what it accomplishes as an instrument of knowledge,*"^X, in a unity of phenomena, which the phenomena must produce out of themselves."

Heinrich Hertz formulated the distinction very succinctly:

> "The images of which we are speaking are our ideas of things; they have with things the one essential agreement which lies in the fulfillment of the stated requirement", [of successful consequences], "but further agreement with things is not necessary to their purpose. *Actually we do not know and have no means of finding out whether our ideas of things accord with them in any other respect than in this one fundamental relation.*"^{XI} [36]

A system of physical concepts must reflect the relations between objective things and their mutual dependency, but, Cassirer argues, this is only possible "in so far as these concepts pertain *from the very outset to a definite, homogeneous intellectual orientation*",^{XII}. It is only within a distinct and specifically *logical* framework that these "images" are significant at all![37]

---

[36] Note the connection to my conception of the "strategic brain"!

[37] Please note the similarity of this situation, as formulated by Hertz and Cassirer, with that I laid out in Chapter 4 for the training seminar. The objects,

He continues:

> The "object" cannot be regarded as a "naked thing in itself", *independent* of the essential categories, (and framework), of natural science: "for only within these categories which are required to constitute its form can it be described at all."
>
> This broadening of perspective, (and it is a *genuine* "Copernican Revolution" in Kant's sense), necessitates and validates Cassirer's conclusion of the innate *symmetry* and *a relativity of interpretations* for phenomena. "With this critical insight ... science *renounces its aspiration and its claim to an 'immediate' grasp and communication of reality*."[XIII]

> "It realizes that the only objectivization of which it is capable is, and must remain, *mediation*," [my emphasis]. "And in this insight, another highly significant" [critical] "idealistic"[38] consequence is implicit. If the object of knowledge can be defined only through the medium of a particular logical and conceptual structure, we are forced to conclude that a *variety of media*", [my emphasis], "will correspond to various structures of the object, to various meanings for 'objective' relations."[XIV] [39]

This is the assertion of symmetry and the foundation for his thesis of "Symbolic Forms".

("images"), in a very real sense, are a function of the calculus. Insofar as they are justified, it is on the conjoint basis of utility.

[38] *Everywhere*, where Cassirer uses "idealism", it must be understood as "critical idealism" in the sense that Kant used it. This is very different from ordinary or Berkelian idealism, and, as I discussed in Chapter 4, is a real misnomer. I have suggested "ontic indeterminism" as a more appropriate alternative, and one I think both Kant and Cassirer would have been happy with. Also compare the "mere X", (below), with my discussion earlier.

[39] Think about Hilbert's "beer mugs" and the "Pythagorean Theorem" discussion in the opening chapters!

"... *Even in 'nature'*,⁴⁰" [my emphasis], "the physical object will not coincide absolutely with the chemical object, nor the chemical with the biological –because physical, chemical, biological knowledge *frame their questions* each from its own particular standpoint and, in accordance with this standpoint, subject the phenomena to a special interpretation and formation.⁴¹ It might also seem that this consequence in the development of" [critical] "idealistic thought⁴² had conclusively frustrated the expectation in which it began. The end of this development seems to negate its beginning –the unity of being, for which it strove, threatens once more to disintegrate into a mere diversity of existing things. The One Being,", [i.e. the metaphysical object], "to which thought holds fast and which it seems unable to relinquish without destroying its own form, *eludes cognition*!"ˣᵛ

It is the *phenomena*, (experience), *not reference*, however, that becomes the fulcrum of, (and reunifies), this relativity of perspectives. The particular symbolic forms do not *refer to* (metaphysical) reality, (their objects are not "*images*" of reality), instead they *organize experience*.⁴³

Metaphysical reality becomes "a mere X"! "The more its metaphysical unity as a 'thing in itself' is asserted, the more it evades all possibility of knowledge, until at last it is relegated

---

⁴⁰ i.e., "science" as opposed to the "cultural forms" –see discussion later.

⁴¹ But even within Cassirer's primary "natural forms" –in physics, for instance, I argue –beyond Cassirer– that the exact parallel obtains. There are arguably alternative Hertzian formulations of the problem. Alternative objects and alternative calculuses, (sic), are possible. Fine suggests that Relativity and Quantum Mechanics may represent such alternatives, and certainly Schroedinger's and Heisenberg's conceptions of quantum theory illustrate the plausibility.

⁴² See prior definitional footnote re: "critical idealism"

⁴³ refer back to the prior quote from Hertz

entirely to the sphere of the *unknowable*[44] *and becomes a*[45] *mere 'X'"*, [my emphasis].[XVI] "It is the realm of *phenomena*, "the true sphere of the knowable with its enduring multiplicity, finiteness and relativity", on which we stand.

It is the (multiplicitous and relativized) *organizations* of the phenomena, *not reference* to a metaphysical origin, which lies at the basis of knowledge.

> "And to this rigid metaphysical absolute is juxtaposed the realm of phenomena, the true sphere of the knowable[XVII] with its enduring multiplicity, finiteness and relativity."[XVIII]

But this reorientation does not destroy either the unity or the coherence of knowledge.

> "But upon closer scrutiny the fundamental postulate of unity is not discredited by this *irreducible diversity"*, [my emphasis], "of the methods and objects of knowledge; it merely assumes a new form. True, the unity of knowledge can no longer be made certain and secure by referring knowledge in all its forms to a 'simple' *common* object which is related to all these forms as the transcendent prototype to the empirical copies." [my emphasis][XIX]

(This latter demand is, of course, the rationale of the dogmatic Naturalist claim of reference.)

> "But instead, a *new* task arises: to gather the various branches of science with their diverse methodologies - with all their recognized specificity and independence - into one system, whose separate parts precisely through their necessary diversity will complement and further one

---

[44] See Chapters 13 & 14 for d'Espagnat's and Bitbol's perspectives on ontic unknowability
[45] (Kantian)

another.⁴⁶ This postulate of a purely functional unity *replaces the postulate of a unity of substance and origin, which lay at the core of the ancient concept of being.*"ˣˣ

This is a profound expansion of the very conception of "relativity" itself far beyond any other ever proposed! (I will shortly propose yet another expansion to a still wider conception.)

Cassirer conceives his "symbolic forms" *functionally*, (and serially), i.e. in terms of the "mathematical concept of function".

> "And this creates a new task for the philosophical critique of knowledge. It must follow the special sciences and survey them as a whole. It must ask whether the intellectual symbols by means of which the specialized disciplines reflect on and describe reality exist merely side by side *or whether they are not diverse manifestations of the same basic human function.* And if the latter hypothesis should be confirmed, a philosophical critique must formulate the universal conditions of this function and define *the principle* underlying it."ˣˣᴵ ⁴⁷

Instead of dogmatic metaphysics, *"which seeks absolute unity in a substance to which all the particulars of existence are reducible"*, he seeks after "*a rule* governing the concrete diversity of the functions of cognition, a rule which, without negating and destroying them, will gather them into a unity of deed, the unity of a self-contained human endeavor."ˣˣᴵᴵ [my emphasis]⁴⁸

---

⁴⁶ Note: This is *Cassirer's* "principle of complementarity" —see Appendix D, (Niels Bohr) —for an argument focused on a profound similarity of their epistemological perspectives.

⁴⁷ This is one of the explicit purposes of my present book.

⁴⁸ Cassirer extends his theory of symbolic forms *beyond "nature"*, (i.e. beyond the sciences), into the "cultural forms": art, myth, religion, etc. -i.e. beyond cognition itself. I will deal with this aspect of his thesis presently, taking a neutral perspective, but first I would like to extend and modify this, his core and scientifically grounded position somewhat.

Perhaps the most succinct overall statement of Cassirer's thesis is found in his "Einstein's Theory of Relativity".[XXIII] Each of the perspectives of scientific knowledge: physics, chemistry, biology, ... (the "cognitive forms"), - and ultimately myth, religion and art, ... (the "cultural forms" which I may perhaps begin to question),[49] are taken as alternative and equipotent *organizational* perspectives on the phenomena. They are in a way like the Hilbertian "beer mugs" and "Pythagorean theorems" of Chapters 2 and 3, or like the alternative models of the universe of the Rosen letter discussed much earler.

> "Each of the original directions of knowledge, each interpretation, which it makes of phenomena to combine them into the unity of a theoretical connection or into a definite unity of meaning, *involves a special understanding and formulation of the concept of reality.*"[XXIV]

Repeating myself, ordinary Naturalism confuses a *particular* organization, (mathematical physics), with the phenomena themselves which are organized. That is the basis of its assertion of reference -and its "scientific realism"[50]. "The "objects", (the organizational primitives -i.e. "images"), of one *particular* form are assumed, (incorrectly), to *reference* ontology - to relate to "an ultimate metaphysical unity".

> "Where there exist such diversities in fundamental *direction* of consideration, the *results* of consideration cannot be directly compared and measured with each other. The naïve realism of the ordinary view of the world, like the realism of dogmatic metaphysics, falls into this error, ever again. It separates out of the totality of possible concepts of reality a single one and sets it up as a norm and pattern for all the others. *Thus certain necessary formal points of view, from which we seek to judge and*

---

[49] I will question but not refute the eventual scope of his vision presently
[50] another misnomer

understand the world of phenomena, are made into things, into absolute beings". [my emphasis]^{XXV} [51]

What these "formal points of view" do, *instead*, he argues is organize, (i.e. "order"), the phenomena. What is consistent under *all* forms, however, are the phenomena themselves.

Yet again, Naturalism confuses a *particular* "frame of reference", i.e. "form",[52] with the *invariant relationality of experience* in the abstract –i.e. under *all consistent frames*.[53] It confuses a specific organization, (and a specific characterization), of experience with the experience itself[54] which is organized. It results, (and I repeat myself), in an improper assignment of unique metaphysical reference rather than a legitimate judgment of empirical, (i.e. experiential), adequacy for the primitives of its theories.

> "Only when we resist the temptation to compress the totality of forms, which here result, into an ultimate metaphysical unity, into the unity and simplicity of an absolute 'world ground' and to deduce it from the latter, do we grasp its true concrete import and fullness. No individual form can indeed claim to grasp absolute 'reality' as such and to give it complete and adequate expression. [my emphasis]"^{XXVI} [55]

This is Cassirer's claim of epistemological cognitive relativism –Cassirer's *own* "principle of complementarity"! But

---

[51] Naturalism, at *whatever* level of sophistication, clearly falls under this injunction.
[52] and assumes that there is only one comprehensive frame *possible* –i.e. that of Naturalism
[53] compare Van Fraassen's "co-ordinate-free descriptions". "Quantum Mechanics: an Empiricist's View"
[54] to include scientific experiment as an extension of ordinary experience
[55] Please see my mathematical ideals discussion of Chapter 9 for a petty lucid explanation of this idea.

his denial of "completeness", and "adequacy", (i.e. *epistemological completeness* and *epistemological adequacy)*, however, is not the same as denying that any individual form can grasp the whole of the phenomena *comprehensively*. I argue that a form can be *both* comprehensive and adequate without being unique![56] This is one of the lessons we got from Hilbert, (and from Quine as well).

Nor does Cassirer speak definitively on the issue of reduction! I will address both of these issues shortly.[57]

> "It is the task of systematic philosophy, which extends far beyond the theory of knowledge, to free the idea of the world from this one-sidedness. It has to grasp the *whole system* of symbolic forms, the application of which produces for us the concept of an ordered reality, and by virtue of which subject and object, ego and world are separated and opposed to each other in definite form, and it must refer each individual in this totality to its fixed place.
>
> If we assume this problem solved, then the rights would be assured, and the limits fixed, of each of the particular forms of the concept and of knowledge as well of the general forms of the theoretical, ethical, aesthetic and religious understanding of the world. Each particular form would be 'relativized' with regard to the others, *but since this 'relativization' is throughout reciprocal and since no single form but only the systematic totality can serve as the*

---

[56] This is the lesson we learned from the King of Petrolia.

[57] If a given form were, in fact, capable of reducing all other theories, and no other could, it would obviously cut against equipotency and "relativization" –i.e. against the whole sense of his thesis! This is the current rationale for dogmatic Naturalism as grounded, (problematically, I believe), in mathematical physics. A likely candidate would be the biologist Maturana's alternative perspective which is one of the progenitors of my own thesis.

*expression of 'truth' and 'reality*⁵⁸", [my emphasis], "the limit that results appears as a thoroughly immanent limit, as one that is removed as soon as we again relate the individual to the system of the whole." ˣˣᵛᴵᴵ

(At this point I would definitely refer you once again to my discusssion of "mathematical ideals"⁵⁹ which is the subject of Chapter 9 and which is a much simpler perspective towards an understanding this idea!)

Cassirer's "Symbolic Forms" is not a *capricious* relativism however; it is a relativism as rigorous in concept as is Einstein's – or as is Hilbert's. Just as Einstein characterized his theory as having removed "the last remainder of physical objectivity from space and time", Cassirer's conclusion removes the last remainder of metaphysical, (i.e. absolute), *reference* from knowledge!

It is based in the essential methodology of science: in its (Hertzian) theorizing *function*! It is the nature of science to construct a form, complete and interdependent between symbols, ("images"), and a "calculus" which acts as a whole.⁶⁰

Under all the forms, (of "nature", at least), Cassirer maintains that what must be maintained are the "invariants" -i.e. that which must be preserved under any consistent form –i.e. the invariant phenomenal relationality. These are not "things" or "images", but rather, (mathematically), that which remains constant under all epistemologically legitimate forms. In the sense which I will expand the notion, I argue that it corresponds to my prior (relativized) definition of "experience".

> "But above all it is the general form of natural law which we have to recognize as the real invariant and thus as the real logical framework *of nature in general*ˣˣᵛᴵᴵᴵ......No sort of *things* are truly invariant, but always only certain

---

⁵⁸ This is the rationale for my later claim that no single form may adequately describe "the interface". See Chapters 8 & 9!

⁵⁹ See Chapter 9

⁶⁰ cf. the "training seminar" of Chapter 4

fundamental relations and functional dependencies retained in the symbolic language of our mathematics and physics, in certain equations." [XXIX] (I will qualify this assertion slightly in Chapter 12.)

I will postpone my critique of Cassirer's thesis for a little. Though I think there are problems and questions which need to be resolved, I would like to make the connection to my own thesis before going into those. In its essence, i.e. the essential relativism of knowledge, and his case against reference, I think his argument is very strong and very fundamental. There are very strong questions and delimitations that I will raise when I return to Cassirer's broader thesis later. They will not, however, question this, his core position.

Chapter 8: Cassirer's "Symbolic Forms"

(The Solution to My OWN epistemological Dilemma –and a *Profound* Change in Perspective!)

Nowhere does Cassirer question the profound *effectiveness* of modern science, however. His orientation is wholly and profoundly scientific. Rather, he preserves the various sciences as *perspectives*,[1] specifically as *organizations* of phenomena. He has, moreover, provided the very tools necessary to resolve the epistemological dilemma created by the combination of my first and second theses!

I therefore propose a fundamental, (and final), "Copernican Revolution" -a profound change in perspective- contrary to that of the Naturalist perspective which I conditionally adopted at the end of Chapter 6[2], (but with perfect legitimacy, I now maintain -as a beginning *relative* stance), and *to* the perspective I will ultimately proclaim as the *ultimate realist perspective*!

This perspective "reduces" the materialist position *itself* to organization (taken quite specifically in Quine's sense as cited in the very Preface to this book), and *not* to reference! I argue *against* ordinary Naturalism, and for a more sophisticated realism, (essentially a *biologically* Kantian-Cassirerian-Maturanian-Freemanian one), consistent with the results of the first two theses.[3] By this, (once again), I do not mean to say that the

---

[1] You might compare my usage here with Neil Bohr's perspective on "complementarity" as discussed in Appendix D

[2] And in the very first chapter of the present MS –in the "Nutshell précis "

[3] Kant's thesis is profoundly difficult to accept admittedly, both intellectually and intuitively -but so was Einstein's. Where Einstein relativized the physical world, Kant sought to relativize the epistemological one. His lapses can be assigned to his deprivation of the examples of modern mathematics and modern science - which subjects were always his primary focus -and which could have corrected

relationality of Naturalism, (or of Naturalist science), is faulty, (on the contrary it is SUPERB[4]), but that its metaphysical reference *as reference per se* is faulty.

My thesis, though built with Naturalist "bricks", does *not* therefore entail the further and unnecessary Naturalist "foundation" of reference. Though it assumes the validity of the Naturalist organization, (at least on the human scale), it does not assume the metaphysical reality of Naturalism's primitive "material", ie. of its "substance". (I have called this "relativized Naturalism").

In questioning our actual, (referential), cognition of metaphysical reality, my thesis is not, therefore, innately self-contradictory! Though stated in Naturalist terms, it can legitimately question the actual (metaphysical) existence, (or even the very *possibility* of knowledge), of the referents of those terms!

Ordinary Naturalism, though it will not say so, is through and through grounded in a specifically metaphysical dogma, i.e. absolute reference, (howsoever "sophisticated" it may claim to be), to *absolute*, (rather than to relativized), "material" which latter it equates with "substance". This is the "material" in "materialism",[5] and was the specific target of Kant's and Cassirer's profound arguments.[6]

As realists, on the other hand however, (and I speak to no one else), we *must* posit the existence of an absolute, external reality.[7] It is, I have argued, an intentionl axiom at the very foundation of realist reason itself! But, I further argue based on Kant, on Cassirer, on the advances of modern physics, on Maturana's penetrating analysis and on the results, (and natural

---

him. That he was two hundred years before his time is surely not an argument against his credibility.

[4] Using Penrose's classification again –his CAPS

[5] as usually conceived -i.e. not in a Cassirerian sense

[6] Again I might refer you to d'Espagnat's "multitudinism" discussion in Chapter 13.

[7] See Chapter 6

concordance), of my first two theses[8], that human cognition does not know, and *can never know* that absolute reality! (There is still "nobody home" in the materialist sense.) I argue we cannot know that *metaphysical* world in itself, even in "sophisticated" reference.

I propose that we stand, *even at the human scale*, in the same relation to ontology that current physics does, (at least as I understand, let's say, Bohr's, Heisenberg' and d'Espagnat's positions[9] to be). I propose that our human scale cognitive world is as much -and *as solely*- a pure algorithm as is the worldview of quantum physics. I argue that it is utilitarian and not referential.[10] But it is an organic, "tactile"[11] algorithm, (a "GUI"), that evolution constructed.[12] This sentence, however, is no longer paradoxical. It must itself now be understood in my larger context, -as the very "evolution" in it is *itself* relativized, (i.e. it is a relative assertion *within* the particular and *relativized* Naturalist form).

The results of my first two theses are therefore consistent under this epistemological rationale. The resolution lies in the scientifically and mathematically, (but most certainly *not arbitrarily*), conceived relativization of knowledge itself!

---

[8] See especially the heading "Turning our Perspective Around —a Model of Process!" in Chapter 4

[9] —and Penrose's which is not far from it.

[10] I will refer you once again to the section in Chapter 4, "Turning our Perspective Around" as I think it is highly relevant to this question. We do incorporate metaphysical "objects", but rather our "objects" are the implicitly defined objects of functioning itself, *not objects of reference*!

[11] See, for instance: http://science.nasa.gov/science-news/science-at-nasa/2004/21jun_vr/ for an article on research into "tactile virtual reality".

[12] This is the implication of my remark in Chapter 4. Let me repeat it here: Ideally instrumentation and control would unify in the *same* "object". We would manipulate "the object" of the display itself and *it* would be the control device. Think about this in relation to our ordinary "objects of perception" -in relation to the sensory-motor coordination of the brain and the problem of the utility – the "why" of naïve realism! We do not *use* our biological algorithm, we *live* in it!

Relational implications, predictive systems, (to include scientific theories), are not, (with Quine), epistemologically determinate. Rather, their essence, (which is their predictivity), can be isolated, (following Cassirer), as relational invariants, (the latter word taken in a specifically mathematical sense),[13] over the field of consistent hypotheses in a sense parallel to that in which Einstein's equations of special relativity were isolated as invariants from the "ether" in which they were originally grounded by Lorentz.[14]

Or, rather, relational implications are invariant, but predictive *organizations*, (i.e. theories and epistemologies), even comprehensive ones, are not![15] They are the (better or worse), "SUPERB" or "MISGUIDED"[16] "forms" which organize those implications.

Whence Cassirer's Thesis:

There is, interestingly, a very real similarity of intent at least, (if not in scope or rationale), between Bas Van Fraassen's "co-ordinate free" and "semantic" approach to modern physics and Cassirer's "symbolic forms".

> "To formulate a view on the aim of science, I gave a partial answer to the question of what a scientific theory is.

---

[13] Discussing Hilbert, Cassirer says:
> "The procedure of mathematics here", (implicit definition), "points to the analogous procedure of theoretical natural science, for which it contains the key and justification. But above all it is the general form of natural law which we have to recognize as the real invariant and thus as the real logical framework *of nature in general*......No sort of things are truly invariant, but always only certain fundamental relations and functional dependencies retained in the symbolic language of our mathematics and physics, in certain equations."

[14] The paradigmal case is, of course, that of Galileo's laws of motion.

[15] Quine and that very wise man, the "King of Petrolia" that we met in Chapter 7 taught us this!

[16] cf Penrose "The Emperor's New Mind" (his CAPS!).

... It does not follow that a theory is something essentially linguistic.

That we cannot convey information, or say what a theory entails, without using language does not imply that -after all, we cannot say what anything is without using language. We are here at another parting of the ways in philosophy of science. Again I shall advocate one particular view, the *semantic view* of theories. Despite its name, it is the view which de-emphasizes language."[1]

"Words are like coordinates. If I present a theory in English, there is a transformation which produces an equivalent description in German. There are also transformations which produce distinct but equivalent English descriptions. This would be easiest to see if I were so conscientious as to present the theory in axiomatic form; for then it could be rewritten so that the body of theorems remains the same, but a different subset of those theorems is designated as the axioms, from which all the rest follow.

Translation is thus analogous to coordinate transformation -is there *a coordinate-free*", [invariant?], *"format as well?*"[II] The answer is *yes* (though the banal point that I can describe it only in words obviously remains)."[III] [17]

Though Van Fraassen ultimately rejects axiomatics, and confines himself to the domain of physical science, his position in regard to modern physics has a very definite resemblance to that of Cassirer, at least insofar as the latter is confined to "nature". Each is epistemologically relativistic,[18] and each is grounded in

---

[17] See Chapter 9

[18] "There are a number of reasons why I advocate an alternative to scientific realism ... One concerns the difference between acceptance and belief; reasons for acceptance include many which *ceteris paribus*, detract from the likelihood of truth. This point was made very graphically by William James; it is part of the legacy of pragmatism. The reason is that, in constructing and evaluating theories, we follow our desires for information as well as our desire for truth. We want

invariants. Van Fraassen rejects axiomatics, (which I believe is the most cogent formulation of the problem), however, on the basis of a *need for meaning and interpretation*, i.e. reference.[19]

Van Fraassen goes on:

> "To show this, we should look back a little for contrast. Around the turn of the century, foundations of mathematics progressed by increased formalization. Hilbert found many gaps in Euclid's axiomatization of geometry because he rewrote the proofs in a way that did not rely at all on the *meaning* of the terms (point, line, plane, ...). This presented philosophers with the ideal: a pure theory is written in a language devoid of meaning (a pure *syntax*) *plus* something that imparts meaning and so connects it with our real concerns."[IV]

My thesis of the "schematic object", (Chapter 4), and the first three chapters of this book are directed *precisely* to that point, (to his "*plus* something that imparts meaning and so connects it with our real concerns[20]), -and denies it! It is precisely my point, (and was Hilbert's as well), that "meaning" be taken in its *mathematical* sense for such a system. A mathematician understands the "meaning" of a term to be *precisely* that which is implied by the syntax, i.e. it is a virtual term "ordering" the *whole* of the system in which it is defined. (Note: This, I have and will argue, is a specifically *structuralist* conception!) But there is no

---

theories with great powers of empirical prediction. For belief itself, however, all but the desire for truth must be 'ulterior motives'." (ibid p.3) Please note the connection to the essential Hertzian perspective. "Information" is concerned with predicting future events; "truth" is something else altogether.

[19] Hilbert's "concept of implicit definition", combined with Cassirer's "Symbolic Forms" is my answer to his objection. See the Moritz Schlick quote about Hilbert re: "meanings" as cited in Chapter 1.

[20] Unless we were to interpret his "something that imparts meaning and so connects it with our real concerns" simply as pragmatic strategy.

"plus" behind or after it. It is the purely mathematical Concept, (and conception), which is our new focus.

If the mind and perception specifically, (the phenomena), is taken in this sense, as "ordering"[21] biological process- if it is taken as an organization, and its terms as metaphors[22] of that organizational process then there is no longer the *metaphysical* question of meaning or of reference —Maturana's "structural coupling"[23] does not allow it! The terms, instead, mean precisely what the syntax implies -i.e. they are virtual terms *only*! (But our artifacts, those "terms" now clearly reflect a "best fit" evolutionary solution[24] for the specifically *strategic* brain.) I maintain these actually *are* our "real concerns".

The deeper problem is the one that Cassirer defined: i.e. that of "experience" itself and how theoretical science relates to it,[25] -and *that* involves a total reevaluation of the problem of reference.

---

[21] i.e. serving as "positions in a structure"

[22] In the sense of mathematical structuralism

[23] See Chapter 6

[24] organizing around "atomic" process rather than around "atomic" information —see Chapter 4: "Turning our Perspective Around"

[25] Theory, (seen as a Hertzian, free construct -as developed in the last chapter), must match, (in *some* sense), the "topology", so to speak, of temporal and spatial consequence in experience. As stated thus far, this idea is, of course, Kantian. Russell however, (in his "Foundations of Geometry"), argued to extend the Kantian frame to projective geometry. I feel it must be broadened again past that -past even topology and into the mathematics of abstract transformations. What is required is that the predicted results of the theoretical system (through *some* transformation!) must match the results of naïve (?) experience, -*and conversely*! That is, the results of naïve experience -through *some* (mathematical) transformation - should match the retrodictive predictions of the theory. But this transformation, (since, in light of W.J. Freeman's findings in Chapter 4, it is *past* topology), it need not preserve "objects". It need not, therefore, preserve *reference*! What its Hertzian premise demands is that it *must* preserve the web of relationality in its most abstract sense. It must preserve the "abstract frame".

Cassirer's epistemology, of course, is firmly grounded in axiomatics. Discussing Hilbert, Cassirer says:

> "The procedure of mathematics here", (implicit definition), "points to the analogous procedure of theoretical natural science, for which it contains the key and justification."[v]

Contra Cassirer: (What are the real parameters?)

Though I accept, (and argue), Cassirer's core position of epistemological relativism specifically as regards cognition and the sciences, (I believe it is absolutely warranted on the very pure and very strong phenomenological grounds[26] wherein he evolved it), I will now question and ultimately refine his extension of its scope and its applicability.[27] What *are* the legitimate forms?

Cassirer's thesis goes beyond "cognition" and science, ("nature") into a symmetry of cultural forms, (to include science as a special case), as well. Van Fraassen does not, nor did Kant, (who remained entirely within "nature" in his core thesis), but this is a question of scope. There is also a question of the identification of the legitimate (primitive) forms –even within "nature" itself.

Before addressing these questions, however, let me first complete my examination of the broadest formulation of Cassirer's thesis –though I will ultimately refine its focus.

Going beyond the "natural forms", (physics, biology, chemistry, etc), he extends his thesis into ground which I must at least begin to question. He proposes that the forms of "nature", of "cognition", are only part of the innate symmetry of perspectives across the whole of the phenomena. They, (the natural forms), represent those forms which relate phenomena directly to a *metaphysical*, (cognitive), framework. He asserts however that the phenomena can be organized[28] on *other* grounds: art, myth,

---

[26] cited in the preceding chapter
[27] See "The Base Issue" section at the end of this chapter.
[28] Please note his usage of "the phenomena *can be* organized" –see later

religion, etc., "*but they achieve this universal validity by methods entirely different from the logical concept and logical law*". (Note: it is his "entirely different" that I will question.)

> "*But again our perspectives widen,*" [i.e. beyond "nature" and into the purely cultural forms], "if we consider that *cognition,*" [itself], "however universally and comprehensively we may define it, *is only one of the many forms* in which the mind can apprehend and interpret being. In giving form to multiplicity it is governed by a specific, hence sharply delimited principle.
>
> All cognition, much as it may vary in method and orientation, aims ultimately to subject the multiplicity of phenomena *to the unity of a 'fundamental proposition.'* The particular must not be left to stand alone, but must be made to take its place in a context, where it appears as part of a *logical structure*, whether of a teleological, logical, or causal character.[29] Essentially cognition is always oriented toward this essential aim, the articulation of the particular into a universal law and order."[VI]

(I disagree with his distinction —I believe that so too can the "cultural forms" embody law.[30] The difference is in the *orientation* –i.e. from *cognition* –to "externality" as world-ground. Any form, even the "cultural forms", will have, (by definition), its own sense of law and logical structure. It is a question of the meaning of "logical structure" –just "structure" by itself, I suggest, is a better formulation.)

> "But beside this intellectual synthesis, which operates and expresses itself within a system of scientific concepts, the life of the human spirit as a *whole knows other forms.*

---

[29] Note: this is a reiteration of his "mathematical concept of function and a reference to its necessary "rule".

[30] I will develop this possibility shortly in an expansion of the possibility of new and different cognitive forms which I will suggest.

They too can be designated as modes of 'objectivization': i.e., as means of raising the particular to the level of the universally valid; but they achieve this universal validity by methods entirely different from the logical concept and logical law.

Every authentic function of the human spirit has this decisive characteristic in common with cognition: it does not merely copy but rather embodies an original, formative power. It does not express passively the mere fact that something is present but contains an independent energy of the human spirit *through which the simple presence of the phenomenon assumes a definite 'meaning', a particular ideational content.*"[VII]

But please note carefully that *all* of Cassirer's "functions of the human spirit" -even his "cultural forms" specifically articulate *phenomena* -i.e. they are not free, "idealistic" constructs, (i.e. of Berkelian idealism) but "...an independent energy of the human spirit through which the simple presence of the phenomenon assumes a definite 'meaning', a particular ideational content". They are, rather, objects of the mind![31]

"This is as true of art as it is of cognition; it is as true of myth as of religion. All live in particular image-worlds, which do not merely reflect the empirically given, but which rather produce it *in accordance with an independent principle.*"[32]

Each of these functions creates its own symbolic forms which, if not similar to the intellectual symbols, enjoy equal rank as products of the human spirit.

None of these forms can simply be reduced to, or derived from, the others; each of them designates a particular

---

[31] Please note his specific use of the word "meaning", and then consider it from Hilbert's perspective!

[32] That is, *in accordance with "a rule"*!

approach, in which and through which it constitutes its own aspect of 'reality'. *They are not different modes in which an independent reality manifests itself to the human spirit, but roads by which the spirit proceeds towards its objectivization, i.e. its self-revelation.*"[VIII] [33]

I will repeat a paragraph from the previous chapter here - the paragraphs above are his claim of epistemological relativism!

But Cassirer's denial of "completeness", and "adequacy" are correctly to be understood as denials of *epistemological* completeness and of *epistemological* adequacy, these being taken at the highest level of philosophical overview.[34] To deny epistemological completeness or epistemological adequacy for any given form is not the same as denying that any individual form can grasp the whole of the phenomena *comprehensively* -for a form can be both comprehensive and adequate without being unique! This is one of the lessons we got from Hilbert, from complex transformations, and which comes from the cross-reductions ubiquitous in modern mathematics. Nor does Cassirer speak definitively on the issue of reduction! I will address both of these issues shortly.[35]

That he claims that "none of these epistemological forms can simply be reduced to, or derived from, the others" seems to provide an essential argument to dogmatic Naturalism.[36]

---

[33] This point will be considered again within my next book which I optimistically hope to complete before I pass over "the great divide"!

[34] I will explore, and perhaps modify my views on this subject in "The Base Issue" conclusions section of this chapter. This is a very deep and subtle problem and demands a precision of language not appropriate right here.

[35] If a given form were, in fact, capable of reducing all other theories, *and no other could*, it would obviously cut against equipotency and "relativization" -i.e. against the whole sense of his thesis! This is the current rationale for dogmatic Naturalism as grounded, (problematically, I believe), in mathematical physics —see Chapters 13, 14, and Appendix D for my counter argument. A more likely candidate, I believe, would be the biologist Maturana's alternative perspective.

[36] But you must understand that he is talking about the epistemological forms t*hemselves* here, not their content.

Conversely, I will argue that it suggests and delimits a more correct extension of Cassirer's solution to the overall problem. I will address these very large problems very shortly.

His meaning must be examined very closely:

> "If we consider art and language, myth and cognition in this light, they present a common problem which opens up new access to a universal philosophy of the cultural sciences.[IX]

> "The 'revolution in method' which Kant brought to theoretical philosophy rests on the fundamental idea that the relation between cognition and its object, generally accepted until then, must be radically modified.

> Instead of starting from *the object*", [my emphasis], "as the known and given, we must begin with the law of cognition, which alone is truly accessible and certain in a primary sense; instead of defining the universal qualities of *being*, like ontological metaphysics, we must, by an analysis of reason, ascertain the fundamental form of *judgment* and define it in all its numerous ramifications; only if this is done, can objectivity become conceivable.

> According to Kant, only such an analysis can disclose the conditions on which all *knowledge* of being and the pure concept of being depend. But the object which transcendental analytics thus places before us is the correlate of the synthetic unity of the understanding, an object determined *by purely logical attributes.* [37]

> Hence it does not characterize all objectivity as such, but only that form of objective necessity which can be

---

[37] Here I think that Cassirer made a distinct mistake: he utilized "logical" in the classical sense instead of viewing it as a manifestation of his *own* "mathematical concept of function" which is much broader than that!

apprehended by the basic concepts of science³⁸, particularly the concepts and principles of mathematical physics. ..."ˣ

Cassirer asserts, beyond this, an absolute "*spiritual*" relativism, (but *always* articulating the phenomena), -i.e. an absolute symmetry across the whole of the "cultural forms", (across the "spirit"), of man.

> "There result here not only the characteristic differences of meaning in the objects of science, the distinction of the 'mathematical' object from the 'physical' object, the 'physical' from the 'chemical', the 'chemical' from the 'biological', but there occur also, over against the whole of *theoretical* scientific knowledge, other forms and meanings of independent type and laws, such as the ethical, the aesthetic 'form'.³⁹
>
> It appears as the task of a truly universal criticism of knowledge not to level this manifold, this wealth and variety of forms of knowledge and understanding of the world and compress them into a purely abstract unity, *but to leave them standing as such.*"ˣᴵ

Though starting from very stable ground, I think that Cassirer ended up in a somewhat ambiguous position. He, like Kant, used words with great precision,⁴⁰ so he must be read very carefully -even technically. "Nature", and "the forms of nature", for Cassirer, are *technical words.*

He defines the "forms of nature" for us -e.g. physics, biology, chemistry. These are some of the "values" of his specific *function*, (his "purely functional unity"), of the human spirit,

---

[38] Please note this specification which is my own as well!

[39] Please note his use of "other forms and meanings *of independent type and laws,* such as the ethical, the aesthetic 'form' as against his prior "But the object which transcendental analytics ...is an object determined *by purely logical attributes.*" cited just previously.

[40] I think it is a necessary concomitant of the very abstract nature of their ideas

(here specifically the cognitive forms). A philosophical critique "must formulate the universal conditions of this function and define the principle underlying it."

We must place this passage in the context of Cassirer's *redefinition* of the formal Concept itself however. We must see it in the context of "the mathematical concept of function" to understand it. The various forms are functional "values" –in a technical mathematical sense –of a definite, and, for Cassirer, *serial* ordering, (and principle). They are the alternative orderings of the phenomena, (defined by a serial function), –and constitute a series of series.

The phenomena, however, remain always the orientation –the focus –of *all* the forms, (even the "cultural forms"). There is in these citations no assertion of comprehensiveness, (and even a seeming denial of it), for any given form however. He seems to argue against reduction,[41] (and therefore comprehensiveness), as well –but against "reduction" and "comprehensiveness" in *what senses?*

It is against *epistemological* reduction and *epistemological* comprehensiveness, taken at the highest level of philosophical consideration that he argues, but not against the comprehensive adequacy of a given perspective as regards the phenomena.

Compare:

(1) "none of these" [epistemological] "forms can simply be reduced to, or derived from, the others",[XII]

(2) "no individual", [epistemological], "form can indeed claim to grasp absolute 'reality' as such and to give it complete and adequate expression."[XIII], and

(3) "each particular form would be 'relativized' with regard to the others, but since this 'relativization' is throughout reciprocal and since no single form but only the systematic totality can serve as the expression of 'truth' and 'reality', the limit that results

---

[41] "None of these forms can simply be reduced to, or derived from, the others"

appears as a thoroughly immanent limit, as one that is removed as soon as we again relate the individual to the system of the whole."[XIV]

What is the sense of Cassirer's "*cannot be simply reduced to or derived from*"? That no individual form can give "complete and adequate expression to reality" and that no form can be "simply reduced" does not necessarily imply that reduction, (i.e. translation / transformation), in a *non-simple* sense, or that *comprehensiveness*, (as a complete accounting for phenomena), is impossible *within* any given form[42]. (3), moreover, seems to contradict (1) and (2).

Consider, moreover, his "*invariants of nature*": though "no sort of *things* [his emphasis] are truly invariant, but [it is the]...fundamental relations and functional dependencies retained ... in certain equations... [which are truly invariant]".

He proposes these, (the functional invariants of these forms), as "the real logical framework of nature *in general*" [my emphasis]. But "nature" is a *pluralistic* word for Cassirer -the "natural forms" are *all the forms of science*!

We have, therefore, an assertion of invariance[43] across *all* the forms of science -and cross-reduction, (i.e. morphisms?), across those invariants.[44] Indeed, this is the only sense in which "invariance" makes any sense at all -i.e. *it is a "coordinate-free" perspective*!

"Invariance", therefore, means invariance across different, (*all the different*), perspectives of nature -and epistemologic relativity. For what *other* interpretation of the "relativization" of (3) is there except as alternative orientations of the *same* phenomena?

Consider also his seeming denial of comprehensiveness. "The 'relativization' [of forms] is throughout reciprocal". "No single form but only the systematic totality can serve as the

---

[42] Quine and the King of Petrolia would argue otherwise!
[43] of functional dependency but not of "things"
[44] See my "Rosen letter" of Chapter 3 for a beginning on this perspective

expression of 'truth' and 'reality'." What he is actually asserting, I argue, is that although multiple forms are legitimate, no single one of them can describe the structure *as abstracted from an orientation*![45] What Cassirer is portraying here is exactly a "*coordinate free*" perspective! It is not, therefore, a denial of *comprehensiveness*[46] that he is arguing, but a denial of the

---

[45] This corresponds to the concept of a "mathematical ideal" which I will present shortly in the next chapter.

[46] Comprehensiveness is, of course, a highly pertinent issue because of the very definite, (and very powerful), claim by ordinary Naturalism for just such an ultimate comprehensiveness for mathematical physics . (I will address this issue presently). This is a very strong claim, and one I think we all actually do accept - at least in principle.

However, if one particular form, (e.g. Naturalism), *is* actually capable of such comprehensiveness, (even in principle), *and no other were*, then this would constitute a very definite objection to his thesis. The question lies in the "no other were" part of the premise. Certainly Quine would argue otherwise. Cassirer believed that the only salvation for the symmetry and relativism he envisaged lay in his extension across the cultural forms:
"As long as philosophical thought limits itself to analysis of *pure cognition*, [his emphasis], the naïve-realistic view of the world cannot be wholly discredited," [I will disagree with this], "The object of cognition is no doubt determined and formed in some way by cognition and through its original law -but it must nevertheless, so it would seem, also be present and given as something independent outside of this relation to the fundamental categories of knowledge. If, however, we take as our starting point not the general concept of the world, but rather the general concept of culture, the question assumes a different form. For the content of the concept of culture cannot be detached from the fundamental forms and directions of human activity: here 'being' can be apprehended only in 'action'." [Note the connection to Maturana!]

I believe the *actual* salvation of his thesis and the guide to its extension lies in the idea of *converse* -i.e. mutual reduction. If his basic conception is right, and I think it is, (on phenomenological grounds), then multiple *cross-reductions* and a true relativism will be possible. The possibility is founded in the conception of alternative axiom systems, (and orientations), in formal mathematics, in the developments in the foundations of mathematics presented at the outset of this paper, and in my extension of Cassirer's reformulation of the formal logical concept into the Concept of Implicit Definition, (C.I.D.)!

Note, December, 2011: A thought just occurred to me. If we were to begin within a genetic, or within a *biochemical* or *hormonal* cognitive perspective, (as I will shortly propose) —as a viable symbolic form, it might

(metaphysical) adequacy and the *sole* truth of any *particular epistemological orientation*, [symbolic form].

It is only in their multiplicity that he believes that they express "'truth' and 'reality'". "The limit that results appears as a thoroughly immanent limit, as one that is removed as soon as we again relate the individual [form] to the system of the whole."[xv] (Again, see my discussion of "mathematical ideals"'" in Chapter 9.)

If the cognitive forms are "the real logical framework *of nature*", (i.e. of the natural sciences), and they are invariant across *all* the forms of nature, then all the forms of nature are, by implication, *cross reductive and comprehensive*! That these forms themselves cannot be "*simply.*", (epistemologically), "reduced to, or derived from the others", does *not* mean, therefore, that their "objects" cannot be reduced *as transformations* or derived *at all*![47] This is what "invariants" are all about.

It is *cross-reduction* and relativistic invariance which tie the forms together and it is only in their totality that they express reality -and experience. The mathematical axiom system will serve to illustrate the case again. That any (adequate) axiom system for a given discipline will be comprehensive is, of course, clear by definition. But to confuse the discipline itself with any one of the particular, (of many possible), adequate axiom systems, is incorrect. Peano's system *per se* is not the same as the positive integers.[48] (A more specific and perhaps a more elegant tool for illustrating Cassirer's conception, lies the mathematical notion of "ideals" in abstract algebra, (when properly considered in this light). (I have referred to this a couple of times and will present it shortly, with illustrations, in Chapter 9.)

Cassirer is asserting alternative functional orientations across the phenomena in his thesis of "Symbolic Forms". Each

---

produce something equivalent to the perspective and the need he is outlining here. But it, in itself, would become a new center, a new symbolic form exploring a wholly new and maybe more important aspect of humankind!.

[47] Think about Hilbert's "beer mugs" and "Pythagorean theorem" comments of Chapter one, and the heading "Another Look at Hilbert" in Chapter 3..

[48] See the prior quote from Van Fraassen

draws different functional, (and serial), perspectives, "diverse manifestations of the same basic human function".[49] This is an explicit invocation of his "mathematical concept of function".

I suggest, instead, an extension of it: that the objects of knowledge are constituted in different, (and alternative), "axiom systems"[50] which "crystallize" the phenomena as virtual, but probably better here as virtual *intentional* objects, (of the feedback loop), under the "concept of implicit definition". (This is certainly consistent with the Hertzian perspective, more so, I believe, than even Cassirer's interpretation.)

But I suggest that it is the phenomena *themselves* which are the actual invariants![51] It is a solution based, not in the mathematics of functions but, as Cassirer suggested often as the true focus of modern thought, -in that of the manifold itself. What results is a *true* epistemological relativity, (in a mathematical sense), and the possibility of multiple, each-truly-comprehensive and cross-reductive independent perspectives.[52]

I will leave the problem of the definition of the *actual* (valid) forms without reaching a definite conclusion at this point. Cassirer's solution is seductive, to be sure -and may very well be correct, but it is outside of the needs for my thesis. What is unquestionable, I think, is his "coordinate-free" orientation to phenomena. Such a perspective on Cassirer's Hertzian stance, narrowed to Van Fraassen's smaller physical perspective, and his case for the "forms of nature" in general, (biology, chemistry, …), will adequately serve my case. But, as Cassirer himself explicitly states, beyond that we leave the arena of "cognition" altogether.

---

[49] Also: "A philosophical critique must formulate the universal conditions of *this function* and define the principle underlying it."
[50] Alternatively, "generators of an Ideal"
[51] Are the phenomena themselves, then, invariant *equations*? No, they are what the equations embody.
[52] See the discussion of mathematical "ideals" shortly for a further elaboration of these ideas.

But cognition is precisely our area *of interest* here. Our context here is precisely that of cognition and metaphysics –and argues against the *specifically metaphysical* claims of Naturalism!

If my area of *interest* were to change –if I chose to look at "the phenomena" artistically, let's say, then this would *no longer* be my orientation, and his broader case might be argued. But then, conversely, I would *no longer* be able to express it in a cognitive context![53]

Note: December 2011: Thinking about it now, I see another possible and highly promising line which does *not* logically separate the cognitive symbolic forms, (cognition), from Cassirer's "cultural symbolic forms", but which *still preserves the invariants* –the *cross-reductivity* of his symbolic forms that I have argued as inherent in his original conception! I think it actually *fullfills* Cassirer's implicit but very clear demand for a view which preserves the *whole* of man's "self-revelation", "*the whole of the human spirit*". But that demand need not obviate the independent and complete legitimacy of cognition. I suggest that there are new possibilities within the cognitive perspective itself which make his overall perspective much cleaner and more symmetrical. If we can think about each of these new forms that I will suggest shortly specifically as independent *organizations*, each still preserving all the invariant relationality of experience, then *each* will still retain its validity over the whole of the perspectives

---

[53] An interesting and important point comes up here, however. If his broader thesis is correct, and my extension of it as well –i.e. mutual cross-reductions and comprehensiveness - then the "invariants", (if there should be such), of *those other forms* will be (reductively) retained as invariants even in the sciences! Thus, if there be absolutes, (invariants), in art, in music, in religion, then they will be *retained* as invariants even in the sciences, (in psychology, for instance). I consider this a very significant scientific conclusion, and running contrary to current social relativism. There *may be* an ultimate scientific decision possible between, let's say, John Cage and Beethoven! –Or between Zoroaster and Jesus!

on the mind –both reductively[54] and independently, (i.e. as *beginning* perspectives)!

(1) The Unicellular Perspective –a New Cognitive Symbolic Form

Think first, about a unicellular organism, (and about its specifically *genetic* inheritance). But its genome represents the incorporation of the genomes of even more primitive unicellulars, e.g. mitochondria). Then think of a unicellular or a system of such unicellulars, (a "tissue type" or subsystem of such[55]), as being subsumed into a metacellular as representing "modules" of reactivity", melded and integrated by metacellulars to ultimately form (reactive) "axioms" of process at the "tactical level".

Now further consider taking Gerald Edelman's line of "massive pruning"[XVI] during the gestation of the embryo for sorting and selection, (i.e. for appropriateness), and therefore producing, let us hypothecize still further, something like W.J. Freeman's "equivalence classes" in the central nervous system. But remember that in this hypothesis the whole process is fundamentally tactical and purely operative, embedded and predetermined, (the latter word taken in its broader sense), in the genome of the metacellular! I think this is a legitimate extension of Maturana's conception of the "autopoietic organism"[56], and *as such* stands -as did Maturana's more limited vision -as a viable symbolic form in itself as I argued earlier specifically for Maturana's particular thesis.

---

[54] "reductively" is used here in a much deeper and different sense than its current common usage, which latter derives from mathematical set theory. It is employed throughout this book, unless specifically or contextually obvious otherwise, with the meaning of "transformationally"!

[55] and histogenesis

[56] i.e. as "second order coupling" It also fits very well, I believe, with Gerald Edelman's core position –sans his "epistemological error" –see Chapter 4. Its compatibility with W.J. Freeman's position is, of course, implicit.

Let me repeat a section from Chapter 4: "Turning our Perspective Around" which mentioned the possibility:

> But it is just the *converse* of the argument made above, (i.e. "The Engineering Argument"), that I propose for evolution however. It is not the *distribution* of function, but rather the *centralization* of disparate "atomic" -*but blind*-biological function into efficacious schematic -and virtual- objects that evolution effected while compositing the complex metacellular organism. (These are clearly just the complementary perspectives on the same issue.)[XVII]

But let's talk about the "atomic" in the "atomic biological function" of the previous statement. There is another and crucial step in the argument to be taken at the level of biology. The "engineering" argument dealt specifically with the schematic manipulation of "data".

At the level of primitive evolution, however, it is modular (reactive) process that is significant to an organism, not data functions. A given genetic accident corresponds to the addition or modification of a given (behavioral/reactive) process which, for a primitive organism, is clearly and simply and *solely* merely beneficial or not. The process itself is *informationally indeterminate* to the organism itself however -i.e. the process is a modular whole.[57]

No one can presume that a particular, genetically determined response is informationally, (*rather than reactively*), significant to a Paramecium or an Escherichia coli, for example, (though *we* humans may consider it so). It is significant, rather, solely as a modular unit which either increases survivability or not.[58]

---

[57] See Maturana's "structural coupling" in Chapter 6 for a rationale.

[58] Note: This is an argument toward a specifically genetic and unicellular symbolic form, but this particular stage of the argument points to yet another level and form: i.e. a biochemical form and perspective.

Let me therefore extend the prior argument to deal with the schematic organization of atomic, (modular), process, rather than of primitive, (i.e. absolute), data. It is my contention that the cognitive model, and cognition itself, is constituted as an organization *solely* of that atomic modular process, an organization designed for computational and operational efficiency as embodied within a simplistic and optimized schematic calculus. The "atomic" processes themselves remain however, and *will forever remain*, informationally indeterminate to the organism! They are locked solely to reactive adequacy, not to knowledge.

And further:

How could evolution organize –as it *had* to organize– the reactive function of this colossus of seventy trillion cells? Even this formulation of the question disregards the yet more profound complexity of the reactivity of the individual cells –also organisms– themselves! It was the overwhelmingly crucial issue in the evolution of complex metacellulars.

That the progressive evolutionary reactivity of this megacollosus occurred under the bounds of real necessity is, of course, a given. It is the basic axiom of Darwinian "survival". But that it could *match* that possibility[59] –i.e. that it could achieve a (reactive) *parallelism* to that bound –i.e. "information!" –is a hypothesis of quite another order and teleologically distinct. It is, I assert moreover, mathematically immature.

This new idea relates the metacellular with the individual cells –but it does it within a specifically genomic and evolutionary conception/form. I believe this line of reasoning yields an entirely new and viable perspective on the evolution of the central nervous system itself –to include its "cognition"– across *all* species!

---

[59] See Chapter 6: Maturana and my arguments on the specific issue of "*congruent* structural coupling" versus raw "structural coupling".

To quote Maturana:

> That is, a recurrent or invariable correlation is established between a perturbed or sensory surface of the organism and an area capable of producing movement (motor surface), which maintains unchanged a set of internal relations in the amoeba."[XVIII]
>
> "This basic architecture of the nervous system is universal and valid not only for the hydra, but also for higher vertebrates, including human beings. ... the basic organization of this immensely complicated human nervous system follows essentially the same logic as in the humble hydra ...the nervous tissue understood as a network of neurons has been separated like a compartment inside the animal, with nerves along which pass connections that come and go from the sensory surfaces and motor surfaces.
>
> The sole difference lies not in the fundamental organization of the network that generates sensorimotor correlations, but in the form in which this network is embodied through neurons and connections that vary from one animal species to the other. ...
>
> But we emphasize: ... this is the key mechanism whereby the nervous system expands the realm of interactions of an organism: *it couples the sensory and motor surfaces through a network of neurons whose pattern can be quite varied.* Once established, however, it permits many different realms of behavior in the phylogeny of metazoa. In fact, the nervous systems of varied species essentially differ only in the specific patterns of their interneuronal networks."[XIX]

The Biochemical Symbolic Form in Cognition

Let us now think about the hormonal level. This system, (dispersed throughout the physical geometry of the metacellular organism of seventy trillion cells), has its own specific role in reactivity –but it must be understood from the standpoint of biochemistry.

You see, I believe that *even Cassirer's* scope for his "symbolic forms" was too small! Consider the possibility of a genetic, biochemical or hormonal symbolic form, standing as purely as Maturana's whose epistemological purity I discussed in Chapter 6, for instance.

How is it possible that we can believe that the brain, or more generally, a nervous system at *whatever* stage of evolution *is purely a rational thing*? Hormonal influence is very clear, (you can supply the obvious examples[60]), as is the powerful influence of street drugs… Are these then *rational*, "objectivist" perspectives, or rather, mighten they not embody –or at least permit the conception of the emotional, aesthetic … "cultural forms" *which affect* the brain's rational or extra-rational functionality[61]? Here lies a possible answer to Cassirer's dilemma.

Does this new (inclusive) cognitive perspective lessen the importance or the status of these new forms in their role as being *primary centers*, as primary and freely independent symbolic forms as well? I don't think so. It does not lessen their importance as *beginning* perspectives, *each of which* "asks its questions", echoing Cassirer's phraseology regarding the various physical sciences.

Each generates an organization of the phenomena, but *different* "objects" obtain, and lay in *different* (structural) positions from the new centers of organization. Ultimately, from our biological perspective, it becomes a question solely of their usefulness, of how well they serve *as function* in the "structural coupling" of the brain. No, let me modify that last sentence to: … how well they serve as function in the "structural coupling" *of the whole of the metacellular* –of Maturana's "second order coupling"!

---

[60] I believe that much, if not most of ongoing human society is driven by it!
[61] Consider: the "intentional" functions contain not only "belief", but "want", "need", … as well! I.e. they include desire as well!

You see, this line of reasoning has a *myriad* of extensions, I have suggested just a few, but all fall within Maturana's scope, and inherit their legitmacy, their "legality" therefrom. These are all, I claim, valid and promising symbolic forms conducive to future research. I will leave this theme temporarily, and then take it up again in the section: "The Base Issue".

These may very well be amongst the "roads by which the spirit proceeds towards its objectivization, i.e. its self-revelation. I believe that they are fully cross-reductive, but, at the same time, they are fully independent and viable symbolic forms in themselves! Cassirer's is a profoundly beautiful and elegant conception, but I feel it was limited in its scope. It is the *extension* of the scope of the cognitive forms which reorients it and allows its full brilliance to shine through. I have said it before, but let me repeat it here –that I think that Ernst Cassirer will eventually be seen to have been the greatest and most significant Philosopher of the 20th century!

The Power of Naturalism:

Naturalism, however –and on the other hand, is a profoundly *comprehensive* theory! Not only mathematical physics, but its reductive incorporation of the *other* disciplines, from biology and chemistry through (purportedly) psychology, philosophy, ethics, religion,[62] presents a purportedly complete (comprehensive) theory of *all* the phenomena.

Quine demonstrates, however, that there are *always* other interpretations of the phenomena, *no matter the level of detail*. Can there be *other* comprehensive forms then? I think the answer is necessarily yes! But need they be *physical* forms?

---

[62] The primitives of some of these forms are distributed and derivative under the reduction, however. Think about "beer mugs" and Rosen again.

The possibility of alternative, *and comprehensive*, other physical forms, certainly seems quite believable though. Heisenberg vs. Schroedinger illustrates the plausibility. Based on the sense of this current writing, I believe that Cassirer's *other* "natural forms": biology, chemistry, etc. are capable of such a legitimate extension to comprehensiveness[63] as well. I believe it is absolutely sound as demonstrated in Chapters 1 through 5 of this book. In fact, the very book you are reading right now is another such an attempt from a biological perspective.

Cassirer wrote in another era,[64] but this does not, in itself, invalidate his conclusions or their possible extension to a broader relativism. On the subject of biology, for instance, he dealt with the issues of vitalism. In modern times, however, there is a much stronger case made on much more rigorous grounds which supports the same, independent case for biology. It is that of Maturana and Varela as presented in Chapter 6. To appreciate it, it is necessary, of course, to effect the same "Copernican Revolution" which Cassirer suggested.

Maturana and Varela's case is made on very pure epistemological grounds. The biology they propound is not grounded *upon* mathematical physics however. Its primitives are not those of the latter, but rather, physics, (and human knowledge in general), is derived as a function of linguistic coupling, (third order structural coupling) –i.e. it is contained as a (non-centralized) theoretical derivative of biology's own epistemological primitives:

> "It is by languaging that the act of knowing, in the behavioral coordination which is language, brings forth a world. ...We find ourselves in this co-ontogenic coupling, *not as a preexisting reference nor in reference to an origin*, [my emphasis], but as an ongoing transformation in the

---

[63] with equivalent distributions and derivativeness of primitives

[64] though not *that* long ago!

becoming of the linguistic world that we build with other human beings."[xx]

Maturana and Varela's thesis does not find its epistemological roots in substance, but drives past its materialist beginnings to find its new epistemological center in "autopoietic unities" and "structural coupling". It ends up[65] questioning the very physical ground from which it began.

In many ways it represents the "Heisenberg" case of biology. It represents an alternative theoretical perspective on experience and on science itself. It *works* because of the purity of its phenomenology. Can other "natural forms" be asserted in this same sense?[66] Could chemistry, for instance, be stated with the phenomenological purity with which Maturana and Varela stated biology? That is the only real issue. This is Hertz's problem, after all, pure and simple. It is also the case I made for the training seminar, (the generic model), in Chapter 4.

I will not profess an *absolute* conclusion on these questions but I think the discussion we have just finished furnishes a new perspective on it —essentially validating —but refocusing Cassirer's magnificent conception. I conclude, (on Quinean grounds), that there must be, indeed, multiple possible comprehensive forms. But my conclusion in its essence, and beyond Cassirer's original idea, is a *fully* relativistic one.

The truly fundamental forms are (necessarily) *comprehensive* forms -i.e. they are fully functional alternative

---

[65] Or, rather, *should end up* —see Chapter 6
[66] Maturana and Varela reveal such an alternative orientation in "structural coupling" and "autopoietic unities". That these other "symbolic forms" *must* encompass the whole of experience, (i.e. the whole of past and future experience -to include scientific experiment), I think is incontrovertible. But they need not encompass it *in the same way* as does physics, for example. They need not encompass it as the primitive and hierarchical ground of their science, (think of Bell's perspective in Chapter 2), but may weave and distribute its relationality into a much less central, (i.e. that particular relationality removed from "axiomatic" status), much less concentrated position in its theoretical structure a la Quine. They need not adopt the primitives of another orientation as their own primitives as the latter may become "theorems"!

"axiom systems"[67] capable of exhausting the phenomena. (Alternatively, "the phenomena" is that which remains constant - i.e. invariant- under all such exhaustive perspectives.) These forms "slice" the phenomena, (*all the phenomena*), from different perspectives. To be fully relativistic, *each* form must be complete however. Though Cassirer seemed to drive towards this complete relativism, I don't think he ever completed it.[68]

But, you might object, must not a comprehensive organization *be categorical,* i.e. must there not be *only one*? (If we could achieve the Laplacean ideal, would it not be unique?)[69]

---

[67] again see later discussion of mathematical "ideals" in Chapter 9

[68] I believe because of the limitation in his formal concept

[69] The Laplacean ideal is not realist by definition.
"In the introduction to his "Theorie analytique des probabilites" Laplace envisages an all-embracing spirit possessing complete knowledge of the state of the universe at a given moment, for whom the whole universe in every detail of its existence and development would thus be completely determined. Such a spirit, knowing all forces operative in nature and exact positions of all the particles that make up the universe, would only have to subject these data to mathematical analysis in order to arrive at a cosmic formula that would incorporate the movements both of the largest bodies and of the lightest atoms. Nothing would be uncertain for it; future and past would lie before its gaze with the same clarity. ...Du Bois-Reymond elevated scientific knowledge far above all accidental, merely empirical bounds...If it were possible for human understanding to raise itself to the ideal of the Laplacean spirit, the universe in every single detail past and future would be completely transparent. 'For such a spirit the hairs on our head would be numbered and no sparrow would fall to the ground without his knowledge. He would be a prophet facing forward and backward for whom the universe would be a single fact, one great truth'." Cassirer, "Determinism and Indeterminism in Modern Physics", pps.3-4

Under a functional logic, (i.e. one not based in the generic concept), there is the possibility of alternative "axiom systems", (organizational perspectives), exposing alternative utilities, (e.g. biology, psychology, etc. -or alternative physical theories). The Laplacean ideal does not, therefore, presuppose a unique theory, (Newtonian, for instance), and reference.

If we were, in fact, to achieve a science, (theory), such that "the hairs on our head would be numbered and no sparrow would fall to the ground without his [our] knowledge", i.e., comprehensiveness, I maintain that it still not need be unique. The Laplacean ideal is not tied necessarily to Newtonian or any other

Or, conversely, could there not *still remain* alternative but still comprehensive predictive organizations with *different* perspectives and *different* utilities[70]?

Under the Aristotelian logic, and assuming comprehensiveness, (i.e. assuming the possibility of a single and complete accounting of all phenomena), there would *have to be* a linear reduction of all true theories to a single substratum of primitives.[71]

Hierarchy, (set-theoretic, type ordered inclusion), is an essential component of the existing Naturalist perspective: i.e. that there is a necessary hierarchy of spatial scale. It argues that this hierarchy is mirrored in the process of the reduction of scientific theories: e.g. biology is a subset of chemistry, and chemistry of physics. (Thus psychology and all the phenomena of experience, of knowledge, and of the "spirit" as well, are seen as embedded in that hierarchical ordering -as biological subsets.)

It presumes that our *naïve world*, (or at least most of it), is hierarchically mirrored in the primitives of any true theory, (i.e. that the "objects" of naïve realism are objects of that true theory as well). It presumes that they can be represented as legitimate and necessary groupings of those primitives. Thus our ordinary objects and the ordinary things they do are, in fact, real and ultimately *metaphysical* objects and happenings. This argument is crucial to the strength of Naturalism and its metaphysical claim!

But scale is not inherent *a priori* nor is it the *only* way to preserve the phenomena, i.e. it *need not* necessarily "cut reality at the joints". If *other* organizations, more effective, (i.e. other schematic organizations), are found, then they are legitimate as well. Our naïve objects, *as objects per se*, are not *necessarily* metaphysical objects!

---

particular theory, but constitutes the basic rule *per se* of determinism and could apply generally. (ibid)

[70] I.e satisfying different intentional goals

[71] See Appendix B: Lakoff and Edelman for a further discussion of classical logic and science

Chapter 8 — The Power of Naturalism

Science, until very recently has supported Naturalism's spatial, (and theoretical), hierarchy -from the macroscopic to the human scale to the microscopic to the atomic, (which, of course, theoretical reduction generally supports -i.e. biology -> chemistry -> physics), -or from cosmology right down through the human scale to the atomic.

At the smallest level of scale, of course, (and at the largest scale as well), the case for hierarchy has broken down in this twentieth century. As an example, let me once more cite Penrose's "most optimistic" view of quantum mechanics, (most optimistic for scientific realism, that is):[72]

> "I shall follow the more positive line which attributes *objective physical reality* to the quantum description: the *quantum state.* ."

> "I have been taking the view that the 'objectively real' state of an individual particle is indeed described by its wave function psi. It seems that many people find this a difficult position to adhere to in a serious way. One reason for this appears to be that it involves our regarding individual particles being spread out spatially, rather than always being concentrated at single points. For a momentum state, this spread is at its most extreme, *since psi is distributed equally all over the whole of space*, (my emphasis),...It would seem that we must indeed come to terms with this picture of a particle which can be spread out over large regions of space, and which is likely to remain spread out until the next position measurement is carried out...."

The particle -this *smallest part of our "object"*- is *not* included, (spatially, reductively), *within* the spatiality of the atom or within the molecule -or even within the *human scale* object of which it is the theoretical (and supposed material) foundation.

---

[72] Also see Chapter 9, d'Espagnat on "multitudinism"!

Naturalism can no longer support, therefore, a consistent hierarchy of scale!

At the human level, of course, it is a very useful tool, and that is just what I propose it is -*constructed by evolution*! Schematism, (and Cassirer's "Symbolic Forms" as well), suggest *other*, non-scaled and non-hierarchical organizations -i.e. they support *any* other truly efficacious organization. It is a simple matter of Hertzian utility.

Naturalism's primitive substratum, (the primitives of mathematical physics), is deemed unique and "true of" == "refers (isomorphically) to" ontology. It is Naturalism's epistemological basis for a claim of reference. But under a functional logic, (i.e. a logic not based in the generic concept), there is the possibility of *alternative* "axiom systems", (different functional logical concepts/theories, -*not* as class abstractions from phenomena or as hierarchical spatial perspectives into the phenomena, but as lines drawn *across* phenomena -as connective functional rules), and a different sort of "reduction", or, rather, using a more correct word instead, i.e. "*translation*", exposing alternative utilities, (e.g. biology, psychology, etc. -or even alternative purely physical conceptions).

So may we consider the new possibility that the relationality of experience, (and experiment), can be entirely preserved under varying (comprehensive) functional perspectives, no one of which stands as the canonical revelation of ontology/experience. The assertion of comprehensiveness for a given reducing theory would not then imply that it would necessarily, therefore, be *the sole and unique* organizational primitive -i.e. that it would be the only one.

This is the sense of my extension of Cassirer's "symbolic forms". I argue, with Cassirer for a relativism of forms which organize the phenomena, but, (disagreeing with him), against reference in any guise.[73]

---

[73] because he *still* incorporates a conception of "presentation", even within his theory of symbolic forms. I believe it was not "the Concept", but rather "the percept" that was hypostasized! My conception of the Concept of Implicit

It is in Cassirer's sense of the organizational, rather than the referential relevance of theories that I propose that the relations of ordinary Naturalism -and my own thesis as well- can be, (must be), retained in a deeper realism. "Experience", our true primitive, (and, I have argued, the second of the primitive and *intentional* axioms of realist reason), is not the same as any particular organization of it. It is not identical with its (legitimate but particular) characterization as "sense impressions" under the Naturalist form, for instance. I have argued a broader -and truly relativistic definition of "experience" as that which remains invariant under *all* consistent and comprehensive worldviews.[74]

What must be preserved is the web of implication of experience in our world, but hierarchy as such[75] *need not be maintained*. A comprehensive theory, ("form"), e.g. Naturalism, stands as an "axiom system" to generate the field of experience. But if *other* workable theories, (forms), and other workable "axiom systems" are found, (and Quine definitely implies their existence), also comprehensive, then the preference is no longer epistemological but utilitarian. Each, however, must fully preserve "experience" -to include the whole body of past (and future) scientific experiment.[76]

Definition, (C.I.D.), makes it possible to move beyond that error, and achieve the totality of his original beginning.

[74] But does "experience" itself absolutely, (i.e. metaphysically), refer to *something else*? My thesis proposes that it does not. I propose, rather, that it is an organization of atomic, (and indeterminate), process. It is, therefore, real and ontic, but irreducible and non-referential.

[75] Remember the work of W.J. Freeman in Chapter 4

[76] An earlier, but I think substantially true footnote: This is the point on which I question, (but do not necessarily deny), Cassirer's suggestions of the *particular* comprehensive "symbolic forms" -i.e. in that I believe that they must each embody the whole as past and future scientific experiment. In defense of his choice, however, that relationality of experiment need not necessarily be maintained as "central" to the organization of a particular form. That is, it need not lie close to its "axiomatic" base, but need only be maintained somewhere and somehow within the form as a whole. Thus biology could stand as such a "form" in Maturana's conception, for instance, wherein the experimental results of science would be maintained within third order structural coupling, for example.

The Base Issue

Here I must ask you to jump over the linear sequence of this book and refer to Appendix D for an expansion of the current dialogue, (I will summarize it below). It will give me the ground to present my final conclusions, both on Cassirer and on Niels Bohr and will allow me to summarize my conclusions on reality itself in the concluding chapters.

I think there is a striking parallelism of perspective between Niels Bohr's very early and personally original usage of "multivalued", purely mathematical functions[77], (as he utilized them in his own final epistemological perspective), and Cassirer's broader and philosphically more deeply based theory of epistemological relativity[78], i.e. his "Theory of Symbolic Forms" which we have just been exploring in depth.

From the transcript of the following audio taped interview[79], we find that Bohr was trying to approach the deepest

---

But how would science be retained in a *mythical* form, for instance? Or *language*? And yet he has touched something very powerful in both of these. That I am, as yet, unable to see the specific relevance of these suggestions does not convince me that they are, therefore, wrong! In the specific case of religion, for instance, however, I believe that Cassirer has misconstrued the problem. Let me make a counter suggestion: that religion, identified not with its ordinary practice, but with its incarnations in the "religious mystics" of the various faiths- exhibits an alternative *biological* form corresponding to the rational form suggested by Quine, i.e., one in which "ordinary objects" are no longer the organizing rationale. (cf. William James "Varieties of Religious Experience").

---

[77] like the multivalued square root function: $F(x) = x^{-2}$ over the non-negative Real line

[78] using Swabey's characterization

[79] Interview of Niels Bohr by Thomas Kuhn and Eric Rüdinger on November 17, 1962. Niels Bohr Library & Archives, American Institute of Physics, College Park, MD USA. www.aip.org/history/ohilist/LINK

problems of our mental world even as a very young man.[80] He built his original mathematical perspective on "multi-valued functions" into his eventual perspective on the deepest problems of the mind –to include cognition and epistemology.[81] He was able to conceive the possibility of *simultaneous* and *equipotent* legitimate, *but alternative*, foundational, (i.e. epistemological), perspectives on the "real world" from this viewpoint:

> **Bohr**: "I took a great interest in philosophy in the years after my [high school] student examination. ... At that time I really thought to write something about philosophy, and that was about this analogy with multivalued functions. I felt that the various problems in psychology — which were called big philosophical problems, of the free will *and such things*[82] ... that one could really reduce them when one considered how one really went about them, and that was done on the analogy to multivalued functions.
>
> If you have square root of x, then you have two values. If you have a logarithm, you have even more. And the point is that if you try to say you have now two values, let us say of square root, then you can walk around in the plane, because, if you are in one point, you take one value, and there will be at the next point a value which is very far from it and one which is very close to it[83]."

---

[80] just after he had finished his high school examinations
[81] Granted that he began with things like "free will", but the course of the interview shows how he broadened his beginnings all the way up to alternative explanations of "real things" themselves –e.g. to biology and to the wave/particle duality of light!
[82] my emphasis As mentioned earlier, he subsequently expanded this range to include cognition, epistemology, biology and the "real world" of physics.
[83] Ibid, my emphasis

COMMENTARY –to paraphrase and clarify: consider the function $f(x) = x^{-2}$ over the non-negative real line and more specifically the related functions:

$$g(x) = +|f(x)|^{84} \text{ and } h(x) = -|f(x)|.$$

Bohr visualizes tracing *continuous* curves within, alternatively, either the range of g(x) or of h(x) which lie, respectively. above and below the X axis. Each of these alternatives serves as an equipotent expression of the logical consequences, ("walking around in the plane" and the "next point very close to it" –roughly equivalent to continuity), within its particular theoretical perspective.

> **Bohr**: "If you, therefore, work in a continuous way, then you — I'm saying this a little badly, but it doesn't matter — then you can connect the value of such a function in a continuous way. But then it depends what you do. If in these functions, as the logarithm or the square root, they have a *singular value at the origin*, then if you go round from one point and go in a closed orbit" [i.e. continuously] "and [if] it doesn't go round [through] the origin, you come back to the same [value]. That is, of course, the discovery of Cauchy.
>
> But when you go round" [through] "the origin, then you come over to the *other* [value of the] function, and that is then a very nice way to do it, as Dirichlet [Riemann], of having a surface in several sheets and connect them in such a way that you just have the different values of the function on the different sheets. And the nice thing about it is that you use one word for the function, f(z).
>
> Now, the point is, what's the analogy? The analogy is this, that you say that the idea of yourself is singular in our

---

[84] i.e. the absolute, (positive), value!

consciousness then you find [85] -now it is really a formal way -that if you bring this idea in, then you leave a definite level of objectivity or subjectivity. For instance, when you have to do with the logarithm, then you can go around; you can change the function as much as you like; you can change it by 2 & pi; when you go one time round a singular point. But then you surely, in order to have it properly *and be able to draw conclusions from it*, will have to go all the way back again *in order to be sure that the point is what you started on*.[86]

Now I'm saying it a little badly, but I will go on. -That is then the general scheme, and I felt so strongly that it was illuminating for the question of the free will[87], because if you go round [through the origin], you speak about something else, unless you go really back again [the way you came]. That was the general scheme, you see".[88]

In his later application of this conception of "multivalued functions" to the specifically physical world as his "Principle of Complementarity" in quantum physics, he was able to comprehend the fundamental wave/particle duality of light. Therein Bohr incorporated that same "multi-valued" and "branching" epistemological perspective within the specific context of physics and Naturalism. But Cassirer, long before him, had gone one step deeper to incorporate such a "multi-valued"

---

[85] Free will, and consciousness are taken as his example and are the only specific referent in this interview of an application of his multivalued perspective. His elaboration immediately after it explains and expands his perspective more fully to the whole of the sciences!
[86] My emphasis
[87] Note: "Free Will" is not my focus here, it was his sole expansion of the core idea. It was his incorporation of "multivalued" functions as viable epistemological tools within the core of physics itself that gained my attention!
[88] ibid, my emphases

viewpoint into a deeper perspective -across the *whole* of the various sciences' differing perspectives on reality itself!

> **Bohr**: (expanding his perspective to the physical sciences): "… If you have such a thing like this, and you go around here, then you certainly are treating things in an orderly manner, *but you gradually get over into some other meaning of the words.* Now, I say it very badly, but that was the kind of interest [I had]. We were later on very interested in the particle-wave problem. I felt also — but not to do anything with it — that it was more so that if one created a photon, then one had made a knot in existence, a knot which was of a very difficult kind to say, and only when that photon was absorbed, annihilated, that knot was untied. … But now we know that these are solved by the non-commutation rules, and therefore, the non-commutation rules are certainly something great. *But in order to understand what they mean -you cannot get over that problem of the particle and the wave. And,* therefore, it is also so nice that this lies in the complementary description."

COMMENTARY:
Consider the meaning of the paragraph above in light of the one preceding it. "… If you have such a thing like this, and you go around here," [i.e. continuously], "then you certainly are treating things in an orderly manner, but you gradually get over into *some other* meaning of the words." But, consider further, that each of the manifestations of the multivalued functions must incorporate "the origin", "the zero" as an integral part of its own domain!

> "But then you surely, in order to have it properly *and be able to draw conclusions from it*, will have to go all the way back again *in order to be sure that the point is what*

*you started on.*⁸⁹ ... That is then the general scheme, ... because if you go round, [through the origin] *you speak about something else*, unless you go really back again [the way you came]. That was the general scheme, you see".⁹⁰

One last excerpt:

> **Bohr**, [Perhaps his most emphatic point made in this session]: "... Does [Einstein] think that, if he could prove they were particles, he could induce the German police to enforce a law to make it illegal to use diffraction gratings or, opposite, if he could maintain the wave picture, would he simply make it illegal to use photo-cells? That was, of course, in all friendliness, but it was the idea to say that this [is a] problem *we cannot get over*, and that means that actually *we got something new in the quantum*. That was the point."

He maintained this perspective continually to the end of his life, and his principle of "complementarity" is seemingly the only perspective that makes sense of quantum physics' fundamental wave/particle duality of light!⁹¹ But this perspective, in a very real sense, is just another, (though a much more delimited), way of looking at Cassirer's "Symbolic Forms" –as Cassirer had initially limited it to the physical sciences.⁹²

---

⁸⁹ My emphasis
⁹⁰ ibid, my emphases
⁹¹ Kumar's book, Kumar 2008, is the best reference I have seen on this point. It is a crucial and essential question at the very foundations of the subject.
⁹² See my headings "Cassirer's Theory of Symbolic Forms" in Chapter 7, and, more pointedly, "Contra Cassirer" in Chapter 8.

Conclusions:

Consider Bohr's schema in its most basic sense. His are multivalued functions which take a shared, unique, (single) value only at the origin, but which branch as equipotent instances beyond it as in the square-root, or in the Riemannian sheet analogies. But even restricting them to the origin, they do not even begin to constitute the basis of –a fecund analogy for– a viable epistemology, only a methodological route *towards* such a conception. But it, ("complementarity"), works as a fundamental explanatory principle in the field of modern physics! It lies at the basis of the Copenhagen Interpretation! But there is no explanatory basis other than successful consequence.

His conception, however, is very much equivalent to and provides a rationale, now found to be workable in the "real world" for Cassirer's earlier and more deeply based and philosophically legitimized epistemological center, ("origin"), of his "Symbolic Forms".[93] Cassirer provides the basis of his largely equivalent conception *for the whole of the sciences* in his considerations of Hertz's beginnings, (as cited earlier), and in his philosophical conclusions which derived from it.

Bohr's perspective, I think however, provides an actual scientific basis for an acceptance of Cassirer's thesis. It is applicable "in the real world". But that thesis, *Cassirer's own* "Principle of Complementarity", as just stated, is much broader than Bohr's –it is applicable, not just within one particular discipline, (i.e. Quantum Physics), *but across each and all of the sciences.* Repeating:

> "...But instead, a *new* task arises: to gather the various branches of science with their diverse methodologies - with all their recognized specificity and independence - into one system, whose separate parts precisely through

---

[93] i.e. Cassirer's multivalued "principle of Complementarity" based in its Hertzian premise as cited above

their necessary diversity *will complement and further one another.* This postulate of a purely functional unity replaces the postulate of a unity of substance and origin, which lay at the core of the ancient concept of being. "

I specifically used the latter as the underlayment for my (realist intentional) assertion of the ontic existence of the "interface".[94] It is this "interface" that corresponds to Bohr's "origin" and to the philosophical center, this "gathering into one system" of Cassirer. This congruence of perspectives between Cassirer, Bohr, and myself I believe confirms my *relativized*, (conditional), materialist perspective of ontic indeterminacy for the functionalist/materialist brain as a *viable* epistemological principle!

Each of these multi-valued functions <u>*must include*</u> "the origin" specifically *as part of* its domain however! This what allows us to transition with perfect *legitimacy* from one perspective to another, (in Bohr's sense)! It allows a *continuous* progression of our reasoning from one perspective to the next. The rationale implicit in Cassirer's "symbolic forms" constitutes the explanatory, (and non-"null"), basis, ("the origin"), of this continuity, and allows the subsequent unity of viable interpretations. It explains the *absolute relativity* of human cognition. It is not "empty", ("null"), because it has a profound and viable philosophical foundation based in science!

A Remarkable Parallelism

This parallelism between Bohr's and Cassirer's conceptions, and the former's crucial –and pragmatically successful– role in "the real world" of modern physics I believe, specifically validates my incorporation of an epistemologically *relativized biology* into the "real world as well! Based in

---

[94] See Chapter 10

Maturana's profound ideas, it becomes a legitimate and primal beginning perspective on the *whole* of the problem of cognition and epistemology. I believe it is the *best* perspective, (utilizing Penrose's and Einstein's aesthetic criteria for the evaluation of theories), –on the *whole* of the problem of "scientific realism" itself![95]

What I suggest, ultimately, is a *blend* of these two, very similar perspectives. Bohr, (here we might use his Riemannian Sheets as a prototype), *provides much of the scope*, while Cassirer *provides the profound philosophical rationale* –as seen, especially, from my biological perspective!

The key issue for me is that I consider Cassirer's "symbolic forms" as *fully* symmetric, and *fully* cross-reductive, and I am unsure as to whether Cassirer himself thought so. I think he (subconsciously) got trapped in the "schools" of academia[96], –in his own and Kant's supposed "critical idealism", and therefore classed himself as within a subset, (*classically defined*), of the set of all *Berkelian* idealists! Otherwise stated, he could not expand his "mathematical concept of function" to include *his own unique and brilliant conception of Symbolic Forms itself* therein. I think this must be one of the reasons for the obliqueness, the non-linearity of his writing in his "Theory of Symbolic Forms" as opposed to the linearity and concise logical flow in the writing of "Substance and Function"!

There is a distinct and more fundamental problem that arises here however. I believe that a Concept or a category *cannot see outside of itself.* That, I assert, is what it *means* to *be* a concept –or rather that is what it means to be a legitimate *fundamental* or *immersive* perspective *per se*. It *always* sees within its own perspective, its own context! But it is therefore *always* confined within that very perspective.

The classical concept is still basically set-theoretic. Philosophers, logicians, and mathematicians keep trying to get

---

[95] See Penrose, 1989, p. 421 Also see my argument in Chapter 14.
[96] See the Niels Bohr dialogue in Appendix D for a referent

around that fact, keep trying to build "upwards" from it –e.g. hierarchies of infinities, power sets, aleph null, aleph one, two, … …..

But that classical Concept *remains* fundamentally a *digital* idea, derived from and confined within extensions of the natural number sequence[97], as opposed to being founded on a continuous or "meaningful"[98] idea. The classical Concept is *still* always set-theoretically conceived and digital. It is basically still a digital idea always trying to compound from the countable rationals to something countable conceived above them. It is still confined within simple digital extensions of the countable Aristotelian "Concept[99]" –which latter was derived from *countable and finite sets*!

Cassirer's Concept which is an *actual enlargment* and a broadening of the classical Concept, retains the classical concept as a special case, but within its *own* context. That classical Concept is "visible" *within* Cassirer's immersive "Functional Concept of Mathematics" –and it is possible to "see" the former's limitations therein. *But not conversely!*[100]

Each of Cassirer's "series" may be ordered by *radically variant* principles however: "according to equality", (which is the special case of the generic concept), "or inequality, number and magnitude, spatial and temporal relations, or causal dependence" - *so long as the principle is definite and consistent.*"

But Cassirer's extension involves *any* kind of consistent rule whatsoever! Causal dependence is the "escape hatch"! His Hertzian foundation provides the rationale and supplies the sole

---

[97] to yield the rational field -which is *still* countable!
[98] "meaningful" taken in Hilbert's sense within his "implicit definition" by axiom systems rather than in the sense of modern logic
[99] See the criticisms of Cantor's conception in Chapter 3 by Poincaré, Weyl, Thurston and Kline in the "Conclusions" section. The Aristotelian Concept was derived from *finite* sets!
[100] When the latter is taken as the overriding context and which it, in practice, actually is so taken!

rule —that *of applicability* —i.e. *does it work?* [101] "The images of which we are speaking are our ideas of things; they have with things the one essential agreement which lies in the fulfillment of the stated requirement, [of successful consequences], but further agreement with things is not necessary to their purpose. Actually we do not know and have no means of finding out whether our ideas of things accord with them in any other respect than in this one fundamental relation." It is the "successful consequences" taken in its broadest and most general sense that moves Cassirer's new concept totally beyond the classical concept which preceded it!

Each "dimension" of his new Concept, in itself, is given *total* freedom —i.e. *any consistent rule*! And yet Cassirer's Concept is *still* built on the natural numbers —on its listing of its *dimensions*! F(x,y,z,..;..). But its "dimensionality" *itself* is still grounded on the natural numbers)! This is the ambiguity and the distinct limitation of Cassirer's Concept. This is where I think the young Hilbert defined a newer, larger Concept built on semantic meaning[102], and within it, a new concept of "meaning" itself!

Cassirer's is a broadening and actual redefinition of the very idea of the Concept, (it is an enlargment of the original classical Concept), but still I do not think that is the *largest Concept* -i.e. the largest *possible* conceptualization, (Concept).

The largest one, I think, is Hilbert's concept of "implicit definition" because it "understands" —i.e. "knows", (a*nd it actually does know!*), that it is self-referential and self-determining *of its own* "objects" *—they are "objects" of the whole of the system.*[103] I think it understands its expansions and its limitations as well —this is why, for instance, all proofs, (even of its very consistency, *must be relative* proofs per se! There is no way

---

[101] It is this "causal dependence", viewed from Cassirer's Hertzian perspective, which provides the "escape hatch" allowing us to go beyond hierarchy!
[102] On the line of "we need only replace 'points', 'lines' etc. with 'tables', 'chairs' or 'beer mugs' …"*and transform our theorems to match*'", [the latter is crucially important to his conception here as I have argued]!
[103] See Hilbert references to "the whole of the system".

to get beyond itself, no way to establish ontological truth, even the truth of consistency itself. This, I believe, is the actual (subtextual) subject of his debate with Frege which I cited at the very beginning of the section dealing with Shapiro in Chapter 2.

To repeat the citation of Moritz Schlick from early on,

> "[Hilbert's] revolution lay in the stipulation that the basic or primitive concepts are to be defined just by the fact that they satisfy the axioms.... [They] acquire meaning only by virtue of the axiom system, and possess only the content that it bestows upon them. They stand for entities[104] *whose whole being* is to be bearers of the relations laid down by the system."[XXI]

As stated before, I think that Hilbert gave us the gift of "meaning" itself within our conceptual world —which is nowhere accessible in the prior two definitions —either in the classical nor even in Cassirer's "Concept"[105]. To conceive it properly, (meaningfully), is essentially impossible *within* those fundamental and immersive perspectives. "Meaning" *for both* of these perspectives is always conceived as "pointing". *But to where? And how?*

In Hilbert's Concept, (and in my C.I.D.), meaning is derived internally, solely as a function of *the whole of the axiom system* —taken *purely mathematically* in the former and *purely operatively* in the latter cases!

Citing Hilbert again:

> "It is impossible to give a definition of point, for example, since only the whole structure of axioms yields a complete

---

[104] Please note his use of the word "entities"

[105] within the classical frame, or even within C*assirer's* frame, (which still incorporates "presentation" from "without", as juxtaposed to the "rule of the series" which is his "new form of consciousness")

definition. A concept can be fixed logically only by its relations to other concepts. These relations [are] formulated in certain statements [which] I call axioms, thus arriving at the view that axioms are the definitions of the concepts."[106]

And yet again:

"I do not want to assume anything as known in advance. I regard my explanation ...as the *definition* of the concepts point, line, plane ... If one is looking for other definitions of a 'point', e.g. through paraphrase in terms of extensionless, etc., then I must indeed oppose such attempts in the most decisive way; one is looking for something one can never find because there is nothing there; and everything gets lost and becomes vague and tangled and degenerates into a game of hide and seek."[107]

To conclude, once you get *inside* of a fundamental perspective, a fundamental (conceptual) Concept, *you really can't see outside of it!* Everything you consider is seen from its exclusive perspective.[108]

Chapter Conclusions:

I have proposed that our ordinary perceptual world –our innate and functional organic naïve realism is specifically an *organization* of primitive "atomic" process constructed by

---

[106] Hilbert via Shapiro
[107] ibid, my emphasis
[108] very early in my thinking about this subject, I used to phrase this as "interior to" a system of relations to refer to it,

evolution for efficiency and viability[109]. At the human scale, scientific Naturalism is an extension of that existing organization -i.e. of that which evolution has given us. But there is clearly no paradox remaining in these statements in light of the prior discussion. My thesis is, therefore, self-consistent and non-contradictory. The epistemological dilemma is resolved!

My thesis is, I believe however, *more* than consistent. Even from a purely Naturalistic perspective, I maintain that it is the only complete and consistent explanation yet offered of what it is we have set out to understand -i.e. the whole of cognition! This is *how* "a machine" could know its "objects". This is *how* a "Cartesian theatre" could exist. This is *how* there could be "meaning"!

The problem of the "Cartesian theatre", (sentiency), for instance, has heretofore either been trivialized and eliminated by ordinary Naturalism, (leading to a sort of linguistic or materialistic "idealism"), or it has been referred, for instance, to epiphenomenalism or emergence. But the latter are little more than an invocation of magic, (they do not *vivify* the ghosts they summon).

On its own grounds, I believe my *scientific* thesis stands well vis a vis its competition -it is biologically, psychologically, logically and teleologically cogent. It is, moreover, far more compatible with the epistemology of modern physics than is any other alternative -it speaks the same language.

It "covers the territory" of mind and mind-brain for the first time and assumes no "magic", (*also* for the first time). But our "ordinary objects", (the objects of naïve realism), *need not* be, (and in fact, *are not*), preserved as *metaphysical* primitives -i.e. as necessary unities within it. Quine acknowledged the possibility:

> "One could even end up, though we ourselves shall not, by finding that the smoothest and most adequate overall

---

[109] as stated in *relative* -but legitimate- Naturalist terms, i.e. within a "relativized Naturalism"

account of the world does not after all accord existence to ordinary physical things......*Such* eventual departures from Johnsonian usage[110] could partake of the spirit of science and even of the evolutionary spirit of ordinary language itself."[XXII]

This is exactly the case I have made. I argue that the "smoothest and most adequate overall account of the world" does not indeed "accord existence to ordinary physical things". My departure from Johnsonian usage does indeed "partake of the spirit of science and the evolutionary spirit of ordinary language itself".

---

[110] Johnson, once again, demonstrated the reality of a stone by kicking it!

Chapter 9: A Simpler Alternative Approach to Cassirer's Symbolic Forms: "Mathematical Ideals":

There is an easier and more intuitive approach to Cassirer's ideas and to my own, especially concerning my characterization of "the interface", (which I will explicitly define soon in the next dedicated short chapter).

That route is by employing the purely mathematical notion of an "ideal". The example given in Birkhoff and Mac Lane's "A Survey of Modern Algebra", is clearly directly applicable, (by its substance), to the immediate problem and should make Cassirer's ideas much clearer and more immediate.

The subject of mathematical ideals illustrates a very different and very concrete notion of "relativism" itself. While actually encompassing a scope much wider than simple geometry, this simple example provides a very clear illustration of what it means to be a truly and scientifically relativistic concept.

The point I will make is that *the very same object*, (in this particular example "the mathematical circle of radius 2", (taken as an illustrative token for just about any mathematical "thing"), and, in general, for I want you to consider it as a standin for human phenomena themselves -baseballs, elephants (and all the things these things do). It will show how these phenomena can be preserved in a *context-free* setting.

Try to envision "the circle C of radius 2" itself throughout the following discussion as though it were an *actual object of perception* –as an "elephant" perhaps -and consider the profound philosophical *consequences* of this conceptual re-orientation!

An Alternative Approach to Cassirer's and My Ideas: "Mathematical Ideals":

Chapter 9 "Mathematical Ideals"

"***The circle C of radius 2***"[1], [standing in place of our "object"], "lying in the plane parallel to the (x,y) plane and two units above it in space is usually described analytically as the set of points (x,y,z) in space satisfying the simultaneous equations:

(16) $x^2 + y^2 - 4 = 0, \quad z - 2 = 0.$

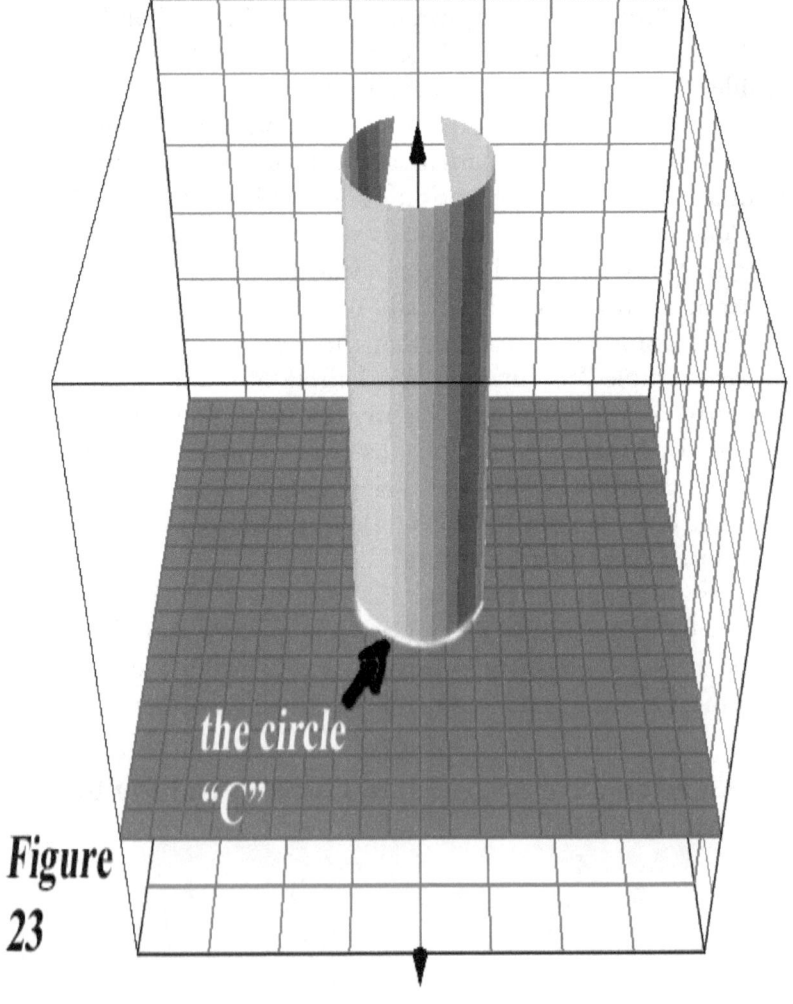

the circle "C"

Figure 23

[1] my italics

These describe the curve C as the intersection of a circular cylinder and a plane.

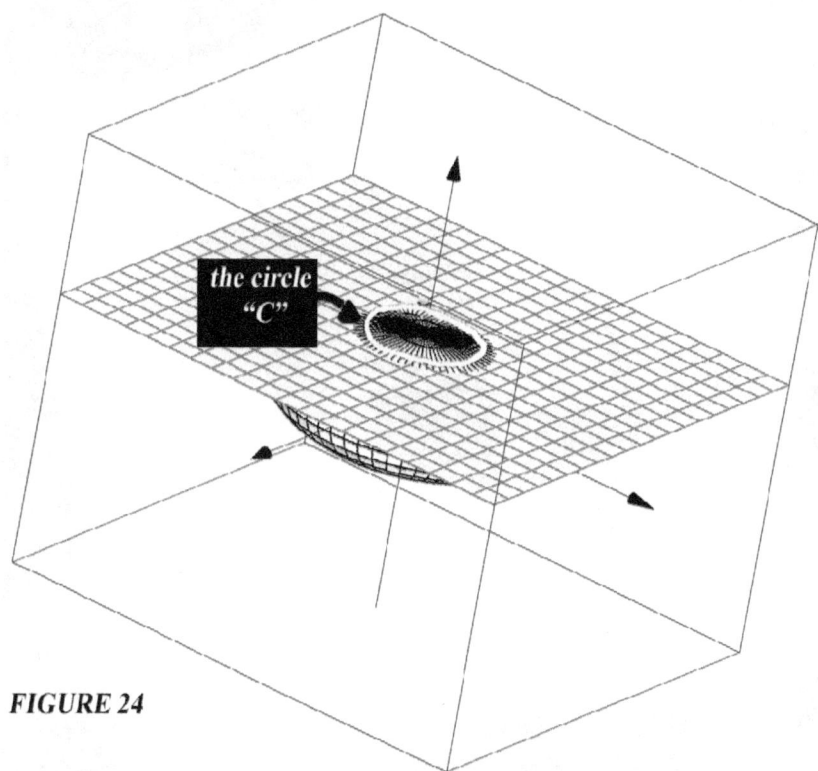

FIGURE 24

But C can be described with equal accuracy", (as well), "as the intersection of a *sphere*", (my emphasis), "with the plane z = 2, by the equivalent simultaneous equations:

(17) $x^2 + y^2 + z^2 - 8 = 0, \quad z - 2 = 0.$

Chapter 9   "Mathematical Ideals"

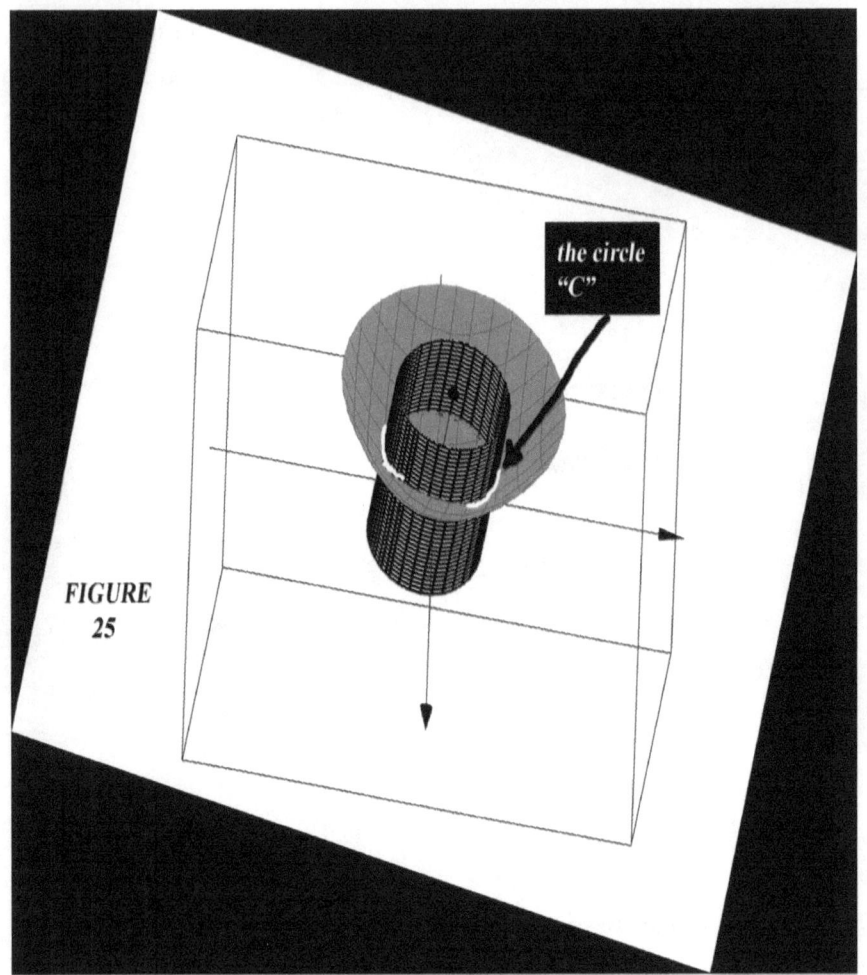

FIGURE 25

Still another description", (my emphasis), "is possible, by the equations

(18) $x^2 + y^2 - 4 = 0, \quad x^2 + y^2 - 2z = 0.$

These describe C as the intersection of a circular cylinder with the paraboloid of rotation:

$$x^2 + y^2 = 2z.$$

Therefore the only *impartial* way to describe C", (my emphasis), "is in terms of *all* the polynomial equations which its points satisfy."

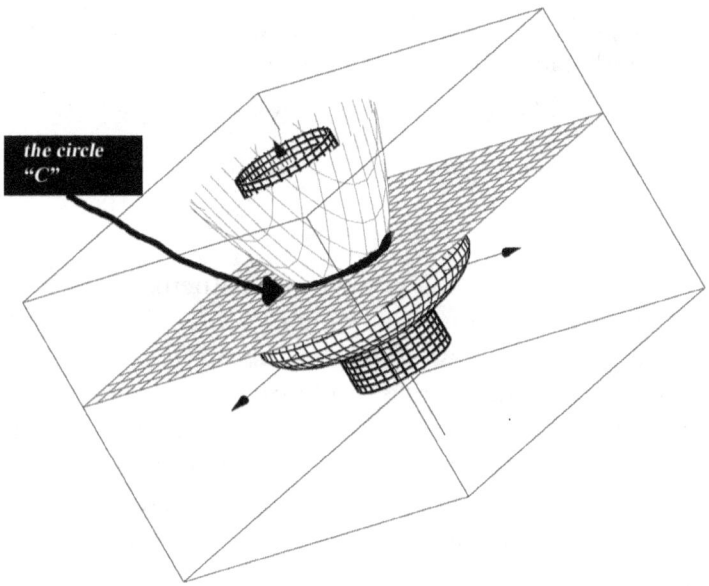

Figure 26

The descriptions above represent just a few of the ways to represent "the circle C" however. But in fact there are an *infinity* of ways to do so!

"But if f(x,y,z) and g(x,y,z) are any two polynomials whose values are identically zero on C, then their sum and difference also vanish identically on C. So, likewise, does any multiple a(x,y,z)f(x,y,z) of f(x,y,z) *by any polynomial a(x,y,z) whatsoever.*", (my emphasis). "This means that the set of all polynomials whose values are identically zero on C *is an ideal*.[2] This ideal then, and not any *special* pair of its elements, *is the ultimate description of C*. In the light of this observation the special pairs of polynomials occurring in equations (16)-(18) appear *simply as generators* of the ideal of all polynomials which vanish identically on C. ...", (my emphases).

"The polynomial ideal determined by this curve thus has various", [actually an infinity of[3]], "bases,

(20) $(x^2 + y^2 - 4, z - 2) = (x^2 + y^2 + z^2 - 8, z - 2) = (x^2 + y^2 - 2z, z - 2)..., ..., ..., ..., ...,$
...................."

An understanding of this concept of mathematical "ideals" opens a door to a better understanding of Cassirer's arguments, and a simpler understanding of my third thesis. It illustrates the conception of a *rigid* invariance –and not a mere, unstructured or trivial relativism! (In no respect is it anything like the conception of "cultural relativism", for instance). *None* of these generators stands prior to any other, nor does it *create* the figure comprehended, but each is comprehensive and exhaustive!

Each stands, rather, as an *equipotent* and relativistic "logical", (i.e. explanatory), basis fully exhausting the actuality of the figure: "**The circle C of radius 2**" which we may consider as a stand-in for a phenomenon –e.g. the elephant. No one of these *organizations* replaces the reality, the "ideal" of the figure itself!

What this says is that the *particular* perspective we begin with in our explanation is not the ultimate determining factor. To quote Cassirer in his "Symbolic Forms": "each asks its own questions" and constitutes a *different* perspective, but the "object" which is described is the invariant "ideal". Descriptions, explanations are not the same thing as the actual "object"

---

[2] My emphasis
[3] Simply concatenate the operations!

described. Ultimately it remains "a mere X"! This will relate later to my Chapter 10 and my conception of "interface"!

We *start* with the phenomena themselves, not with theories and explanations, (orientations, organizations). Theories must *validate* the phenomena, not the converse. (But we must incorporate Merleau-Ponty's input-output loop –his "intentional arc"- to truly understand the relationship).

"The circle" cited here would stand in place of Cassirer's "phenomena", (sic), for my "percept" or for *the perceptual elephant!* It is the *invariant* component of perception that we must needs preserve. It is focused as an invariant under varying perspectives relativistically but rigidly. This is *how* we can preserve the actuality of our phenomena, (evolutionarily fixed, I argue), as relativistic invariants of our scientific symbolic forms and gain an understanding of the rationale of those forms themselves! Percepts are not created by, nor are they dependent upon any particular frame of reference. If they reference ontology, then they do it as a composite ideal, and not in their particular frames. This was the sense of Galileo's profound insight long, long ago.

But we must consider the "ideal" within the larger context of mathematics. Not only can such descriptions be relativized in relation to a fixed coordinate system, but the very coordinate systems themselves stand in like case. Axes need not be orthogonal, nor need they be rectilinear, (e.g. polar coordinates are possible). Nor need they be fixed.

They may be in translation –e.g. relative motion, (which correlates to conditions of special relativity for instance), and they need not be Euclidean, (nor Hyperbolic nor Spherical). Bertrand Russell, for instance, argued that our descriptions of phenomena might even be based in projective geometry.

But need they be *even spatial?* Can we not conceive of such explanations being framed as abstract transformations,[4] which latter are *not* defined on spaces, but on raw and unstructured

---

[4] Morphisms?

abstract domains as suggested in my illustration for brain function in my first hypothesis, and the raw and unspecified "externality" of Maturana's "structural coupling".[5] Abstract domains, however, fall naturally within the scope of axiomatics which ground Cassirer's "Symbolic Forms", Hilbert's "Implicit Definition", and my own thesis of the "strategic brain".

Cassirer's Theory of Symbolic Forms:

Cassirer's "symbolic forms" is the broadest, and, I think, the deepest conception of truly scientific relativism yet expounded as it is grounded in invariants themselves. Cassirer concludes that the actual metaphysical "object" is a mere "x", which can be interpreted from a myriad of perspectives, but always preserving the *relations* of phenomenology. This is a perspective analogous and similar to the effect of combining my prior citations from Benacerraf and Bell:

> "It [logic] remains the tool applicable to all disciplines and theories, the difference being only *that it is left to the discipline or theory to determine what shall count as an 'object' or 'individual.'* [Benacerraf, 288]

> "There is an evident *analogy* between mathematical frameworks and the local coordinate systems of relativity: each serve as the appropriate *reference frames* for fixing the meaning of mathematical or physical concepts respectively. [Bell]

But Cassirer's conception confirms that there is no *canonical* context in which to view reality. Repeating myself

---

[5] Let me repeat Bell's comment from our Structuralism discussion –it is very pertinent here: "...it becomes natural, indeed mandatory, to seek for the set concept a formulation that takes account of its *underdetermined* character, *that is, one that does not bind it so tightly to the absolute universe of sets with its rigid hierarchical structure.*"

once again, it is the confusion of (the "objects" of) a *particular* form, (e.g. mathematical physics), with the invariant relationality of the phenomena which it organizes, he argued, which leads to an unwarranted assertion of metaphysical reference for its objects. His genuine "epistemological theory of relativity" is, I argue, "coordinate free", (and non-referential), in Van Fraassen's and Bell's sense as well.

My third and final hypothesis, (in Chapter 10), will be epistemological -an extension of Kant's, and ultimately of Cassirer's epistemology. Its purpose will be to supply a plausible answer to the "what" of mind.

I have argued an essentially Kantian position —greatly deepened by Maturana's insights and consistent with Cassirer's to reduce the de facto *metaphysical* presumptions of naturalism to their legitimate and necessary minimum.

This, surprisingly, leaves room for the *actual existence* of a "substance" of mind for which I will propose a specific and plausible answer. Ultimately I argue that our mental reality comes down to intentional *strategies that work*, and not to certain knowledge. But this is all we will ever need!

There remain, of course, significant problems -the most obvious of which still remains "reference". But I argue that there is a categorical difference between *metaphysical* reference and the internal, *model/model* automorphisms, (transformations), of what I maintain is our logically closed human cognitive world. (cf Quine). It is the latter which constitute the problem of science, and I have suggested a particular kind of automorphism between the brain and the world.[6]

The Substance of Mind:

Here is another excerpt from my earlier book. Hopefully you can now understand it in the mathematical context of the current paper:[7]

---

[6] See Iglowitz, 2005
[7] from Iglowitz, 1995, Chapter 5 on "the substance of mind"

"Though I have argued against the "material" and the "substance" *of Naturalism* as metaphysical existences, there is a deeper -and truly metaphysical sense of substance that I *do* wish to maintain. It is embodied in our's, (and Kant's), minimal *realist* assumptions -in the axioms of externality and of experience, [stated formally in Chapters 6 and 7 of the current text]

Though Cassirer argues for a broad range of symbolic forms, there is *another form* implicit in his thesis, (roughly equivalent to the whole of the specifically natural forms), -and innate in Kant's as well. It is the *metaphysical form*, i.e. the whole of the metaphysical context of the problem itself. (It was as a "Prolegomena to any Future Metaphysics" that Kant himself characterized his work, after all.)

This metaphysical form is the proper context for any conception of cognition, (and realism), but, precisely because of Kant, Cassirer and Maturana, it is necessarily severely restricted and analytic.

Inside of the *form* of metaphysics, (wherein we are now framing the problem), we are constrained by Kantian parameters - i.e. the fundamental, (rather than the historically limited), parameters discussed earlier. These *abstract* limits, the axioms of externality, and of experience, and the relativity of perception to the (human) instrument whereby it is effected, dictate a necessarily general, relativized and abstract solution to the problem.

Always implicit in Kant, however, was the assumption of *some connection* between our cognition, and the reality which is perceived, (metaphysical reality), -and that connection was assumed to be reflected in experience, ("intuition").

Always implicit in Kant is the relationship between the absolute external existence which he affirms and the modifying, coupling relationship of cognition itself. Kant's is very much a modern mathematical conception. He argues that we cannot separate the facts of our "instrument", (our cognition), from that which it "measures", (cognates).

The *relationship* between that cognating entity and its object, however, is understood in a very profound and sophisticated sense –very much in the sense of modern algebra. [e.g. "structuralism"] His concept of intuition, (experience), is a relativistic one. The connection is seen as a limit concept -as the most abstract possibility- conceived relativistically to the complete "X" of metaphysical reality.

Alternatively, we might today characterize this connection as the most abstract reinterpretation of Maturana and Varela's "structural coupling", (which I addressed in Chapter 6), but removed from its strict Naturalistic (metaphysical) formulation. I think the most natural characterization of it is, simply and abstractly, "interface"!

This concludes the epistemological argument. In the next chapter, I will complete my solution of the mind-body problem with a statement of my third thesis which will supply the "what", the "matter of mind". All the hard work has already been done, however, so the chapter will be brief. The *problem* is not so hard; it was our presuppositions which made it seem so!

Chapter 10: "The Interface"

This "interface", this connectivity, (or, rather, the mathematically conceived "ideal" of the same - I think Dennett would use the words "heterophenomenologically conceived"[1]), between cognator and that which is cognated, is assumed in *any* realist conception of reality, (most definitely to include Kant's itself).

It is implicit in materialism, in dualism ...; it is implicit in behaviorism, and identicism ..., in "memes" and in neural process. I mean it to be the *minimum* intersection, (the limit, i.e. *the most abstract mathematical conception*), of all of these realist, (i.e. non-idealistic –taking that word in its historic philosophical meaning here), possibilities. It is the invariant commonality, the "mathematical ideal" of all materialistic interpretations of the sensory boundary, and it is therefore a necessary and legitimate (intentional) realist ontological *existence* postulate in itself!

I mean it to be taken in the sense of a "mathematical ideal", (see Chapter 9) –in the sense of a strict *mathematical relativism*- in the same sense we take the relativism of the equations of Special Relativity. This minimum conception of interface is then, (by definition), necessary and apodictic to any realist position –to mine, and to yours as well. *Realistically*, (i.e. - therefore Realists *must* so believe), it does *metaphysically* exist, (-whatever it "is"!)

Here follows my Personal Metaphysical Assertion!

This "interface" is the metaphysical reality that Kant does not name, but which is implicit in his and any other realist position. *As a realist*, I claim it therefore to truly *metaphysically* exist, -it is a fundamental postulate of realist intentionality -and I call it relativized "substance".

---

[1] Or "co-ordinate-free" as I understand Van Fraassen

This is not the "substance" of materialism however, but an *analytic* conception –i.e. it is the metaphysically and mathematically minimal logical necessity of realist cognition. It is a some-*what*, not a some-*thing!*[2]

That there is something more, some *other* "substance", some externality other than the interface, is also apodictic to realism –it is presumed in the "axiom of externality"[1] –and I confirm it as well. Kant and Maturana have stripped the latter of all knowable determinate form, (but not of existence), but it is the former, ("interface"), with which I wish to concern myself here.[3]

---

[2] See Chapter 4: "Turning our Perspective Around"

[3] Note: I also believe in "other minds". My problem, however, is that I have no idea what they might "look like"! If my perspective is valid, then it opens a whole new perspective on my fellow man. One may have the attribute, (under some viewpoint), of blackness, or whiteness or beauty or oldness, or ugliness, or "crippled-ness", without *being that!* What I am left with is humanity, not bigotry or zealotry. "Is" is the trap of limited minds.

Chapter 11: The Last Hurdle

There remains one last difficulty with my (Naturalist) hypothesis of Chapter 4. From the standpoint of my original claim of a *complete* solution to the mind-body problem, "mind", (at the stage of Chapter 6 –and even at the stage of Chapter 9), remained conceivable only in a reductively materialist, (alternatively: an *organizational*), sense. It remained only process and without "awareness" except as the latter was itself considered reductively.

What is "mind" and *where* is it? How could it *be*? The answer is that *it is*! It *must* "be"!

For it is the apodictic metaphysical realist intentional presumption of the *ontological "substance"*[1] of the interface *itself,* (as just affirmed *as an innate realist posit* in the previous brief chapter), that I propose is the substance of mind. The reality, the metaphysical presence of this interface is the immediate and necessary consequence of the synthesis of our two realist intentional ontological fundamentals: externality and experience. It is the *relativistic* equation between a cognitive entity and externality, (and you may correctly interpret this from Maturana's perspective). This necessary presumption of the *actual existence* of connective "substance" supplies the last remaining element for the complete solution of the mind-body problem.

The Third Hypothesis:

Please spend the time to truly *contemplate* the import and meaning of this very brief chapter. It is a very deep idea, though you may not think so at first. It is like "the sound of one hand clapping", and to understand it you must "contemplate your navel" for a reasonable amount of time for a true understanding of Cassirer's and of my perspective. This, I believe, is the deepest

---

[1] Taken "heterophenomenologically"

possible conception of a scientific relativism, -and of reality itself - and it needs some "quiet time" to enable comprehension.

Swabey described Cassirer's "Theory of Symbolic Forms" as a genuine epistemological theory of relativity. Take awhile to digest this characterization. It characterizes a *relativism* of epistemologies *themselves*, (and specifically of all the scientific disciplines), and proposes that there is more than one totally legitimate way of viewing the world.[2]

But to describe it as "a *genuine* epistemological theory of relativity", says something more. I am forced to interpret it, (and I think the nature of Cassirer's own history of thought forces this conclusion), -as a *truly scientific* relativism in the sense of Einstein, rather than in the trivial senses of social relativism, irenic relativism, ..., et al. The "genuine relativism" of Swabey's characterization reflects the necessary incorporation of invariants, (in the sense of *mathematical* invariants),[3] across all the viable epistemological forms, and it is these invariants themselves, (as distinct from any particular -necessarily "localized"[4]- description of them), which allows us to define the "interface", (a la Maturana's "structural coupling"), *in an abstract sense*, but which prohibit us, at the very same time, from definitively grounding it within any single one of the particular forms of knowledge.

To repeat an earlier reflection: With Cassirer, I argue that the essential flaw in the referential conception of knowledge, ("scientific realism"), lies in its confusion of a *particular* "frame of reference", i.e. "symbolic form", (and its assumption that there is only one comprehensive frame possible). It is confused with the *invariant relationality of experience in the abstract*, (i.e. under all consistent frames).

---

[2] See Chapter 13:d'Espagnat for a further development of this argument and a confirmation in the "real world" in Appendix D: The Epistemological roots of Niels Bohr"!

[3] See my Rosen discussion Chapter 3 for a partial idea of the kind of perspective I intend.

[4] Please review Bell's perspectives on "local mathematics" in Chapter 2, and Chapter 9 in its entirety.

This, Cassirer and I both argue, is the heart of the issue. It results in scientific realism's confusion of a specific *organization* of experience[5] with the experience itself, which is organized.

A formal statement of my third hypothesis:

The Axiom of the Interface:

Given that the interface, (as just defined —whatever-it-is), (1) metaphysically and *actually exists*, ("heterophenomenologically", to use Dennett's term again), and given further (2) *that it is structured in parallel with the structures as postulated in my first and second hypotheses*, and *this* is the specific formal statement of my third hypothesis —the (intensional) "Axiom of the Interface", then (3) it internally and necessarily defines our objects and what they do -and they *too* exist! And, as demonstrated by my arguments in Chapters one through eight, it *knows* them!

All the problems of structure, all the problems of logic have been dealt with in the previous hypotheses, and a plausible Naturalist rationale is in place. All that remained was *existence*. It is the sole further (realist intentional) assumption of the *metaphysical existence* of the interface *itself* which supplies the reality and the existence of sentiency!

Mind is the "unified concept", (the rule), of this interface, (and of the brain) —seeing it from Cassirer's[6], (via Hilbert's), and Maturana's perspective. Under the combination of my three hypotheses, then, mind becomes quickened, becomes "aware", becomes "live".

We *do* know, we *are* aware, we *are* real. What we are sentient and aware *about* however, is *not* metaphysical externality.

---

[5] I.e. mathematical physics
[6] as a "rule" in the sense of Cassirer's reformulated "concept", and further, in the sense of the "ordering" of my "concept of implicit definition"

Rather, from my biological perspective, it is the metaphorical organization of primitive process *by which we deal with the latter*. But that is quite good enough. It works! And it is a *genuine* miracle in all its glory!

A (crude) Graphic Overview follows: (see technical footnote! [7]

---

[7] Repeating a previous Note: Freeman's use of the words "spatial integration" in the following diagram is somewhat confusing and misleading. His use of "spatial integration" refers to integration over the *physical* space of the brain but which actually accomplishes a *divergence* in the mapping of the "data" itself. (Please note the diagram itself which illustrates the mapping of parallel data into a specifically non-parallel distributive mapping!) An afterthought: Both sides of the feedback loops pictured here, considered together, seem to furnish a fairly lucid rendition of Merleau-Ponty's "intentional arc" –"by which experience derives from the intentional actions of individuals that control sensory input and perception".[W.J. Freeman 1994]

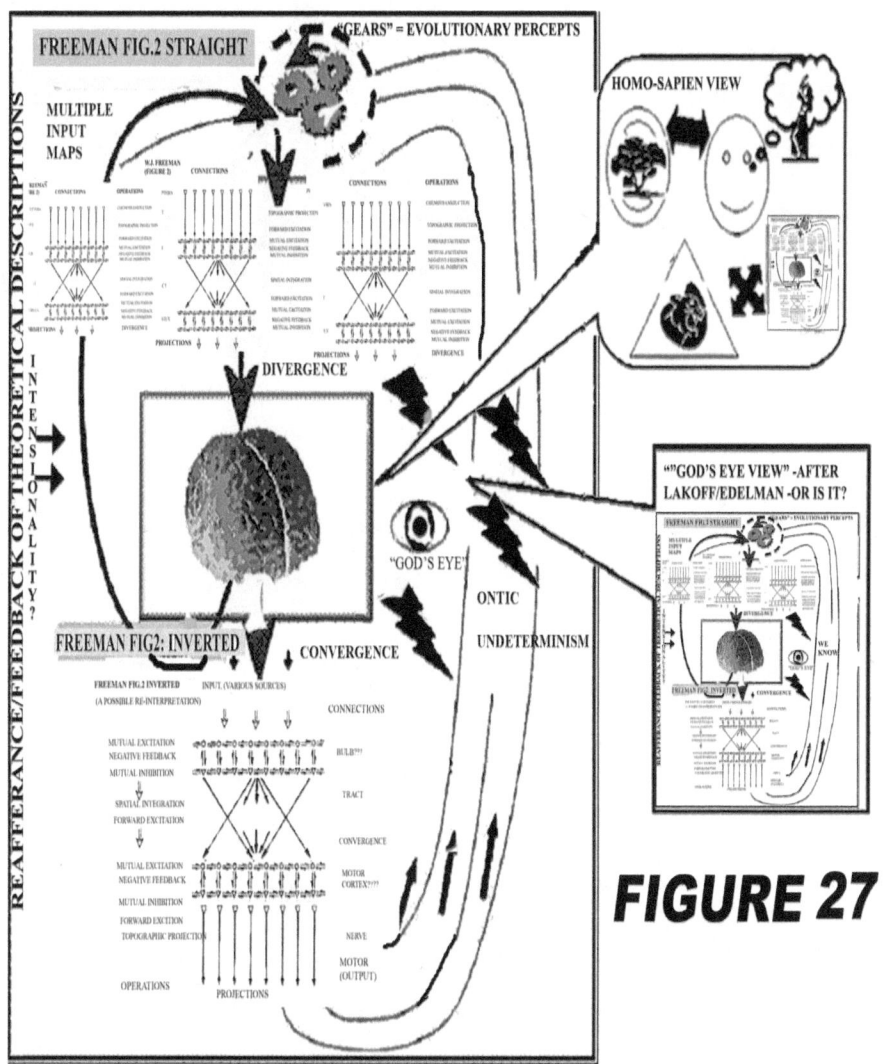

FIGURE 27

The Strategic Brain: a crude graphic overview

In the next chapter, I will answer two relatively modern critiques of my conception. Hopefully it will answer some of your questions and objections as well. The chapter after that will deal with Scientific Realism from the standpoint of modern physics and reach a surprising conclusion.

Chapter 12: Two (Relatively) Contemporary Realist Criticisms of My Conclusions

In this chapter, I am just going to present two different (negative) viewpoints on my perspective. One of them is Will Durant's —who was actually criticizing Kant's huge idea, but, by implication, my own as well as they are quite similar in the aspects he addresses. The other one is from a "Journal of Consciousness Studies" anonymous reviewer commenting on a submission of a piece of my own work some years ago. Hopefully they will raise and answer some of your own problems with this very radical interpretation of reality.

I'll start with Durant. He does a pretty nice description of Kant's conception but he doesn't "buy" it however, (in the sense of William James' "cash value of an idea"). He prefers instead to accept James' very specialized perspective (which is a psychologist's perspective, and which, in fact, makes a great deal of sense as such —purely as a psychologist's perspective). I frankly don't think any biologist would ever accept it however[1]. But then Durant inherits James' problem: i.e. of conversely legitimizing his own perspective to a biologist.

Durant instead adopted James' specifically psychological perspective: i.e. that we get relations right along with our objects as epistemological *primitives*. I think any biologist would ask the question "but how?" But Durant never answers that and neither does James. They just assume they exist *as primitives* and build their worlds from there!

Durant on Kant:

"[Kant's] *Critique becomes a detailed biology of thought*, an examination of the origin and evolution of concepts, an analysis of the inherited structure of the mind. This, as Kant

---

[1] (In Durant's words, conversely, it would have no "cash value" to a biologist!)

believes, *is the entire problem of metaphysics*", (my emphasis)...."¹

This is *my* "take" on Kant as well —I think Kant saw the problem just as a biologist would see it, and as I still see it myself as well! When I was a very young man, I was a student in a biology laboratory. An idea had occurred to me from my own ruminations and I wanted to run it by the lab supervisor, (a Ph.D candidate, I think). I asked the question: "Is it possible to view a multicellular organism merely as an *assemblage* of unicellulars?" Without a moment's hesitation, (and I have had the highest respect for the philosophical abilities of biologists ever since), he replied: "Sure. There are sponges which can be forced through a sieve and dispersed into individual cells, but which then come back together to become a metacellular once again all by themselves."²

This has always been my perspective on multicellulars, and you and I *both*, (you must surely admit), fit that characterization – we are metacellular organisms. Mine is a *biological* perspective, and I think it is clear that it is Kant's perspective as well.

> "... it [the brain] is an active organ which moulds and *coordinates* sensations into ideas, an organ which transforms the chaotic multiplicity of experience into the ordered unit of thought.ᴵᴵ .... But let these various sensations group themselves about an object in space and time —say this apple; let the odor in the nostrils, and the taste on the tongue, the light on the retina...unite and group themselves about this 'thing'; and there is now an awareness not so much of a stimulus *as of a specific object...* "ᴵᴵᴵ ³

---

² He asked me my major and when I replied "philosophy", he said he thought I would be good at it.

³ My emphasis. Note: this is my interpretation of the function of the cortex, and its "a/d converters"

> "But again, was this passage, this grouping, automatic? Did the sensations of themselves, spontaneously and naturally, fall into a cluster and an order, and so become perception?
>
> Yes, said Locke and Hume; not at all, says Kant."
>
> "For these varied sensations come to us through varied channels of sense, through a thousand 'afferent nerves' that pass from skin and eye and ear and tongue into the brain; what a medley of messengers they must be as they crowd into the chambers of the mind, calling for attention!"

This is very reminiscent of Maturana's comment[4] that the input to the brain is "like an animated family discussion with all the members talking at once!"[IV]

> "And left to themselves, they remain rabble, a chaotic 'manifold', pitifully impotent, waiting to be ordered into meaning and purpose and power..."
>
> "Observe, first, that not all of the messages are accepted...a storm of stimuli beats down upon the nerve-endings which, amoebalike,[5] you put forth to experience the external world: but not all that call are chosen; only those sensations are selected that can be molded into perceptions *suited to your present purpose*, or that bring those imperious *messages of danger* which are *always* relevant."[V]

Please note the connection of this passage with the issue of the intentionality of the brain —and to its organizational prioritization of *danger* explicitly. This was my focus also in my first thesis —it lies at the roots of my "schematic model", "interface", and at the root of my argument for a necessary

---

[4] cited in Chapter 6
[5] This is relevant to my conception of the multicellular as an assemblage, a "society" of unicellular organisms.

violation of "hierarchy" to preserve urgent priorities in reaction as well.

In the terminology of computer languages, "danger" may necessitate a "Go To" command which can absolutely violate the "structure"/hierarchy of a program to go elsewhere —even outside the program itself by reason of urgent necessity! [VI]

And finally, one last quote:

> "Consider a system of thought like Aristotle's; is it conceivable that this almost cosmic ordering of data should have come by the automatic, anarchistic spontaneity of the data themselves? See this magnificent card-catalogue in the library, intelligently ordered into sequence by human purpose.
>
> Then picture all these card-cases thrown upon the floor, all these cards scattered pell-mell into riotous disorder. Can you now conceive these scattered cards pulling themselves up, Munchausen-like, from their disarray, passing quietly into their alphabetical and topical places in their proper boxes, and each box into its fit place in the rack, -until all should be order and sense and purpose again? What a miracle-story these skeptics have given us after all!"

What a wonderful metaphor. It represents beautifully the most succinct argument for Kant's thesis. Durant thinks that William James supplies the answer to Kant's objections in his "Radical Empiricism" however.[6]

Durant's brief coverage of James' perspective is quite different. James' is a psychologist's viewpoint, and, lacking Cassirer's insight of "Symbolic Forms", James rejected Kant's fundamental conclusions out of hand. (I think Kant would clearly have rejected his as well.)

---

[6] It will take Cassirer's "Symbolic Forms" to mediate between their conceptions, and which will eventually reconcile them.

First of all, let me say that I have a harder time in dealing with William James' philosophy than I do with Durant's criticism of Kant because I think, under a certain perspective, it makes some sense. I think that James proposed an almost pure epistemology, (form), grounded in a *psychological* perspective and very much equivalent to the sense in which Maturana, (as we saw in Chapter 6), proposed a *biological* epistemological form. As such, James' is a real candidate for incorporation within Cassirer's "Symbolic Forms".[7]

The relationship of these alternative worldviews is again roughly equivalent to Hilbert's "beer mugs and Pythagorean theorem" interpretation of the mathematics we examined earlier. This is what Cassirer's "Symbolic Forms" is all about!

Now let us turn to Durant's brief summary of William James's ideas. He sees James' as a more rational alternative in the modern world and as providing an escape from Kant's fundamental "error":

> "… and if he", (James), "begins with psychology it is not as a metaphysician who loves to lose himself in ethereal obscurities, *but as a realist* to whom thought, however distinct it may be from matter, is essentially a mirror *of external and physical reality.*[VII]

> …And it is a better mirror than some have believed; it perceives and reflects not merely separate things… but their relations too; it sees everything in a context; *and the context is as immediately given in perception as the shape and touch and odor of the thing.* Hence the meaninglessness of Kant's 'problem of knowledge', (how do we put sense and order into our sensations?) –the sense and the order, in outline at least, are already there." [VIII]

---

[7] Paraphrasing Cassirer: "each asks its *questions* each from a particular perspective…"

I think that Kant, as a biologist, would ask the legitimate question "*How* is it 'already there'? Give me a biological rationale!" Within my own (biological) perspective, "things" and "relations" would be specifically "already accounted for" under the rationale of the "schematic model" whose facile relations, "the calculus", I have proposed, was the very purpose of the model itself!

James assumed his personal perspective, I believe, under the "realist imperative" mentioned earlier[8], and, since we *all* possess it, we are inclined to agree from the beginning. This inclination in itself guarantees nothing logically however.

> "Consciousness is not an entity, not a thing, but a flux and system of relations; it is a point at which the sequence and relationship of thoughts coincide illuminatingly with the sequence of events and the relationship of things. *In such moments it is reality itself, and no mere 'phenomenon' that flashes into thought;* for beyond phenomena and 'appearances' there is nothing. ... *the 'noumenon' is simply the total of all phenomena,* and *the 'Absolute' [is] the web of relationships of the world."*

(It would have been interesting to have asked James for his own *specific* meaning of the word "relationships".)

Shifting perspective somewhat, Durant now goes on to develop James' notion of "radical empiricism":

> "To find the meaning of an idea, said Peirce, we must examine the consequences to which it leads in action; otherwise dispute about it may be without end and will surely be without fruit.
>
> [James] tried the problems and ideas of the old metaphysics by this test, and they fell to pieces at its touch..."
> [Pierce's] "simple...test led James on to a new definition of truth. Truth had been conceived as an objective

---

[8] see Chapter 4

relation, ... now what if truth" [itself] " ... were ... relative to human judgment and *human needs,* (i.e. productivity)?"

" ... 'Natural laws' had been taken as 'objective' truths, eternal and unchangeable ...and yet what were these truths but formulations of experience, convenient and successful in practice; not copies of an object, but correct calculations of specific consequences? Truth is the 'cash-value' of an idea."

(This might almost be a paraphrase of my arguments for my "schematic model"[9], - or of Hertz's characterization of the "objects" of science[10] -but lacking its implicit biological rationale.)

"...The true ... is only the *expedient* in the way of our thinking ... The true is the name of whatever proves itself to be good in the way of belief', (productivity, consequences). "...Truth is a process and 'happens to an idea'; verity is verification.

Instead of asking whence an idea is derived, or what are its premises, pragmatism examines its results; it 'shifts the emphasis and looks forward'; it is the 'attitude of looking away from first things, principles, 'categories', supposed necessities, and of looking towards last things, fruits, consequences, facts'."

"Scholasticism asked, What *is* the thing, -and lost itself in 'quiddities'; Darwinism asked, What is its origin? —and lost itself in nebulas; pragmatism asks *What are its consequences?* –and turns the face of thought *to action* and *the future.*"

"...Men accept or reject philosophies, then, according to their needs and their temperaments, not according to

---

[9] See Chapter 4
[10] See Chapter 7, "Cassirer's Theory of Symbolic Forms"

'objective truth'; they do not ask, Is this logical? —they ask, What will the actual practice of this philosophy mean for our lives and our interests? Arguments for and against may serve to illuminate, but they never prove."

As an independent symbolic form I think James' perspective makes sense. And it's a perfectly legitimate form, I believe. I think it's capable of being just as rigorous for psychology as Maturana's was for biology, for instance, (see Chapter 6), but it's a totally different worldview. It does not, however, fill the needs of a biological perspective.

Durant ends up accepting James' generalist rejection of Kant and then basically falls down to an argument "ad populum" which is fundamentally just an appeal to *everybody else's* prejudices. It's not a very good refutation.

Durant Critiques Kant:

Cutting to the chase, here is what I believe constitutes the core and the essence of Durant's criticism of Kant's conception. It is a (naïve) realist's simplistic and absolute dismissal:

> "the annual elliptical circuit of sun by earth [is] independent of any perception whatever; the deep and dark blue ocean rolled on before Byron told it to, and after he had ceased to be…[or] when we see an insect moving across a still background…" "a tree will age, wither and decay, whether or not the lapse of time is measured or perceived." [IX]

The problem, as I see it, is that Durant was unwilling to consider the deepest implications of the existance and of the possibility of the mutual agreement of "other minds", "other brains", (which I think both he and I accept), and which has deep implications to this problem.

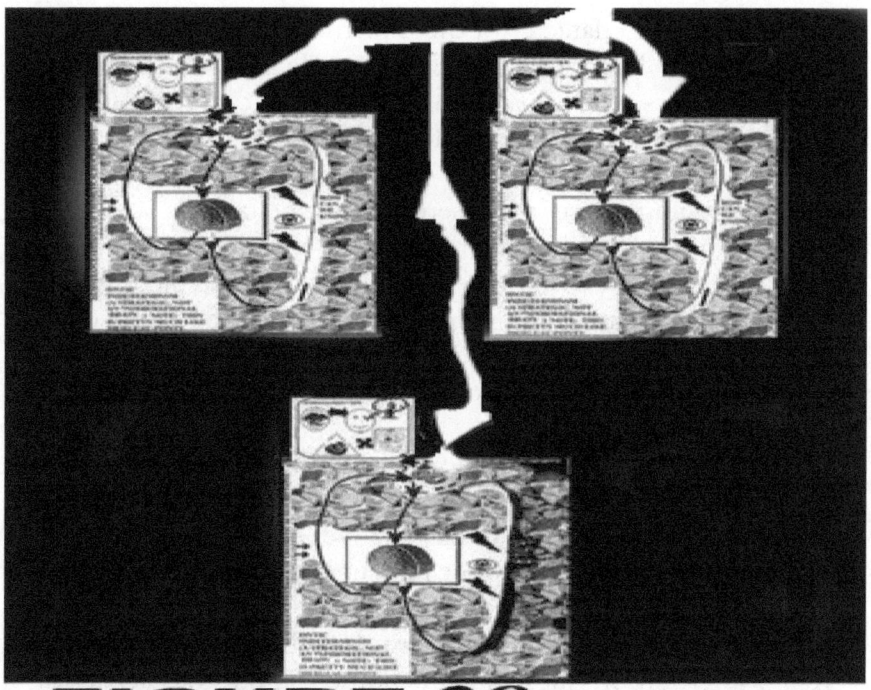

FIGURE 28

This is the model I propose for human reality, but it is lived through the "gears and levers" of our evolutionary artifacts, the latter being understood from the perspective of Biology, itself being just one of Cassirer's multitudinous "Symbolic Forms". This, I believe, is the homo sapien cognitive reality!

If another mind, another brain sees and acknowledges the same "facts of reality" that I do, could it not merely mean that it too has processed these in the same way that I would, that our naïve worlds are similar precisely because our *brains*[11] are so —

---

[11] But what then are "brains"? I will repeat my very early injunction that in any theory of deep metaphysics *all* terms should be assumed to be in quotes –i.e. they should be taken heterophenomenologically – at least until the final conclusions. This allows a "context-free" discussion in the sense of Van Fraassen. I have supplied an actual answer to this problem in my third thesis.

irrespective of the character of the underlying "substantia phenomena"?[12] Durant's argument, a simple appeal to popular agreement[13] does not really address the substance of the issue. That "a tree will age, wither and decay, whether or not the lapse of time is measured or perceived" is a certainty within our worlds, but the very substance of the assertion must be taken heterophenomenologically!

The viewpoint embodied in Figure 28 is the picture I suggest as an alternative and whose substance will be clarified shortly. I believe in other minds, (I get to have beliefs too), but these minds, I believe, see through the exact same "gears and levers" that I do.

That our conclusions about reality should agree does not surprise me. We all see with the same indeterminacy that my later figure of chap 11, (reproduced above), shows *but through the same parameters*! This is our human linguistic and cognitive world: we speak the same language!

That there is something more, (i.e. *somewhat*), that is real is Kant's assertion of "substantia phenomena", but the "what" of it is precisely at issue, and Durant did not debate the substance of Kant's claim against knowledge other than in a casual reference to James' "Radical Empiricism"[x] which I think is a poor answer. He went on against Kant's categories, ethics, et al. I have definite problems with these latter as well, but I think that Kant had the basic problem precisely right.

---

[12] Think about the possibility of two minds with alternate primitive conceptions of physical reality, (just as, for instance, Benacerraf conceives of two minds with alternate conceptions of set theory). Each might see "the deep and dark blue ocean [that] rolled on before Byron told it to".

[13] an example of an "argumentum ad populum"

The JCS Review [14]

The JCS reviewer did a much better job of critiquing my own conceptions, I think, than Durant did of Kant's. He raises some fairly substantial issues, but I think they're answerable. But to repeat once again, mine, like Kant's, is a *biologist's* perspective.

The JCS reviewer raises two questions. There are two fundamental problems that he's looking at –but he mixes them together. One of them is his "my world". He provisionally accepts my viewpoint and then he asks the question "what is my world?" Well what does that mean?

> "No dent has yet been made, however, on the problems of consciousness as they are likely to be perceived by the readers of JCS (including myself). The adoption of a non-representational position shifts the locus of these problems. Given that I have a world consisting of perceptual and conceptual objects (i.e. operational constructs) which I denote by terms such as 'trees', 'despair', 'redness', 'brains' etc, is it possible to envisage how [the] last mentioned of these constructs", [the brain], "could itself embody a world analogous to that of my own?" (Please note that his primary problem is that of envisioning even the *possibility* of an answer: "how...*could*... 'the brain' ...embody a world analogous to my own?"
>
> My answer lies in Cassirer's "Symbolic Forms" –i.e. it lies in the conception of an epistemological automorphism.)
>
> "If, however, it is admitted that what I call a 'brain' might itself form operational constructs and this might thereby

---

[14] This is an anonymous reviewer's commentary on a submission to JCS which encompassed essentially just my first hypothesis as presented in Chapter 4 of the current writing.

explain what I am myself doing all the time, then are we not back in almost exactly the same place as we started?"

"Namely, we have to explain how it is that a pattern of neuronal firing can have the attributes that I designate by the word 'tree'."

(The attributes would be "implicitly defined" in just Hilbert's sense of chapters 1 and 2 by the operative process of the brain. This again is a problem of envisioning possibility.)

"The only gain is that we no longer have the additional problem of hooking it onto a postulated external "real" tree, and this gain may in fact be a loss for those who hold that the real tree may play a role in establishing the qualia of our percepts."

Before answering his objections more fully, let me note emphatically that something very new has slipped apparently unnoticed into his equation, (*besides* his "only gain"): i.e. *an explicit and constructive biological rationale for "consciousness" itself.* And this is *not* a small thing. It was the very purpose of the dialogue.

His first question, (stated last): the reviewer's "real world" and the possibility of the "real" tree's supplying qualia. That part of it I think is addressed in Maturana and Varela's "structural coupling" which is a coupling between the autopoietic organism and externality. It's a very general thing, defined conceptually at the outset and I think it is best understood within the context of the "mappings" or "morphisms" as I understand they are conceived within category theory – between the unspecified realms "domain" and "codomain".

I think this is about as far as a Darwinian analysis can go. It's very, very general. What we're talking about here is some kind of a mapping that only preserves "adequacy".[15] We are not talking about (James') "goodness" or "truth" but *just* adequacy,

---

[15] See also the Gleick reference shortly

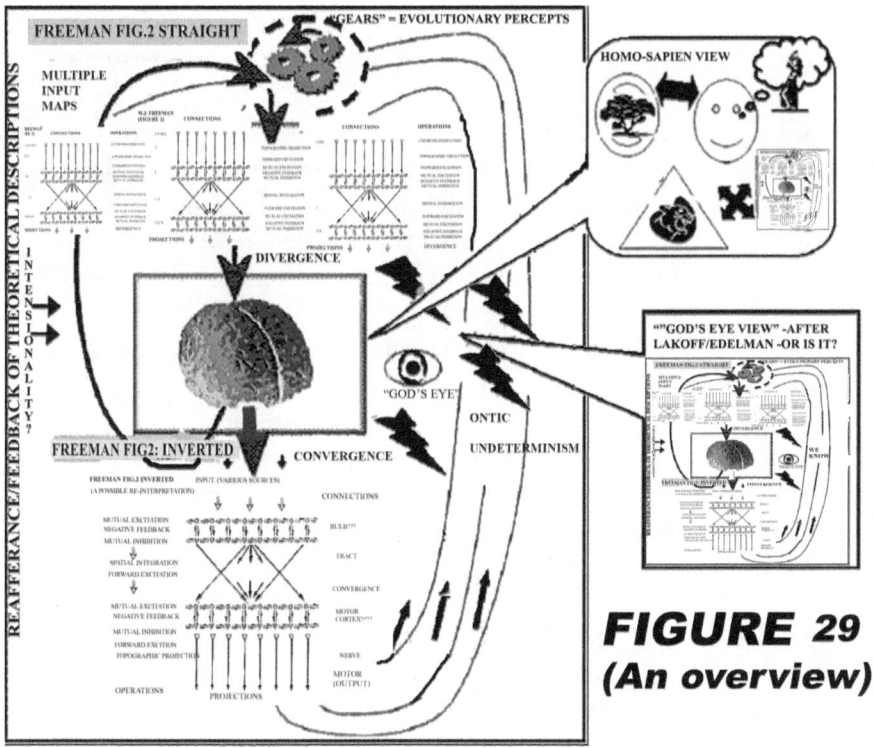

FIGURE 29
(An overview)

mere adequacy. (See my illustration Bounds and Limits in Chapter 4.)

To get a broader picture however, let's look first of all at my God's Eye explanatory diagram from the "Freeman Appendix" of Chapter 4 again. Let me reference and try to explain more fully what this model signifies.

First we have to define our fundamentals and refine our characterizations. The fact is that even this very model, (above), exists *inside of* the closed and bounded cognitive framework of man, (as discussed in Chapter 6) – of me, and of you. We must start from there and we can never really get outside of it no matter how far we concatenate our reasoning.

The question of the "embodiment", the "analogy"[16], (automorphism), of my reviewer's dilemma exists *within* our closed, (but very effective), cognitive model, but the question of possibility *per se*, i.e. what *kind* of possibility results from his limited appreciation of the scope of transformations! Consider, for instance, the mappings from the domain to the range of the logistic difference equation of chaos theory, or Mandelbrot sets. James Gleick gives a pretty good introduction to the whole of the possibility. (Gleick, 1988) These mappings explicitly break hierarchy. W.J. Freeman's experimental researches demonstrate it directly!

(In my early work, I had always understood sets as "unstructured manifolds" inside of my interpretation of Cassirer's ideas.)

Ours must be an automorphism" in the *general* sense rather than the specialized sense that is invoked here as the fact is that we are truly *"blinded"*, (ontologically incompetent, to state it baldly -in the "real" ontological sense), at the periphery of the "GOD's EYE!" map shown above.

---

[16] "How could ['the brain'] itself embody a world analogous to that of my own?"

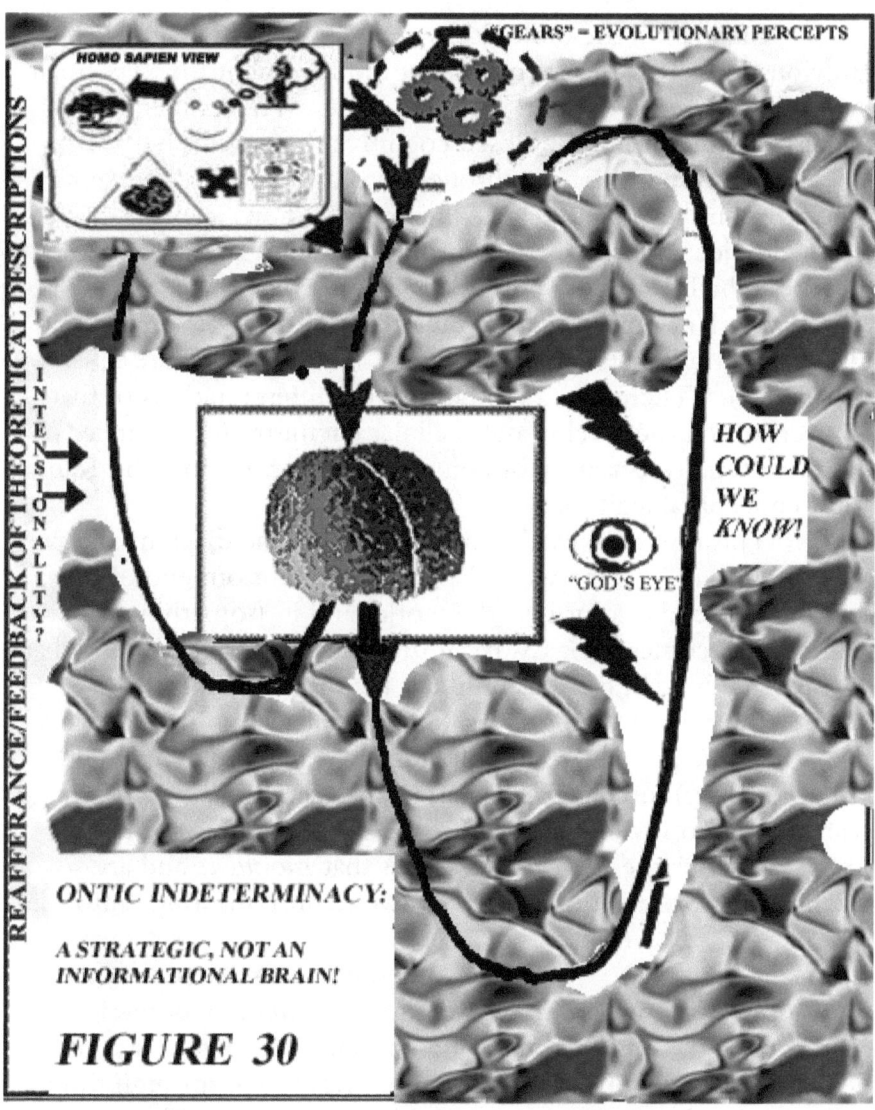

FIGURE 30

This diagram, (directly above), is the ontic model I ultimately propose for cognition and it corresponds pretty well with Merleau-Ponty's. For even the previous conception itself, (my Freeman's God's eye conception), exists *within* the cognitively closed human world! The answer I was attempting to propose to JCS was just too big to fit into that limited journal format.

But *within* the biological symbolic form implicit in the previous model, (i.e. in terms of that model —call it the "Freeman Model"), we receive input[17] into the brain. (Remember that this "brain" is still within our closed cognitive framework.)

This input is passed through into the "objects"/percepts, 'the gears and levers", (i.e. the implicitly and evolutionaryily defined objects), of the brain —through these organizational nexuses which I propose are the "a/d converters", (hierarchical/nonhierarchical converters), of the cortex and distributed[18] as per Freeman's diagram, (Freeman's Figure 2, Chapter 4). (There may be other, deeper interpretations of that conversion on the level of metacellular chemistry for instance, –I don't pretend to certainty or completeness nor, I think, did Kant, –just conceptual legitimacy).

This input travels downward through the diagram and we ultimately "act into the world"[19] at the very bottom end of this loop. We send output into the world, and it, (something or somewhat), comes back. What it does and how is absolutely, that is, ontically hidden from us until the next re-entry loop.

But within that loop reafferent feedback, (which I believe references the particular symbolic form we are employing —and intentionality?), our specialized and particular viewpoint/worldview is "measured" against that transformed input. It is our theoretical hypotheses that *modulate* and are modulated by our actions —by what we're promulgating "into the world".

Every time you or I do something, every time we "act into the world", we are implementing a theoretic hypothesis which may, in fact, turn out to be wrong! It is corrected or at least linearized when it "triggers" input right back through those very same a/d converters employed in the first place I believe.

---

[17] I could, I am sure you realize, employ "scare quotes" almost everywhere in this discussion

[18] or "centralized" depending upon perspective- see the "telescope" reference in Chapter 4.

[19] Using Merleau-Ponty's phrase

But we're still dealing here with the implicitly defined and evolutionary "objects" -the "gears and levers" of the mind/brain. It's a continual loop. This cycle goes on forever and ever and ever.

What is really and truly "out there"? I don't know, you don't know, and nobody *can* know! But inside of, interior to our model itself, this conception is legitimate. This is "*ontic indeterminism*". It is *not* strategic indeterminism!

You might object to this concept of cognition on the grounds of an infinite regress. Sure it's an infinite regress. It exists within the *closed* cognitive framework of the human mind/brain and there's no way it could be anything else. But I think it works there. It works in much the same manner within the presumptions of naïve realism or within any of the scientific theories, so it is really not such a new idea.

"The brain" *per se* can, in fact, be the focus and a starting point of a different but legitimate symbolic form. And why not? Maturana began his conception from much the same grounds, and I think it is legitimate. But again, this is just the sort of thing that transformations do.

Within this context, we can now address my reviewer's "automorphism requirement": i.e. "Namely, we have to explain how it is that a pattern of neuronal firing can have the attributes that I designate by the word 'tree'."

If we take Cassirer's "phenomena", or James' web of relationships of the world[20] in the sense of my axiom of experience; if we identify them with "that which remains invariant under all consistent worldviews" as I proposed in Chapter 7, then *epistemological* automorphisms in the sense of

---

[20] "Consciousness is not an entity, not a thing, but a flux and system of relations; it is a point at which the sequence and relationship of thoughts coincide illuminatingly with the sequence of events and the relationship of things. *In such moments it is reality itself, and no mere 'phenomenon' that flashes into thought*, for beyond phenomena and 'appearances' there is nothing. ,,, *the 'noumenon' is simply the total of all phenomena*, and *the 'Absolute' [is] the web of relationships of the world.*"

Cassirer's "symbolic forms" make sense and "the brain" can form the focus and origin of one specific independent symbolic form.

Quoting from Chapter 7:

> The fable, (in concert with Quine I maintain), helps us to see that "experience" *as such* is not, (a priori *or* a posteriori), identifiable with any of its organizations or orientations. Rather, it must be identified with the *invariant relationality* -i.e. with that which remains fixed- under *all* global, comprehensive and consistent orientations.
>
> "Experience", (tentative working definition), is that for which both the king and the technician *must account* in some manner![21] It is not *itself* an orientation, however. It is, rather, *that* ("thing") *which must remain fixed*, and I argue that it is a *primitive* of reason. [It is a logically primitive invariant!]
>
> Scientific experiment extends, (generates), experience and thereby bounds (and shapes) the scope of consistent theories. It adds new invariant relationality to the abstract frame, (and the history of abstract frames). Following Quine however, it never *determines* them."
>
> But our "objects" and "the things they do" are exactly what I propose as being the implicitly defined "objects" –i.e. the primitive, implicitly defined invariants of the brain and the schematic (naïve) model which embodies them. If this were true, if these were, in fact, the invariants of the various symbolic forms, if these constituted the basis of the phenomena themselves, then the conception of mutually valid automorphisms *over* these "objects" is not problematic. This is exactly the sort of thing that automorphisms do.[22]

---

[21] This identifies, I propose, a viable and legitimate -and theory independent- working definition of experience.

[22] Automorphisms need not preserve operations – i.e. addition could go to multiplication, etc.

These automorphisms are alternative and equipotent Cassirerian symbolic forms. In fact, automorphisms are probably the easiest way to understand Cassirer's "symbolic forms" –i.e. as *epistemological* automorphisms! (See Chapter 9).
From Cassirer's standpoint, (and within the "naturalist forms"), we always maintain the equations; we always maintain the invariants of experience.

> "The naïve realism of the ordinary view of the world, like the realism of dogmatic metaphysics, falls into this error, ever again. It separates out of the totality of possible concepts of reality a single one and sets it up as a norm and pattern for all the others. *Thus certain necessary formal points of view, from which we seek to judge and understand the world of phenomena, are made into things, into absolute beings.*[my emphasis]"[XI]

> "But above all it is the general form of natural law which we have to recognize as the real invariant and thus as the real logical framework of nature in general......No sort of *things* are truly invariant, but always only certain fundamental relations and functional dependencies retained in the symbolic language of our mathematics and physics, in certain equations." [XII]

Where Cassirer and I Fundamentally Differ:

At this point, I think I must differ with Cassirer. I agree that "it is the general form of natural law which we have to recognize as the real invariant and thus as the real logical framework of nature in general." But I differ with his assertion that "no sort of *things* are truly invariant".

I agree with Cassirer that our specifically *theoretical* "objects" are not fixed, -that they are no more invariant than our theoretical *hypotheses* are invariant, ("No sort of *things* are truly invariant"). Cassirer is saying that "the laws of nature" *per se* are invariant –and I agree with that.

The part I differ with is his assertion that "no sort of *things*" [*per se* –i.e. *whatever*] "are truly invariant". I believe that our naïve "things", meaning specifically our perceptual, naïve realistic things, are evolutionarily created, (as seen within the Naturalist form), as an organization of process and that this picture, (form), "objects" and "calculus" *combined* is, in fact, invariant. This is the "realist imperative" that I discussed in Chapter 4.

The composite of a *theoretical hypothesis* plus its concomitant plastic, (i.e. non-invariant) "objects" which it can conceivably distribute differently[23] than directly to our naïve "objects" must match against the sum total of our perceptual objects -the evolutionary "objects and the things they do" of the naïve realistic form. They are what hold the different symbolic forms together and constitute the source and the target of these automorphisms.

This is my hypothesis. If they do match, then we have a successful theory to whatever level it has been tested to. And I think, as biological/mechanical entities that that is all we can ever have. I think, moreover, that it is all we will ever *need to have!*

The strong parallel of quantum physics reasserts itself once again here. Within its model we have a "state equation" which is some solution we've made to the whole, (or the applicable part), of reality. It lasts until we do the next experiment, ("action into the world"), in which case the whole of the model is recast.

This is a very similar situation that I postulate within my model. We act into the world but what's going to come back, we can't know. The new input has to be reintegrated into a new "state equation" which generates new hypotheses. I think the parallel is very, very strong, and I think it works.

This parallel is interesting because it makes the functioning of the brain very much like the functioning of quantum physics. It establishes that we're adopting the same strategy that physics did at the very small and the very large scale. But this is at the middle, biological scale and it deals with algorithms fundamentally.

---

[23] This goes directly back to the schematic models of Chapter 4.

As I stated much earlier, I think that "the mind" is the brain's *rule* of structural coupling. But this "rule" must be understood in Cassirer's sense of the logical rule of a concept –and in its extended sense of the rule of the concept of implicit definition.

It has been argued that Cassirer's "symbolic forms" departed from the tradition of Kant, specifically in the issue of innate categories.[XIII] I don't think this criticism would have bothered Cassirer particularly no more than I believe it would have bothered Kant. Cassirer quoted Kant, (paraphrasing), as acknowledging that his ideas were a beginning, not an end, and that change and development were inevitable.

The reviewer's second question challenges the existence of any possibility of an answer to the dilemma he proposes. In terms of the reviewer's "'objects' that we construct -including the 'brain' which mirrors everything he, (I), do", it's a different issue than discussed above.

We're talking here specifically about the possibility of an *automorphism* that maps from the worldview that the reviewer has, (essentially that of naïve realism), into a specific worldview that orients the whole thing in a different way. (This is Cassirer's "Symbolic Forms" of Chapter 7.) It's a specifically *epistemological* automorphism, another symbolic form. This symbolic form starts from the brain as its central organizing point, ("asks its questions" from that beginning), and builds outward to include all of the things he does. I think it's perfectly legitimate as an automorphism. (Maturana began much the same way.)

In the "Freeman diagram", you see output into the world and coming back through our primitive "gears and levers" and with a concomitant reafferent feedback besides which latter embodies, (and corrects), our (intentional) theoretical hypotheses. In these terms I think it makes a great deal of sense.

I think it works for what we need it to do. I think the perspective of "the brain" as such can be oriented that way and that it is a legitimate biological and cognitive symbolic form.

It's superior to the reviewer's own naïve realistic worldview that he starts with and which he is advocating essentially unchanged in the end. It supplies a specific rationale for "consciousness" as well -which for him is innately impossible![24] I think my perspective is legitimate and answers the basic *biological* question, and I don't have any deep problems in his critique. The biggest remaining problem that I have is the one from organism to externality and I think that Maturana and Varela have framed the essential problem very, very well.

There remains one last fundamental objection to my thesis which I have long considered and which is exposed throughout this dialogue: why then, does it work so well? I have thought this over deeply, and perhaps the best answer that I can make is of a "hive of bees" completing their hive. That is, I think good science is a self-fulfilling prophecy.

The Anthropic Principle

Or, to put it in a more respectable setting, I think it may be the ultimate fulfillment of the concept that I first saw in Penrose's book: of the "anthropic principle".

But the usage I imply here is a deeper sense and meaning of the words. It is not that "if the world were not as it is, then we would not be here to see it", (Penrose, paraphrase), but rather in a sense where "our seeing it that way" allows an algorithmic interaction with a nameless reality. Put more simply, we can only see what we are "designed" = "configured" to see. It is not a matter of external existance, but rather one of "structural coupling"!

We are, however, allowed to extend and expand that vision. But this must be combined with a biological perspective under Cassirer's "Symbolic Forms" to attain the full vision.

---

[24] Save in the "quiddities" of dualism, for instance.

Cassirer and God's Eye:

Durant[25], as well as my JCS reviewer exhibit a fault common with just about all the epistemological philosophers, (to include even Kant[26] himself). They always posit "a God's eye view"[27].

The only plausible scientific alternative to this, the traditional, absolutist approach to epistemology that I can conceive, (to repeat a section of Chapter 6), is in a relativism, (albeit a *rigid* relativism), of epistemology itself.

Cassirer supplied just such a relativism in his "theory of symbolic forms", and mathematics, in its conception of "mathematical ideals" confirms its essence. But Cassirer's thesis, at its bottom is conceptual; it is not based in classes or "objects". It deals instead with perspectives. It deals with abstract "domains". It deals with the (unstructured) "manifold". It is a conceptual, (rather than a set-theoretic), scheme.

---

[25] For instance, when Durant says: "The tree will wither and die whether or not anyone sees it or not." This is a statement that says we really, (ontologically), know! This is an ontological assertion.

[26] E.g. where do the "mind" and the "brain", i.e. cognition actually exist? This was Kant's and Maturana's flaw as well.

[27] They always make statements such as "This" is "a brain", or "The mind does such and so", for instance. These are statements with purportedly absolute ontological meanings.

Repeating Hertz:

> "The images of which we are speaking are our ideas of things; they have with things the one essential agreement which lies in the fulfillment of the stated requirement, [of successful consequences], but further agreement with things is not necessary to their purpose. *Actually we do not know and have no means of finding out whether our ideas of things accord with them in any other respect than in this one fundamental relation.*"[Hertz]^{XIV} [28] (Note: there is an echo of James in this.)

It is just Cassirer's theme —as modified with Maturana's and within my structuralist perspective of the "schematic model" of Chapter 4 that I pursued en route to my third thesis.

It is the only philosophical perspective that allows us to use ordinary descriptive, i.e. naïve realistic language "heterophenomenologically" using Dennett's term. It allows us to use such language without an absolute ontic commitment and allows the employment of a "relativized naturalism"[29] as well –i.e. one that allows us to describe reality in our normal, "natural" terms.

> "Each of the original directions of knowledge, each interpretation, which it makes of phenomena to combine them into the unity of a theoretical connection or into a definite unity of meaning, *involves a special understanding and formulation of the concept of reality.*"^{XV} [Cassirer]

Repeating myself yet again, ordinary naturalism confuses a *particular* organization, (mathematical physics), with the phenomena which are organized. That is the basis of its assertion of reference –and "scientific realism"[30]. "The "objects", (the

---

[28] but there can even be no "things" at all –they may even be "whats".

[29] As developed in Chapter 7

[30] another misnomer

organizational primitives –i.e. Hertz's "images"), of one *particular* form are assumed, (incorrectly), to *reference* ontology –to relate to "an ultimate metaphysical unity". Scientific realism thinks it can salvage its strange entities with "hierarchy" and "emergence", but my objections as stated in the very preface to this book, as well as the whole current effort to reground mathematics beyond set theory effectively counters that claim, I believe. (See my discussion in Chapter 1: Introduction and New Précis: In a Nutshell which I think is conclusive).

> "Where there exist such diversities in fundamental *direction* of consideration, the *results* of consideration cannot be directly compared and measured with each other. The naïve realism of the ordinary view of the world,[31] like the realism of dogmatic metaphysics, falls into this error, ever again. It separates out of the totality of possible concepts of reality a single one and sets it up as a norm and pattern for all the others. *Thus certain necessary formal points of view, from which we seek to judge and understand the world of phenomena, are made into things, into absolute beings.* [Cassirer, my emphasis]"[XVI] [32]

What these "formal points of view" do, instead, is *organize* phenomena. What is consistent under *all* forms, however, are the phenomena themselves.

It results, (and I repeat myself again), in an (improper) assignment of (unique) metaphysical reference rather than a (legitimate) judgment of empirical, (i.e. experiential), adequacy for the primitives of its theories.

> "Only when we resist the temptation to compress the totality of forms, which here result, into an ultimate metaphysical unity, into the unity and simplicity of an absolute 'world ground' and to deduce it from the latter,

---

[31] but see the prior discussion of naïve realism as a biological algorithm
[32] Naturalism, at *whatever* level of sophistication, clearly falls under this injunction.

do we grasp its true concrete import and fullness. No individual form can indeed claim to grasp absolute 'reality' as such and to give it complete and adequate expression.[my emphasis]"[XVII] [XVIII]

In Defense of Kant:

I have said it elsewhere, but I think that Immanuel Kant may be the most misunderstood, misconstrued, and unfairly trivialized thinker in the history of the mind-brain problem. This is quite understandable from the perspective of my earlier comment about the necessity of an inbuilt realist imperative in the human brain.

I argued[XIX] that from a biological perspective it is not important that the "operator" of such a complicated process knows what it is, (specifically), that he is doing. It is important *only that he does it well*. It *is* crucially important that he does it *diligently*, however. It is imperative that he be locked into the loop of his virtual reality -that he "pay attention". This introduces the necessity of an *inbuilt realistic imperative* -i.e. a mechanical guarantee of his dedication, (see P.S. Churchland / Hume).

The universal and dogmatic belief in the simple reality of our natural world is thus itself a consequence of my thesis -and the greatest obstacle to its acceptance!

Durant ultimately dismissed Kant for his "subjectivism" – in Durant's words "'the world is my idea' *as Schopenhauer honestly[33] put it*". Bertrand Russell,[XX] though initially a Kantian, expressed his absolute joy at getting his "objects" back once he had renounced those ideas. Scientists and philosophers have spent two centuries trivializing Kant's brilliant insight. I think it's time to turn our perspective around. Science will demand it! *The science of Mind* will demand it.

---

[33] Durant's characterization

Kant, by his own words characterized himself as a "critical idealist", and I think that may be the biggest mistake he ever made! As I noted in Chapter 6, there is a footnote to this however. Kant himself was never satisfied with "critical idealism" but was forced to retain it for historical reasons.

Kant, I think, was not an "idealist" in *any* sense at all - not even a "critical" idealist as the references to his own words above clearly proves. He was rather what I have termed an "ontic indeterminist" which I think is more descriptive of his actual perspective.

But this is still a "realist" in the most essential sense of the word! Kant was very much a realist about the *existence* of externality. His question instead was what it, i.e. externality, in fact actually *was*.[34] But this is the question that physical science continually asks.[35] Kant's work has severe historical limitations to be sure, but he never claimed his program was completed. This was his implicit sanction for the subsequent evolution of the neo-Kantians, of whom Cassirer was perhaps the most outstanding.

---

[34] Kant reduced externality to a "something". Maturana reduced it to a "somewhat". I have reduced it to the "axiom of externality". It is an intentional *axiom* of realist reasoning.

[35] There is a great similarity between the relationship of Schrödinger's equations and the act of measurement and the reafferent aspect of "acting into the world" and the feedback generated by it as noted in my Freeman Appendix. Neither really has presumptive "objects" before the fact.

I have moved beyond Kant, I believe, but I accept him for the genius he was.

Chapter 13: Discovering Bernard d'Espagnat and Manjit Kumar: On Quantum Physics and Reality. (Bohr, Penrose, and Einstein, of course, were already present in the room!)

This may be the most important chapter in this book, -- along with Appendix D on Niels Bohr, as I believe it confirms that modern physical science has come to largely the same conclusions as has my own thesis which originated from the *completely independent perspective of biology!*

If you were to come to agree with this conclusion then it in itself would constitute an empirical validation of my foundational incorporation of Cassirer's thesis of "Symbolic Forms" -which I will argue in Appendix D is a close parallel to Bohr's "complementarity" principle[1]-*as an actual invariant of epistemology and of cognition itself.*[2] It works "in the real world"!

Upon discovering d'Espagnat's book[1], I think the greatest shock for me was the thought that I have unknowingly been doing work on the "quantum brain" for the last 50 years and not just on the mind/brain relationship![3] For I now think that was exactly what I was doing. Our theses, (and Bohr's as well[4]), are so *very* close in their essential perspectives on reality. [5]

---

[1] though, I believe, a deeper and more philosophically viable conception

[2] Or, equivalently, as I argued in Chapter 9, to the algebraic conception of "mathematical ideals".

[3] I had read Penrose early on in my journey, but saw him as confirmatory rather than as *foundational* to my approach!

[4] Note: it is also argued within Appendix D, ("On Niels Bohr and Cassirer"), that Cassirer's "Theory of Symbolic Forms" is the generalization *across the whole of the sciences* of Bohr's more delimited and specifically physical epistemological principle of "complementarity".

[5] Note: March, 2011: I had never heard of Bernard d'Espagnat till about a month or so ago when he was recommended to my by one of my correspondents as being very much in agreement with my conclusions. In the very opening chapters, I saw that I *must* pursue it. It is absolutely brilliant and

D'Espagnat's profound book has been exceedingly difficult for me however because it is, in seeming contradiction, *so very intimately compatible* with my own conceptions! Each concept, each explanation must therefore be filtered through the *whole* of my own ideas and that is a huge task.

Why Insert Physics into a Discussion of the Mind/Brain?

Modern physics, and especially quantum mechanics is a subject we cannot ignore. It has given us so much of our current technology and our critical life parameters, (good and bad), that it is impossible to dismiss. This chapter is meant to introduce the conclusions of contemporary physics on the nature of reality, -and on the nature of *realism itself,* -into our discussion. Its specific conclusions are precisely relevant to an understanding of the mind and the brain.

I think those conclusions are highly supportive of my own thesis and specifically violate the 19th century presuppositions of most contemporary neurophilosophers. As Roger Penrose might put it: we no longer live in a Newtonian "billiard ball" universe!

This chapter, I believe moreover, actually "rounds off" my conception and original purpose in writing this book. It points to clues for the completion and the creation of the *actual* "calculus" of the mind and the brain which was my original target. It points to a methodology allowing a specific, (rather than a "shotgunning"), intervention into the mind via physical means! I think, moreover, that this chapter in conjunction with Appendix D, (the Niels Bohr appendix), actually confirms my thesis in almost all of its basic aspects!

---

highly relevant to my ideas. I think furthermore that our conceptions fit together like a hand and a glove! I found Manjit Kumar at about the same time. The latter holds degrees in Physics and in Philosophy and is perhaps the most lucid writer I have ever found on the specifically *conceptual* developments of physical science.

Preamble:

First let me say that I claim no expertise in Quantum Mechanics except in the most general sense. I am frankly and profoundly humbled before the intense mathematical and experimental substrate of quantum theory —which I suspect is also the case for most of my readers. This is why I argue from the level of abstraction of the *conclusions* of its greatest thinkers. I am admittedly unqualified to address its problems on their own ground. But, conversely I think, so are *they* unqualified to address the actual problems of biology, and specifically of *cognition*[6] at the same level! I do, however, claim to have some insight into quantum theory's problems from my own perspective which I believe incorporates the crucial imperatives of biology.[7]

Bernard D'Espagnat (and, I suspect the preponderance of quantum theorists as well), treats "observation", "perception" and "experience" as "black boxes", given completely and immediately as wholes. I do not! I see them from the standpoint of biology. I see them from the *operational* standpoint of "structural coupling" and "the interface" as expounded in this book. *At its very core and beginning* my thesis supplies a rationale for concepts largely equivalent to quantum physics' foundational concepts of "decoherence", "entanglement" and "complementarity"! That rationale lies in the independently conceived evolutionary "schematic artifacts" as postulated in chapter 4!

And yet it is the probability of "observing", "perceiving", and "experiencing", i.e. "what we would *expect to* observe/see/experience" of those exact *same* "perceptual objects", (as diametrically opposed to "what is really and truly ontologically there!"), that for d'Espagnat constitutes the real and

---

[6] E.g. "observation" *itself*—which term they incorporate at a foundational level in their own theorization is obviously clearly coupled with a conception of cognition.

[7] You might consider this assertion in analogy to QM's usage of "complementarity"!

actual *core* of modern physics! This, (the probability of observations), is clearly distinct from conceptions of "actual reality itself"[8], (i.e. ontic reality = "substantia phenomena" using Kant's word).

I think my perspective adds a new dimension to the specifically physical problem of "Scientific Realism" *per se* as well, and from which I will presently suggest possible biological answers for some of the quandaries of quantum physics itself. I will make these humbly and tentatively[9] in the course of this discussion, inviting comments by competent physicists. Reading d'Espagnat's book triggers continual flashes of lightning in my mind piercing and illuminating the "swamp gas" of metaphysics!

Preface:

The world of modern physical science and of quantum theory specifically, is perhaps the strangest, and the most important "place" you will ever have to explore if you are seeking the "real truth" about existence. It defines your conception of the actual world you live in!

We adulate, (and believe), Einstein probably, sad to say, because he discovered the theoretical basis for the atomic bomb – which, unfortunately, (though specifically relevant to this exact problem), is a fact of life we *cannot* ignore or deny![10] True, he did other things, and some have called him the smartest man in the world. And we therefore purportedly believe in the kind of world he believed in.[11]

---

[8] Does the Moon exist even when nobody is looking at it, as Durant so pointedly asked in Chapter 12? D'Espagnat's answer is that is a question beyond human reasoning. See d'Espagnat 2006, p.391

[9] Some will say: "presumptuously"

[10] I have personally been to both Nagasaki and Hiroshima.

[11] I have come to believe, I am sorry to say, that most people don't even *care* what scientists believe. That is, they don't care so long as those scientists continue to deliver to them new "toys" – a new modern cellphone, a cure for cancer, a cure for aging or death itself, a snazzier and faster sportscar, less time

But is this true —thinking it over, I don't think so. In my reading, especially of neurophilosophy, and in my occasional perusals of some of the supposedly more intelligent forums on the internet on various subjects, (which reflect the opinions of "educated" laymen), it is astounding how very ignorant the philosophical, physical, and materialist positions expressed therein are of modern science. The most modern ideas that most contain are no newer that Newton's!

Surprisingly we do *not* adulate Niels Bohr in the same way that we do Einstein. But what, in fact, *did* Niels Bohr give us? For one, most of modern chemistry is based on his new rendition of the periodic table which is based in his revolutionary conception of the atom itself. From this came much of modern chemistry —and all the fruits thereof which are enormous!

For another, the cell phone or the computer you are probably using right now —or the Pacemaker that is keeping you alive- is based on the transistor and other components that came directly from Bohr and his co-believers.

But perhaps we should adulate him most of all because he was the man who upheld quantum physics even from the brilliant and ingenious intellectual challenges to it by Einstein that the latter made in support of ontological realism —i.e. in support of your normal "multitudinous" conception of reality, (aka your

---

required to do a job, a pheromone to make them more sexually attractive..... the list is endless.

But those scientists might as well be housed in an insane asylum, practicing the worship and contemplation of the great god "Mumplesquant". When the latter walk to the locked gate with a new and working camcorder, let's say, passing it through the bars to you and saying "here is another proof of the existence and wisdom of Mumplesquant.", we ignore their words or blushingly prevaricate and then accept and use the new toy without a further thought, (except that we will most probably return to that gate!).

When these scientists cease to produce these new toys, we will probably find another asylum which might worship the arcane art of paper cutting. Again we will not pay attention to what they say, but will pay close attention to the quality and quantity of the "proofs" they offer us. We are not committed to their theology you see, but we are most definitely committed to the artifacts they produce!

own normal "billiard ball" world). Recent physical experimentation[12] seems to have actually validated Bohr's position over Einstein's however. Bald "materialism"/"multitudinism" *just does not* seem to work!

I think Bohr was the intellectual equal of Einstein, and that is saying a lot. We *must* take him seriously.[13]

According to Roger Penrose, Quantum Mechanics is the most "SUPERB" theory[14] yet attained by science in that *not even a single instance* of its inaccuracy has *ever* been found![15]

---

[12] More shortly

[13] Here is a plug for a book —but an unpaid and very serious one. Manjit Kumar's "Quantum: Einstein, Bohr and the Great Debate about the Nature of Reality", 2008 It describes the 50 year intellectual duel between these two brilliant minds. It is well worth the reading —essential, I would say —if you *really* want to understand the world you live in.

[14] his classification and his CAPS

[15] Repeating a prior footnote: Penrose gives a nice summation of the accuracy of scientific theories:
" 3. *Accuracy*: need not be perfect, but *extremely* accurate over many orders of magnitude! (*degree* of accuracy is a *value criterion*, however, and is a decision factor in deciding *between* theories.) The degree of accuracy of the "SUPERB" theories is astounding:
A. Euclidean geometry: "over a meter's range, deviations from Euclidean flatness are tiny indeed, errors in treating the geometry as Euclidean amounting to less than the diameter of an atom of hydrogen!" (p. 152)
B. Galilean and Newtonian dynamics: "as applied to the motions of planets and moons, the observed accuracy of this theory is phenomenal -better than one part in ten million. "the same Newtonian scheme applies here on earth - and out among the stars and galaxies -to some comparable accuracy". (p.152)
C. Maxwell's theory: "Maxwell's theory, likewise is accurately valid over an extraordinary range, reaching inwards to the tiny scale of atoms and subatomic particles, and outwards, also, to that of galaxies, some million million million million million million times larger!" (p.152)
D. Special relativity: "gives a wonderfully accurate description of phenomena in which the speeds of objects are allowed to come close to that of light -speeds at which Newton's descriptions at last begin to falter." (p.153)
E. General relativity: "Einstein's supremely beautiful and original theory ...generalizes Newton's dynamical theory (of gravity) and improves upon its accuracy, inheriting all the remarkable precision of that theory...in addition, it explains various detailed observational facts which are incompatible with the

But it is a theory so at odds with our naïve world conception, and at odds even with the Newtonian world, (that "billiard ball world" using Penrose's characterization again), which is still the conceptualization of reality held by most modern philosophers. Even Einstein could not conceive of the legitimacy of Quantum Mechanics, though he helped to form it![16]

Our ordinary realist view of the world starts out with a plurality of objects, localized in space, and, as d'Espagnat characterizes it, obeying the "inverse square law". The latter says that as separation distances double, mutual influence between objects is decreased by a factor of four. (At significant distances, it becomes almost irrelevant!) And Einstein's Relativity adds the postulate: –in which no influence travels faster than the speed of light![17]

D'Espagnat traces the history of this "multitudinous" and "localized" approach in science from Aristotle through its more modern proponents –those of the latter essentially deriving from the foundational approaches to physical reality of Galileo and Descartes up through Newton and finally to modern times. That it has yielded profound and important results D'Espagnat readily admits –this building up from simple, localized initial components through laws to explain huge parts of our scientific and of our ordinary world as well. But it was *already* being contravened, he

---

older Newtonian scheme. One of these (the 'binary pulsar'..) shows Einstein's theory to be accurate to about one part in 10 to the 14th power." (p.153)
F. Quantum mechanics: explains "hitherto inexplicable phenomena...the laws of chemistry, the stability of atoms, the sharpness of spectral lines...the curious phenomenon of superconductivity.. And the behavior of lasers are just a few amongst these." (p.153) "no observational discrepancies" (*at all*) "with that theory are known." " Penrose, 1989

[16] Quantum theory in its "standard interpretation", (Neil Bohr's), was essentially validated at the Solvay conference in Belgium, (1927), by an assemblage of the greatest minds in physics to include Einstein, Bohr, Planck, Marie Curie, Schroedinger,....of whom 17 out of 29 were or were to become Nobel laureates. Einstein disagreed with their consensus, and spent the rest of his life in attempting an alternative solution.

[17] Yes, I'm aware of the recent experiments which question this.

argues, in the very foundations of Galilean relativism itself, and subsequently in the positing of real existence to "fields", and most certainly in Einsteinian relativism.

But in the field of quantum physics, d'Espagnat argues that multitudinism and localization *in themselves* —must be seriously reconsidered. He argues that *demonstrable* and experimentally *reproducible* empirical results[18] force a retreat from the multitudinous, localized conception of reality itself!

But *how* can we change the very way we view our external world?

We are ready to modify our innate naïve realism with the fruits of science —even with the fruits of quantum theory – transistors, nuclear power, et al, but we must seemingly always retain some infinitesimal scale at which "objects" *per se* exist in absolute reality and persevere even when we are not looking at them.

Requoting Durant's assertion, (and implicit demand), from Chapter 12: "the annual elliptical circuit of sun by earth [is] independent of any perception whatever; the deep and dark blue ocean rolled on before Byron told it to, and after he had ceased to be...a tree will age, wither and decay, whether or not the lapse of time is measured or perceived".[19]

Equivalently here is Kumar citing John Stewart Bell himself on the same theme but with the opposite consequence, (it was Bell who had actually framed the context for the Aspect

---

[18] E.g. the Alain Aspect experiments showing a violation of Bell's inequalities which I refer to shortly!

[19] Here is a quote from the closing pages of Kumar's "Quantum" citing a similar remark by Einstein: "'does the moon exist only when you look at it?' he asked Abraham Pais in an attempt to highlight the absurdity of thinking otherwise. The reality that Einstein envisaged had locality and was governed by causal laws that it was the job of the physicist to discover." Kumar, 2008

experiments[20] which severely tested that belief through an actual physical experiment), -a remark made *after the fact* of the experiments themselves:

> "Bell derived the inequality from just two assumptions. First, there exists an observer-independent reality. This translates into a particle having a well-defined property such as spin before it is measured. Second, locality is preserved. There is no faster-than-light influence, so that what happens here cannot possibly *instantaneously* affect what happens way over there. Aspect's" [experimental] "results mean that one of these two assumptions has to be given up, but which one? Bell [himself] was prepared to give up locality. [Bell] 'One wants to be able to take a *realistic view of the world*, to talk about the world as if it is really there, even when it is not being observed', he said'".[21]

We can actually measure the masses and half-lives of particles from a cyclotron for instance -which match Einstein's predictions -and still think of them as persevering "objects". In fact, we can even imagine that a softball sized and shaped piece of metal, which does not seem very formidable as a weapon even to break a plate-glass window, can incinerate an entire city and all its inhabitants —provided that the metal is a specific isotope of the metal uranium! ($E = MC^2$) D'Espagnat argues however that this conviction —of locality and perseverance- is not supportable from modern and reproducible quantum theoretical experimentation which he argues *has already proved it wrong*!

D'Espagnat frames the problem of realism from the standpoint of modern physics -but a very philosophical one, -and he is a *very* deep philosopher. Surprisingly many of his physical

---

[20] See later —these are the actual physical experiments which arguably disprove such a conception!

[21] Kumar, 2008, directly citing Bell in the final sentence, my emphasis.

conclusions are an almost perfect match for my own biological conclusions!

I will not be able to deal with the whole of d'Espagnat's conception in a single chapter[22] – it is too huge, and the parallelism with my own ideas, (up to a certain point), is overwhelming.[23] I will sketch therefore just a few of the correlations I have seen so far.

Introduction: About Bernard d'Espagnat:

Bernard d'Espagnat was a Senior Staff Member of the European Union's CERN project for many years and is a recognized expert on quantum mechanics. Quoting from CERN's website,

> "CERN, the European Organization for Nuclear Research, is one of the world's largest and most respected centres for scientific research. Its business is fundamental physics, *finding out what the Universe is made of and how it works*."[24]

Here is a pretty good capsule summary of Bernard d'Espagnat's background and achievements that I found online[25] and which summarizes his credentials pretty well:

---

[22] See also Appendix D which deals with Niels Bohr's philosophical and epistemological perspectives.

[23] Let me therefore recommend the book itself. It is hard reading but well worth the effort.

[24] My emphasis –please note this "statement of purpose", and specifically the part I have italicized which is incompatible with Bohr's, d'Espagnat's and my own conclusions!

[25] From Mathew Iredale , TPM, The Philosopher's Magazine, Issue 54

[Iredale]

"His interest in quantum mechanics having been awoken [sic] by de Broglie, it was only fitting that it was under de Broglie that he should study for his PhD. Thus began a career which saw him work with some of the greatest physicists of the last century, including Enrico Fermi at Chicago, Niels Bohr at Copenhagen and John Bell at CERN in Geneva."

"D'Espagnat's thinking was very much influenced by his work concerning quantum mechanics, especially the work he carried out on a set of mathematic expressions known as Bell's inequalities.

[D'Espagnat commented:]

"I worked with Bell when he discovered his inequalities. We were at CERN together. ... But we were different in that his intuition was on the realist side, on Einstein's side. He really thought that he had found, with his inequalities, *a test between* [physical] *realism and quantum mechanics*[26] and he really thought that the experimental answer would be that realism is right and quantum mechanics is wrong." [my emphasis]

---

[26] in Bohr's and d'Espagnat's interpretation

[Iredale]

"John Bell formulated his inequalities as a way of testing local realism. Local realism is the commonsense idea that results of measurements are predetermined by the properties that objects carry prior to and independent of observations (the reality part) and that these results are independent of any action (the locality part). Put simply, there exists an external reality independent of observation in which nothing travels faster than light. This was Einstein's view.

In order to determine who was right, Bell and d'Espagnat realized that they had to test Bell's inequalities experimentally. Either the experimental results would obey Bell's inequalities, and thus exhibit a failure of quantum mechanics, or they would violate Bell's inequalities, and force scientists to reject Einstein's (and Bell's own prior) local realist view.

[D'Espagnat]:

"I had the luck to discover in my university a young physicist, Alain Aspect, who was looking for a thesis subject and I suggested that testing the Bell inequalities might be a good idea. I also suggested that he go and talk to Bell, who convinced him it was a good idea -*and the outcome of this was that quantum mechanics won*!".

My comment: -and, equivalently, - *that local realism had lost!*

Chapter 13   d'Espagnat

[Iredale]

"This 'win' for quantum mechanics has had far reaching consequences, leading to a clear confirmation of the phenomenon of 'non-local entanglement'[27] ...

It also meant that physicists had to abandon, once and for all, the concept of local reality.[28] And this raised once again the problem of interpretation: just *what* is quantum mechanics describing?"

[d'Espagnat]:

"I think that quantum physics is most easily interpreted precisely as a tool that enables us *to describe human experience*".[29]

Comment: consider the *meaning* of d'Espagnat's words: "human experience" from the perspective of this book, (from Chapter 4 onwards). This is an area where physicists are blind as they have no way to consider the issue from their own perspective other than a circular one. Unless, I will argue, they were willing to admit the possibility of a biological underpinning to their very own science! (See the discussion of this issue later in this chapter.)

---

[27] See later discussion, and especially the heading: "D'Espagnat Provides Possible Solutions to Some of my own Problems" as regards 'non-local entanglement'.
[28] But see my later comments from the perspective of biology!
[29] my emphases

[d'Espagnat]:

"That is, the questions I raised earlier about uncertainty relationships really arise because intuitively we believe in an *ontological* reality and we believe *that we are able to describe it* and that science can describe it. But quantum mechanics describes *not what really exists* but what we see or what we *would* see", [as "human experience"], "in such and such circumstances."[30]

" 'In other words, the question 'Is an electron a particle or a wave?' is the wrong question to ask as it presupposes ontological reality.[31] Rather, in the light of quantum mechanics, one should say, under certain experimental conditions electrons exhibit wave-like behaviour, and under other experimental conditions, particle-like behaviour. Anything more is pure speculation.'

[Iredale]

For d'Espagnat, quantum mechanics is a *predictive* theory", [of future observations], "rather than a *descriptive* theory", [of actual, i.e. *ontological* reality].[32]

D'Espagnat never claims that his is the only possible interpretation of Quantum mechanics[33]. But he does claim that almost any *other* interpretation violates the actual experimental results of the recent Alain Aspect experiments and the logical results of Bell's theorem.

I think that my particular thesis can actually supply certain alternative perspectives –from the perspective of biology- which I think d'Espagnat might eventually be willing to consider.

---

[30] My emphases

[31] i.e. the existence of an absolute physical and realist world ground –ontology.

[32] Ibid Please note that this is specifically an *operative*, rather than a *representative* perspective!

[33] i.e. his Realist interpretation of a "Veiled, [unknowable] Reality"

Chapter 13   D'Espagnat:Solutions for my Own Deepest Conceptual Difficulties

As I have said, D'Espagnat treats observation / perception as a kind of instantaneous "black box". *But I don't!* I see it instead in terms of the feedback loop of the brain/mind which is the subject of this book and which integrates a much more fundamental, pure and non-representative *strategic process*. I think it makes possible a new and deeper interpretation of this problem than even d'Espagnat's.[34]

I think it exposes a new possibility however, allowing for the re-introduction of a (broadened –and relativized) Einsteinian realism *into* quantum formalism as I will discuss presently.

"Observation" is a foundational part of d'Espagnat's, (and Bohr's), conception, (and of Quantum theories in general), but none of them *ever* examines the fundamentals of "observation" *itself.*

Let me enlarge the thought process which leads me from here onwards. It was one of the first "flashes of lightning" that occurred in my beginning reading of d'Espagnat's brilliant book.

D'Espagnat Provides Possible Solutions to Some of My Own Conceptual Difficulties:

For instance, on "The Static Problem", (introduced at the very beginning of the Dennett Appendix –written long ago), d'Espagnat's book generated a specific inspiration.

Here is a brief excerpt from the Dennett Appendix for reference: "Perhaps the hardest hurdle for my conception is what I call "the static problem". The axiom systems of current mathematics tend to create uniform, "static" fields of objects: the integers[35], for instance, or the real numbers. True, there are special, unique objects, [singularities] within them: pi, or e, or 1 for instance, but these are not promising for the kind of usage we will need to see for viable", [multitudinous], "mental objects…"

---

[34] You should probably review the section "Turning our Perspective Around" of Chapter 4 to follow the sense of my argument.
[35] i.e. the Integral Domain

Chapter 13 D'Espagnat: Solutions for my Own Deepest Conceptual Difficulties

This problem might be productively approached from d'Espagnat's "anti-multitudinous" perspective -that all "samelike manifolds" may be expressed as "quantum states", $(0,1,2,3,\ldots 1,000,001\ldots)$, of the *same* entity![36]

Here is d'Espagnat speaking from the perspective of quantum field theory:

> "To form a broad idea of the general guiding lines of the latter" [quantum field theory], "let us begin by observing that the notion of [*particle*] *creation*", (my emphasis), "is not a scientific one: We do not know how to capture it, and even less quantify it. It is therefore appropriate to try and reduce it to something we can master. Now we do master the notions of a system state and changes thereof.[37] We know how to calculate transition rates from one state to another. And the brilliant idea, the breakthrough, just came from this. It consisted in considering that *the existence of a particle* is a state of a certain 'Something'; that *the existence of two particles* is *another* state *of this same 'Something'*", (my emphasis), ", and so on. Of course, *the absence of a particle* is also a state of this 'Something'. Then the creation of a particle[38] is nothing else than a transition from one state of this 'Something'[39] to another, and therefore we may hope to be able to treat it quantitatively. It is just as simple as that! (In practice – believe it –the matter is appreciably more complex.)..." (Bernard d'Espagnat[40], my emphasis) [41]

---

[36] i.e. of d'Espagnat's "Something"/Veiled Reality or of my "Somewhat"/Externality

[37] Note again: this specifies an operative, rather than an informational perspective

[38] This looks almost like straight-on Cassirer! Is this the comparable place where "the particle" *per se* arises within the strategic brain?

[39] = "ontic indeterminism"?

[40] "On Physics and Philosophy", Princeton University Press, 2006

Chapter 13 D'Espagnat: Solutions for my Own Deepest Conceptual Difficulties

This explanatory perspective could explain the existence of a multiplicity of "protons", for instance, (which under quantum theory *are all "states of" exactly the same particle*), as a solution to the analogous "static problem", (of multiplicity), within the calculus of the brain.

> "...experimental data...demonstrate a complete lack of individuality of particles of the same species lying in the same quantum state."[II] And again: "...the idea that each one of the electrons in an atom is individually in one definite quantum state, (lies on one definite 'orbit'), is just simply erroneous. (According to the only operationally nonmisleading picture we have, every one of them lies simultaneously on all the 'allowed' orbits)".[III]

There need be then, in analogy for the problem of the brain and the mind, *just one* primitive "proton-like-entity", (i.e. a singularity likened to the mathematical "e" –the base of natural logarithms in my prior framing of the problem), or "electron-like-entity", (a singularity likened to the mathematical "π"), varying, analogously, with the "state" of Quantum mechanic's "something", (or of my own "somewhat"). But this singularity immediately implies, from my own perspective, the analogous expansion for the existence within the brain of an analog of quantum physics' idea of "non-local entanglement"[42] within the mind/brain *from the very beginning*! In the adoption of d'Espagnat's perspective on this issue, then from the standpoint of the implicit defintion of the mind, these "objects" comprise aspects of a *single*, connected "entity"! All "objects" / "particles" of the same type would be totally entangled from the very beginning!

---

[41] See Penrose 1989 for another lucid and non-technical introduction to quantum theory. I would also highly recommend Cassidy, 1992.
[42] See later and also d'Espagnat's earlier comments on the Aspect experiment.

Chapter 13   D'Espagnat: Solutions for my Own Deepest Conceptual Difficulties

An Excerpt from Appendix C

Here is an excerpt from my own (recent) Appendix C that I just "borrowed" from there that I believe is relevant to this discussion. If this model is, *in fact* evolutionarily derived as I claim, then it would make absolutely no sense to have just the "primitive model" with its translatory, ("a/d converters"/ "hierarchical/non-hierarchical converters"), "objects"/percepts *by themselves*! *From the very beginning*, intentionality must have been incorporated, so that *action* on those "objects" would have been possible and useful—at whatever early evolutionary stage we assume. The primitive (converter) model and the intentional model therefore must have been *co-evolved* simultaneously! But this then directly implies that the "objects" of the primitive model must *always* have been what I will, (metaphorically only), call "fuzzy objects". I am not talking about truth values on a zero to one scale, but something that is defined, but *loosely defined*, (contiguously, I believe), from its process!

This begins to illustrate the interplay between our intentional functions, (and theorizing is one of them), and our primitive translatory "objects". It makes a case for the "fuzziness" in the primitive contiguity of the latter. Evolution must have constructed them so in complement to the also evolving primitive and specifically *intentional* faculties of the brain! As I suggested in Chapter 1, perhaps they are the objects of a "higher level language" utilized by the intentional faculties.

Quoting, (relevantly), from Appendix B of this book:

Without even considering the deeper implications of QM or of relativity, one need only consider results of the "twin slit" experiment or the implications of its multiple execution to see the point. Not even *cardinality* is preserved!

In answer to a question I asked on this point long ago, a physicist correspondent of mine replied that:

"Yes, you can have many slits one after another, (it is better with Mach-Zehnder interferometers than slits, with the same result that one doesn't know if the photon went

Chapter 13   D'Espagnat: Solutions for my Own Deepest Conceptual Difficulties

> through or was reflected by a mirror.... We can say that one photon may be in an arbitrary number of places at once."[IV]

My point was that even the *cardinality* of this basic object, (the photon), was *purely arbitrary* –it could be 1 or 2 or 3 or 1,000,001 or ..., depending on the branching structure of successive slits and the design of the experiment. But innate cardinality is perhaps the *most basic* "property" we ascribe to ordinary objects, so I think the conclusion is significant.

Similarly, reconsider Penrose's "most optimistic" view of quantum mechanics, (most optimistic for objectivism/naturalism, that is):

> "...I shall follow the more positive line which attributes *objective physical reality* to the quantum description: the *quantum state*.
>
> "I have been taking the view that the 'objectively real' state of an individual particle is indeed described by its wavefunction psi. It seems that many people find this a difficult position to adhere to in a serious way. One reason for this appears to be that it involves our regarding individual particles being spread out spatially, rather than always being concentrated at single points. For a momentum state, this spread is at its most extreme, *since psi is distributed equally all over the whole of space*, (my emphasis),...It would seem that we must indeed come to terms with this picture of a particle which can be spread out over large regions of space, and which is likely to remain spread out until the next position measurement[43] is carried out...."[44]

The particle –this *smallest part of our "object"*- is *not* included, (spatially, reductively, nested), *within* the spatiality of

---

[43] Again note the usage of "measurement".
[44] Penrose, 1989

the atom or within the molecule -or even within the *human scale* object of which it is the theoretical (and supposed material) foundation. Naturalism/objectivism can no longer support therefore *even a consistent hierarchy of spatial scale*!

At the human level, of course, these are very useful tools, and that is just what I propose they are -constructed by evolution into our "interface"! Science and logic suggest *other*, non-scaled and non-hierarchical organizations -i.e. they support *any* other efficacious organization as Cassirerian "Symbolic, [operational] Forms". It is a simple matter of utility and theoretical "shape"[45].

Back to d'Espagnat:

> "What most basically differentiates quantum mechanics from classical physics is not, (as often believed), the fact that its axioms involve intrinsic probabilities. It is the fact that is *not descriptive*" [of ontological reality], "but essentially *predictive* and, more precisely, *predictive of outcomes of observations*. It can therefore be claimed that its 'hard core' *reduces to a set of rules!*"

Comment: (Could these basic and profound rules of its "hard core" be those "implicitly defined" within the "atomic"[46] intentional functions of the brain?)".[V] [47]

---

[45] See Chapter 14: "Cassirer's Symbolic Forms" on epistemology and the aesthetics of theorization.

[46] See the heading "Turning Our Perspective Around" in Chapter 4 for my meaning of "atomic" here!

[47] His discussion of the Von Neumann's "Moveable Cut" and "decoherence" are particularly elucidating in this regard -the latter of which I will examine shortly.

Chapter 13   D'Espagnat: Solutions for my Own Deepest Conceptual Difficulties

Here is Erwin Schroedinger[48] on the same subject:

> "The observations, the individual results of measurements, are the answers of Nature to our discontinuous questionings. Therefore, perhaps in a very important way, they concern not the object alone, but rather the relation between subject and object. For philosophers that is an old truism, but it now gains a heightened significance. It is thus no longer so obvious that repetition of observations must lead ... in the limit to an exact knowledge of the" [ontological] "object. When we interpolate the actual measurements by the best possible means, they are imbedded in continua ... that do not represent the natural object in itself, but rather the relation between subject and object.
>
> ... The wave functions *do not* describe Nature in itself, but rather the knowledge that we possess at any given time *of the observations* actually carried out. They allow us to predict the results of future observations not with certainty and precision *but with just that degree of unsharpness and*

---

[48] Schroedinger was the discoverer of *the fundamental equation of quantum physics* – the "Schroedinger Equation"! Concerning Werner Heisenberg, the equivalent "lynchpin" of quantum mechanics, let me cite a web source: "David Hilbert ... Professor of mathematics at the University of Gottingen ... suggested to Heisenberg that he find the differential equation that would correspond to his matrix equations." [Another source says that the existence of the one necessarily implied the existence of the other!] "Had he taken Hilbert's advice, Heisenberg may have discovered the Schrödinger equation before Schrödinger. When mathematicians proved Heisenberg's Matrix Mechanics and Schrödinger's Wave Mechanics equivalent, Hilbert exclaimed, "physics is obviously far too difficult to be left to the physicists ..."
From http://www.valdostamuseum.com/hamsmith/heishist.htmL

probability with which observations *actually made of the object permit predictions about it.*"[49] [my emphases]

Please note the parallelism with the Hertz quote in Chapter 7 which is part of Cassirer's argument for his Symbolic Forms! But those predictions, as is the case with ordinary, normal cognition could be wrong! Hence, for instance, the actual necessity for the performance of subsequent physical experiment – e.g. Aspect's –see later. This is an instance of D'Espagnat's "the reality which refutes!"

To Return the Favor –a Countersuggestion on the Issue of "Decoherence":

"Decoherence" was *always* implicit in my own theory *from the very beginning.* It describes the re-entrant input of Merleau-Ponty's circular causality through the operative artifacts/metaphors of our primitive, evolutionary and operational "calculus", (i.e. the "calculus" of the brain itself –see Chapter 4). I assert that these artifacts themselves actually *constitute* our "macro" objects, -but they are the (virtually) "concrete" and (virtually) "tactile"[50] translatory metaphors of process itself,[51] (as is the "me" which experiences them)! They are "positions in the structure" of our naïve "calculus"! This is why *all* experiment – be it rigorous scientific physical experiment, or our day to day, minute by minute experiment with ordinary naïve experience must filter through them.

---

[49] Excerpt from a Munich lecture (1930) by Schroedinger - cited in Moore, 1989

[50] such things, e.g. tactile VR, are currently being actively explored and developed in the field of virtual reality! See, for instance: http://science.nasa.gov/science-news/science-at-nasa/2004/21jun_vr/

[51] See W.J. Freeman's Figure 2 in Chapter 4 which is re-iterated shortly.

A Profound Parallelism

Here is a reiteration of part of Chapter 1 of this book which parallels, beautifully I believe, d'Espagnat's perspective:

From the physicalist perspective,[52] what I propose is that "mind" is specifically a function of the organization of behavior itself, and not a function of knowledge of ontology. Loosely stated, I propose that the brain/mind is the evolutionary result (by a multicellular organism) of an optimization of process. It is the result of the specifically *self-organized* evolutionary optimization – but [it is an organization and] an optimization of *blind behavior per se* and not one of knowledge![53]

In that process, I maintain that our naïve perceptual "objects" are non-representative, purely behavioral, i.e. organizational and virtual*)*, *artifacts*, but stable ones. (This, though biologically plausible, is a *very* radical hypothesis, but I believe it is the only viable scientific pathway to the solution of the *other* leg of the problem –i.e. my second hypothesis.)

I propose that these artifacts/"objects" are *re-used* in the "intentional arc", (re: Merleau-Ponty), to test our (behavioral) hypotheses –i.e. both scientific and non-scientific. They are the ground for the whole of cognition.

---

[52] I.e.: Within the specifically Realist and Physicalist biological Intentional Form which is just one amongst the many Cassirerian "Symbolic Forms". Please recall my arguments for a relativized materialism in Chapter 6.
[53] See Chapter 4 heading: "Turning our Perspective Around –a Model of Process" for the rationale for this radical change of perspective.

Chapter 13   A Profound Parallelism

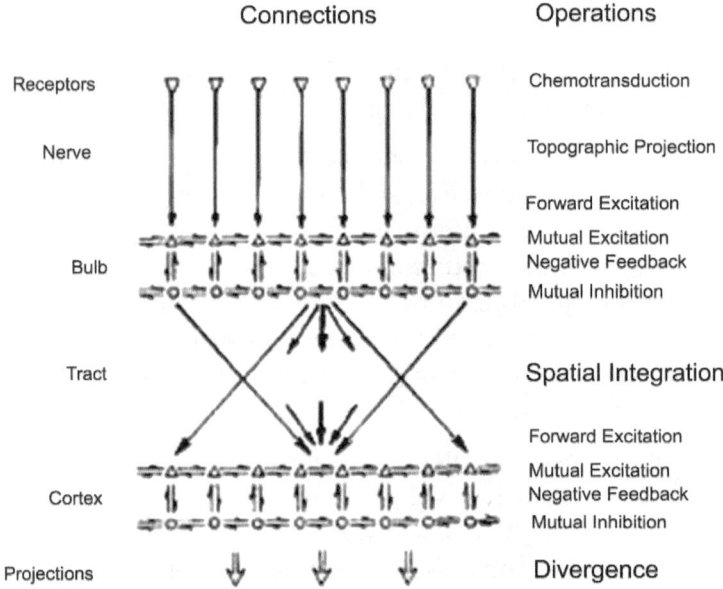

FIGURE 6, (Freeman's Figure 2)

But these artifacts, (our naïve objects), need not correlate hierarchically to absolute [ontic] reality, (see W.J. Freeman for instance, -Freeman's fig. 2—my Figure 6- reproduced above wherein he reveals a specifically *non-hierarchical* mapping into, (not *onto*), the cortex).[54]

It is necessary only that these "objects" be locked into the re-entrant loop between action and perception which passes we know not where!

(Note again how closely this perspective of biological "circular causality" fits with modern quantum theory -i.e. in the

---

[54] I think that Freeman's fundamental role in the foundations of cognition will be found to correspond very closely with Max Planck's role as the discoverer of "the quantum" in Quantum Physics! Remember that Max Planck's original job was to make a better lightbulb! (See Kumar, 2008)

Schrödinger equation vis a vis "measurement"![55]) Repeating Penrose:

> "But there is something very odd about the relation between the time-evolved quantum state', (the Schroedinger equation), "and the actual behaviour of the physical world that is observed to take place. From time to time —whenever we consider that a 'measurement' has occurred – we must discard the quantum state that we have been laboriously evolving, and use it only to compute the various probabilities that the state will 'jump' to one or another of a set of new possible states." [VI]

But each new instance of a measurement causes yet another "loop"!

But What *is* "Measurement" for the Brain?

The mind, I assert, is a similar looping and circular probability machine -in this case utilizing the feedback/intentional aspects of the brain. It must countenance each "measurement" against our biologically innate, (and stable), evolutionary objects/artifacts and then recompute its overall picture and strategies. This is what cognition is.

Re-iterating from Chapter1:

I maintain that our mental, ["naïve"] "objects" are the evolutionary yardstick we carry. They function to crystallize and organize our input, and to crystallize and organize our output. But they must be *rigidly* maintained as the "working gears", (alternatively the "A/D converters", or, better still, as the hierarchical/non-hierarchical converters), of perception. I argue that they are organizational artifacts only!

This is the answer to the question of how a non-hierarchical mapping, (e.g. Walter Freeman's chaotic dispersive

---

[55] You might also consider it in the light of the Raichle discussion of Chapter 3.

mapping, or Edelman's non-topological "global mapping" —or d'Espagnat's or my own), could specifically function in cognition. I think it also gives a very pointed clue to Penrose's problem.[56]

> "In particular, Maurice Merleau-Ponty in 'The Phenomenology of Perception' conceived of perception" [itself] "as the outcome of the "intentional arc", by which experience derives from the intentional actions of individuals that control sensory input and perception. Action into the world with reaction that changes the self is indivisible in reality, and must be analyzed in terms of "circular causality" as distinct from the linear causality of events as commonly perceived and analyzed in the physical world." W.J. Freeman, 1997 ,(as cited previously)

This particular thesis, (the first of my three hypotheses), supplies the necessary perspective of biology and the brain. It is our very own "cave of shadows", (Plato), -but it need not even be projective, (as a "shadow")! I propose that[57] *it is the evolutionary result of a self-organized and virtual optimization of pure response*! It is instead as a GUI, (graphic user interface), rather than as a "shadow" or a "projection" that it functions. And GUI's actually have the potential for this as I concluded in Chapter 4.[58]

---

[56] Note 2011, (again): re "the flow of time"!
[57] from the perspective of relativized materialism, -see Cassirer's "Symbolic Forms", Chapters 7 through 9, —and from the *relativized* biology argued within it —Chapter 8
[58] I think W.J. Freeman's experimental results form a close parallel with the Alain Aspect results of Quantum physics! They force the conclusions of Freeman's own "Bell inequality" of information transfer!

The Crucial Test of Realism: d'Espagnat on Bell's Theorem:

> "To make use of" [Bell's Theorem] "right now, we need not resort to its full explicit formulation, nor to the whole array of the notions it involves. It is enough that we should take into account one of the main points it establishes, which is that *no local hidden variable theory*[59] *is capable* of reproducing in detail all the quantum mechanical predictions *that the Aspect-type experiments actually showed to be correct.*"[60]

> "Since, to repeat, the experiments in question *did corroborate* all these predictions, they prove, in view of the theorem, that no local hidden variable theory is compatible with the data. In other words, they disprove the local hidden variable explanation that looked so attractive at first sight. It is this very disproof that constitutes *the* significant point here.

> It is important to realize that this is in no way a rejection grounded on some a priori philosophical conception but, quite on the contrary, a disproof based on plain facts." [61]

Comment: A Question for d'Espagnat: Just what, precisely, could the expression "plain facts" possibly mean to him?

---

[59] Here is a citation from Kumar which clarifies the concept of a hidden variable: "Einstein's explanation of Brownian motion in 1905 is an example where the 'hidden variable' is the molecules of the fluid in which the pollen grains are suspended. The reason behind the erratic movement of the grains that had so perplexed everyone was suddenly clear after Einstein pointed out that it was due to the bombardment by invisible, but very real, molecules." Kumar, 2008

[60] my emphasis

[61] D'Espagnat, 2002

[He continues:] "Clearly, this result is of considerable generality. Within the realm of a world view consistent with objectivist realism, *neither one* of the two alternative assumptions that wave functions are real or that local hidden variables exist makes it possible for us to keep our instinctive way of understanding correlation at a distance.

Yes, but —some will ask- perhaps it still could be saved, at the price of somewhat watering down a few of our intuitive claims about particles and so on. To this question the answer is 'no'. We shall see that, according to its explicit formulation, the Bell Theorem implies that the outcome of the Aspect-type experiments *is incompatible with a conception of reality involving a "locality" principle* (a principle roughly stating that influences *are not* propagated at infinite speed)."[62]

"Hence, reasonable as it may look, this principle is violated in nature (at least if this word is taken in its usual sense). It is this violation that is called *nonlocality* when it is considered all by itself, independently of any theory. But of course its strong link with quantum theory can hardly be ignored.

Let us remember our analysis of the two-photon experiment. There we introduced the non-separability notion. In its broadest sense nonseparability implies that we must either accept nonlocality ... or, as quantum mechanics forcefully suggests, agree to change our concepts to a greater extent than we should care to imagine, indeed, *to the extent of giving up the view that*

---

[62] My emphasis

human-independent reality is embedded in a normally structured space (existing independently of ourselves) and", [itself], "made up of parts the mutual interactions between which decrease as distance increases."[63]

Comment: So we end up with a "somewhat" of externality contradictory to the materialist/multitudinous realist conception of our ambience/externality!

Thoughts about Einsteinian Realism:

I think, however, that "scientific realism", (which, contrary to what it purports to be, is a specifically *intentional* stance[64]), is itself actually *demanded* in my overall thesis in some *relativized* form because of a requirement for a preservation of invariants *across all the possible, viable Symbolic Forms* within my "theory of relativity" of the mind.[65] This "scientific realism" does not imply *knowledge* of the world we live in however, but it *does* imply an assumption equivalent to Kant's "substantia phenomena"! It does imply an "Axiom of Externality"[66] *taken in its most abstract sense*![67] It implies, moreover, "some conception which will bind together the observed facts" as well:

> "It is existence and reality that one wishes to comprehend. ... When we strip the [this] statement of its mystical elements we mean that we are seeking for the simplest possible system of thought which will bind together the observed facts." (Einstein 1934, Pps. 112-113)

---

[63] ibid, My emphases
[64] because I have argued a relativized "materialism" itself—see Chapter 6. Also please remember my definition of the "Axiom of Experience" in Chapter 7 as that "thing" which remains constant under all possible, viable phenomenal explanations of reality, the latter being taken in Quine's broadest senses.
[65] It is the actual basis of my own "Axiom of Experience".
[66] See Chapter 6: The Axiom of Externality
[67] See the Bitbol/D'Espagnat dialogue later in this chapter.

But the "simplest possible system of thought which will bind together the observed facts" for me *must* include the inherents of human cognition and perception. It must include the specifics of the mind/brain relationship and the bases of "perception" and "observation" themselves!

Quantum physics becomes then a profound revelation and an instantiation of Maturana's "structural coupling" –but *minus* the "congruent"! I think that Maturana made the same mistake in "*congruent* structural coupling" that Kant did in imposing his "categories" on ontology. Both made these moves to account for, (paraphrasing) "the extraordinary success of our human enterprise in the 'actual' world" = ontology! They are both wrong in this instance, and both right in their more basic concepts of (mere) "structural coupling" and (mere) "substantia phenomena". "Ding an Sich", ("the thing in itself"), does not follow as a necessary consequence of either of them!

To cite Kumar: (paraphrasing) "What would Einstein have said in response to the Alain Aspect experiments?" Einstein, according to Kumar was prepared towards the end of his life to give up "locality", (i.e. "multitudinism"), to preserve "realism". But this willingness to give up "locality" translates into an acceptance of Kant's *raw* "substantia phenomena", to Maturana's "ambience" (via *bare* structural coupling), to d'Espagnat's "something", and to my (relativized)[68] biological "somewhat". It is the realist intentional "axiom of externality" in its broadest –and *most minimal* sense[69] that is specifically demanded here![70]

Translating Einstein's statement, (above), through my specific conclusions yields essentially *the same results* that both d'Espagnat and I arrived at independently. This would be a

---

[68] Remember that I proposed a relativization of materialism itself, (and the biology within it), as one of the steps towards Cassirer's "Symbolic Forms" in Chapters 6 & 8.

[69] E.g. Damasius "pante aporeton" see later

[70] See Chapter 6

broadened Einsteinian realism, but one which would take in the results of biology along with those of physics. In deference to this profound and *profoundly open* mind, I will continue to name it so.

Why then, and how then does the human enterprise achieve its efficacy. My "hive of bees" metaphor? I don't know, but perhaps my earlier answer of the extension of the arthropic principle might work as well. Ultimately, I think they are both legitimate –I think they are "complementary" in Bohr's usage of that word![71]

Is Mine a *Simple* Solution?

Is mine a *simple* conclusion? Obviously it is not,[72] but it does bind together all the observed facts, including, but more than just those of physics itself. I believe, however, that my conception "is the *simplest possible* system of thought which will bind together those observed facts" –*all of them*![73]

For the mind, I have concluded, there must be some sort of "localization"[74] in our "fuzzy objects"[75], but I think it lies closer to the perspective of the mathematician John L. Bell's purely mathematical perspective of "local", [and relativized], "mathematics" as expounded in chapter 2, than to Einstein's ontological *physical* perspective.

---

[71] See Appendix D: Bohr's Epistemological Roots
[72] I do maintain that it is cohesive and complete however!
[73] See an expansion of this discussion in Chapter 14: "Overall Conclusions and Opinions" under the heading: Cassirer's Symbolic Forms.
[74] Because of the need for a preservation of *invariants* to meet the requirements of symmetry across the multiplicitous "Symbolic Forms" of Cassirer. It is this invariance that led to my definition of "experience", ("intuition" in Kant's terminology), as our second and necessary axiom of realist intentionality.
[75] because of what I conceive as their evolutionary origin –but *this* particular conception is *itself* a "Symbolic Form" utilizing Darwin's masterstroke of simplification. But it *still* remains just one of a multitude of possible, alternative symbolic forms and incorporating a relativized materialism!

What could "entanglement" mean from the standpoint of the brain? If our "objects" are really metaphors of a process which is fundamentally strategic rather than representative, then it says that it is our *strategies* themselves that are linked together!

But they link through the fuzzy "converter" objects of the primitive model! This then throws us back to Hilbert's original conception as it is presently being mathematically reexamined from the perspectives of category theory and structuralism!

About "Other Minds"

Quoting from Chapter 12: Durant was unwilling to consider the deepest implications of the existance and of the possibility of the mutual agreement of "other minds", "other brains", (which I think both he and I accept), and which has deep implications to this problem.

If another mind, another brain sees and acknowledges the same "facts of reality" that I do, could it not merely mean that it too has processed these in the same way that I would, that our naïve worlds are similar precisely because our *brains*[76] are so – irrespective of the character of the underlying "substantia phenomena"?[77] Durant's argument, a simple appeal to popular agreement[78] does not really address the substance of the issue.

---

[76] But what then are "brains"? I will repeat my very early injunction that in any theory of deep metaphysics *all* terms should be assumed to be in quotes –i.e. they should be taken heterophenomenologically – at least until the final conclusions. This allows a "context-free" discussion in the sense of Van Fraassen. I have supplied an actual answer to this problem in my third thesis. See Chapters 7 – 11.

[77] Think about the possibility of two minds with alternate primitive conceptions of physical reality, (just as, for instance, Benacerraf conceives of two minds with alternate conceptions of set theory). Each might see "the deep and dark blue ocean [that] rolled on before Byron told it to".

[78] an example of an "argumentum ad populum"

FIGURE 28

This is the model I propose for human reality, but it is lived through the "gears and levers" of our evolutionary artifacts, the latter being understood from the perspective of Biology, itself being just one of Cassirer's multitudinous "Symbolic Forms". This, I believe, is the homo sapien cognitive reality!

That "a tree will age, wither and decay, whether or not the lapse of time is measured or perceived" is a certainty within our worlds, but the very substance of the assertion must itself be taken heterophenomenologically! (See my Figure 28 reproduced above).

(Please note again how closely Figure 28 resembles the picture of philosophical idealism! But the "black space" is *not* non-existence; it is ontic *unknowability*.)

This, I assert, is the reality of our human linguistic and cognitive world: we all speak the same language, but we are all equally *ontologically blind!* Therefore the totality of our dialogue must be interpreted heterophenomenologically, (using Dennett's word again).

The viewpoint embodied in Figure 28 is the picture I suggest as an alternative ... I *believe in,* (but I do not in fact *have ontological knowledge of the existence of)*, other minds -I get to have beliefs too![79] But I believe as well that these other minds see through the exact same "gears and levers" that I do.

---

[79] Schroedinger's "Personality Principle"

> "...states that one's fellow human is a being like oneself. This principle makes science possible, because no person in isolation can create scientific works, and any person who embarks upon such works must know that he or she can use the contributions of other persons.

> Yet the other person is not exactly like oneself. He cites the 'scientific report on a crucial experiment' given to him by a three-and-a-half year old child: When you pinch yourself, it doesn't matter to me, but when I pinch myself, it hurts.' 'Take note', he says, 'of the wonderful 'it', that recalls the famous 'it' of Lichtenberg, with which he corrected the Cartesian 'I think, therefore I am' (which should say 'it thinks, therefore it is'). [Schroedinger] argues that science must accept the personality principle. Otherwise 'it must render all its own activities meaningless'. ...The naïve statement of the child leaves science in a painful dilemma, pushing it towards an acceptance of a multitude of egos which must have some incomprehensible, completely unexplained relation to the individual human bodies. 'Yet the most wonderful and most sublime of all teachings, the Brahman doctrine that the all equals the unity of consciousness, culminates in a mystical victory over this dilemma, the words so obscure to the understanding, so close to the intuition *Tat twam asi*, ["That Thou Art."] ...But it would be a vast error to believe that science knows any better or clearer answer concerning these things.' Moore, 1989

That our conclusions about reality should agree does not surprise me. We all see with the same indeterminacy that Figure 28 shows *but through the same parameters*! This is our human linguistic and cognitive world: we speak the same language!

D'Espagnat contributes another perspective:

D'Espagnat's discussions about the power and the central role of the "realism of the accidents" argument as perhaps the most fundamental underlayment to a belief in "Scientific Realism" is entirely relevant here. But, surprisingly and conversely, "the realism of the accidents" might be taken as a specific argument for my own "schematic model"! (See the heading: "The Realism of the Accidents" in Chapter 4.)

His argument –taken at the highest level- matches in detail my assertion of the *permanence* of our fixed evolutionary "objects"[80]. I assert that these *metaphors* and *nexuses* -and the concomitant naïve (schematic) "calculus" to which they are subject constitute the very essence of our consciousness!

His discussion is almost precisely a restatement of my argument for the schematic model and supplies the reason for our innate biological commitment to naïve realism that I have argued. We require an *absolute dedication to the algorithm* for our continuing existence! It supplies a "higher level language" to enable the use of our intentional, (and at the deepest level, of our ultimately "atomic"), faculties.

---

citing Schroedinger P.252 Note: Please compare the very similar conclusion of my Chapter 13!

[80] These "A/D /hierarchical/non-hierarchical" converters in the cortex seen in the context of W.J. Freeman's diagram –See Chapter 1, Figures 6 & 7, and Chapter 4, Freeman Appendix

How Then Do We get to the Aspect Results?

I originally had this problem all wrong![81] Upon reflection it became clear to me that the consequences of the Aspect experiments became a *confirmation* of my thesis, not a problem for it! That is, it became clear once I adopted d'Espagnat's clue in my earlier reference to his work. (See the earlier reference in this chapter to "D'Espagnat Provides Possible Solutions to Some of My Own Conceptual Difficulties" regarding the "static problem".)

He provided the crucial clue to my own prior dilemma of the "static problem" which I had addressed in the very opening paragraph of Appendix A: "the Dennett Appendix. Following d'Espagnat's lead, non-locality then became a *necessary* and *simple* consequence of my thesis as shown in that prior discussion. Non-local entanglement and action at a distance are necessarily implied as simple and direct consequences of my adoption of d'Espagnat's perspective on this issue.[82] As he says the Aspect experiment compels us to renounce the notion that we are living in a "*normally structured space ... made up of parts* the mutual interactions between which decrease as distance increase"!

Repeating a prior citation:

> "In its broadest sense nonseparability implies that we must either accept nonlocality ... or, as quantum mechanics forcefully suggests, agree to change our concepts to a

---

[81] It led, I think, to d'Espagnat's bewilderment when he perused an earlier draft of my book I had sent him prematurely in my initial enthusiasm. Though he agreed with my fundamental presuppositions, I believe he was disoriented and confused with this specific section. I have since reconsidered this particular problem and realized the error of my earlier, incorrect perspective. See "d'Espagnat Replies" cited shortly.

[82] See again my heading: "Turning our Perspective Around –a Model of Process!" in Chapter 4 –but in conjunction with the "static problem" of Appendix A: the Dennett Appendix

greater extent than we should care to imagine, *indeed, to the extent of giving up the view that human-independent Reality is embedded in a normally structured space,* (existing independently of ourselves), and made up of parts the mutual interactions between which decrease as distance increases."[83]

I suggest that we live, instead, in a "space" of predictions and process –i.e. intentionality. The strategic brain counsels *only* strategies and feedback, it does not deal with "things"! It gives us "what we would expect to observe! This yields an almost perfect parallelism between the results of D'Espagnat's and my own vastly different conceptions and beginnings!

The potential of the "GUI" and "the interface" *per se* was a specific target of my argument in my early paper: "Why: Mind- the Argument from Evolutionary Biology, (Virtual Reality -A Working Model)". It culminated in my discovery and interpretation of the experimental neurological researches of the noted neurophysiologist Walter J. Freeman.

Repeating Figure 7 below:

---

[83] d'Espagnat, 2002. P. 57, my emphasis

Chapter 13 How do we deal with the Aspect Experimental Result?

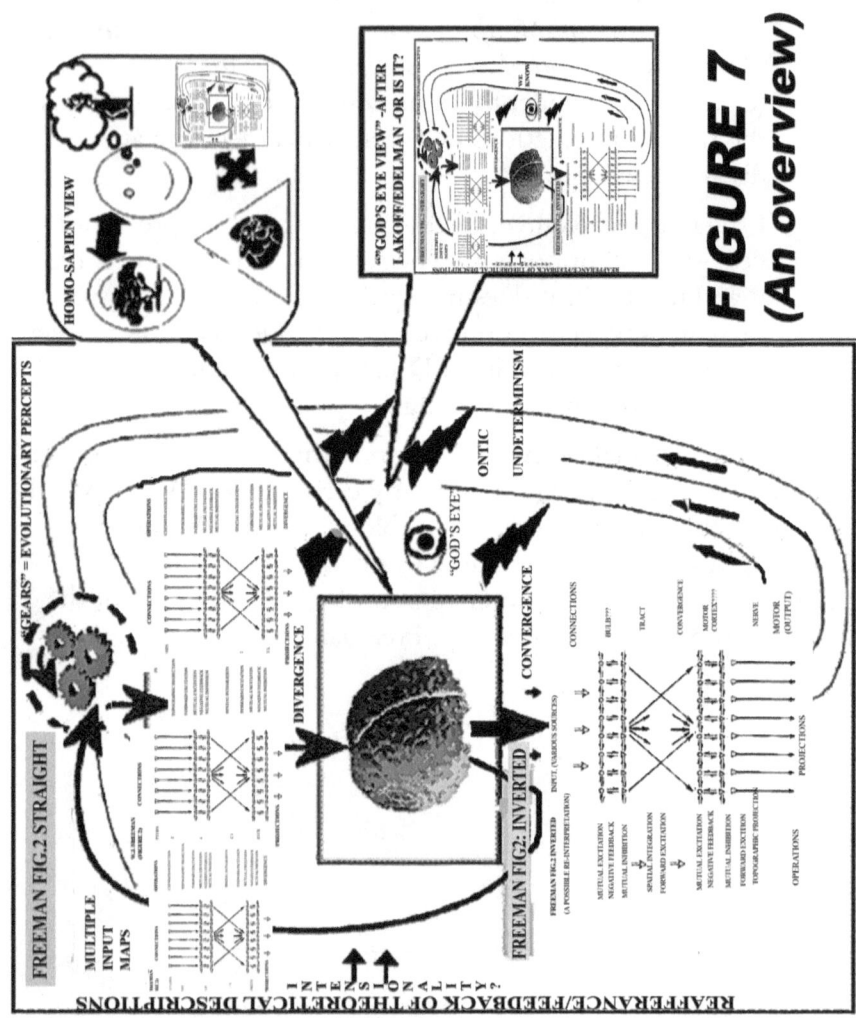

FIGURE 7 (An overview)

Another Idea Inspired from Reading D'Espagnat

D'Espagnat maintains that his three main branches of the quantum formalism: Dirac's Sea, Feynman's computational calculus, and Quantum field theory are *precisely equivalent* (computational) interpretations of the quantum formalism, so Feynman's, specifically here, is computationally equivalent to d'Espagnat's choice which is quantum field theory. There is nothing that can be explained in one that cannot be explained in the other!

> "...it so happens that the mathematical formalism", [of Quantum Mechanics], "yields not one but three distinct theories, all of them grounded on the general quantum rules, yielding essentially the same observational predictions, but widely differing concerning the ideas they call forth. They are called the 'theory of the Dirac Sea', 'Feynman graph theory', and 'quantum field theory.'"[VII]

Think about this: though d'Espagnat veers away from Feynman's "objects" as being absolutely uninterpretable as "real objects", think about those computational "objects" themselves. Think about them instead within the context of this book -in the context of the primitive beginnings of our evolutionarily derived intentional "calculus" of brain process which, I have come to conclude, *co-evolves* with the primitive A/D model of process.[84]

> "Let it just be mentioned that this sham ontology", (as he characterized the attempts at a metaphysical interpretation of Feynman's work), "spontaneously emerged on the basis of the findings of just one (but outstandingly brilliant!) physicist, the theorist Richard Feynman. Outwardly, the formalism Feynman elaborated ...quite appreciably differs from the quantum field theory one. In it, for example,

---

[84] see Appendix C: Some Random Further Thoughts about the Proposed Model

there is no question any more of a strange 'something' whose state changes account for apparent particle 'creation.' But the two theories merely predict observational results, and they are quite strictly equivalent in this respect.

The advantage of the former is of a purely computational nature. However, this advantage is enormous, due to the fact that, in Feynman's formalism, most complex calculations may be pictured visually. More precisely, these calculations involve long mathematical formulas that are essentially products of several complex factors. But it so happens that the way the latter combine with one another may be represented by means of graphs, constituted of straight line segments, each one of which corresponds to one of the said factors.

Feynman had the simple but most effective idea of mentally associating with each such segment the motion of a particle that either is identified with one of those actually taking part in the process under study or is just a fictitious one. For example, if the graph has the form of the letter H the 'horizontal' segment in the middle, which has no free ends, corresponds in the formula to a factor the nature of which is that of a denominator. It is mentally associated with a fictitious particle called 'virtual'. And as for the other segments, those with free ends, they are mentally associated with the two particles whose 'collision' the calculation is meant to study.

This procedure, which amounts to replacing a long and abstract formula by a picture, summons up in view of the planned operation *the (considerable) visualization abilities of the brain.*" [Please take note of this phrase!] "It thereby makes the calculations easier to such an extent that it enables physicists to perform highly complex ones *that could never have been done without this help.* Under such conditions, it is not surprising that the physicists in question should, on this basis have invented something akin to a pseudo-ontology. It consists in, for example, interpreting the H diagram by saying: 'One of the incident particles (one 'leg of the H) proceeds to a certain point, where it emits a virtual particle (the horizontal segment) and continues on its way, while the other one (the other 'leg' absorbs the virtual particle."[VIII] [my emphases]

D'Espagnat finds this pseudo-ontology impossible and incomprehensible however!

"Knowing that virtual particles are but symbols for denominators present in formulas yielding calculation recipes, we may deplore that thinkers of great ability devoted so much effort to the task of understanding to what 'category of beings' virtual particles belong. But it must be said in their defense that the physicists' current language could not but incite them to quite seriously consider this question."[IX]

But *why not?* From the perspective of my interpretation of the brain and the mind, I think there exists a definite possibility here. If our brains are just huge, specifically strategic mathematical probability machines with feedback, and our realist "objects" are *themselves* "just metaphors i.e. artifacts", ("utilizing the considerable visualization abilities of the cortex", then *why not!* Why not think of Feynman's "virtual particles" as the "x-ray image" so to speak, showing *the mathematical "skeleton"* of primitive or extremely precise intentional objects at the fundamental level of brain process? Please think about this —I think it is a very pretty and pregnant idea!

I think mine is a very plausible rendition of the strategically predictive and intentional mind *per se* in contrast to its being considered as a representative rendition of ontological reality! As d'Espagnat says "the problem arises because we ask whether the electron exists as a wave or a particle —that is we are asking questions of ontology that we can never answer" (paraphrase)

D'ESPAGNAT –repeating: "That is, the questions I raised earlier about uncertainty relationships really arise because intuitively we believe in an *ontological* reality and we believe *that we are able to describe it* and that science can describe it. But quantum mechanics describes not what really exists but what we see or what we would see", [as "human experience"], "in such and such circumstances." [My emphases]

"In other words, the question 'Is an electron a particle or a wave?' is the wrong question to ask as it presupposes ontological reality. Rather, in the light of quantum mechanics, one should say, under certain experimental conditions electrons exhibit wave-like behaviour, and under other experimental conditions, particle-like behaviour. Anything more is pure speculation. For d'Espagnat, quantum mechanics is a predictive theory"[85]

My comment is that it is a predictive theory by and of the *specifically human* mind –but at a particular moment in time![86]

Quantum Physics and "Normal" Cognition:

Quantum physics predicts "what we would expect to see/observe" if a carefully controlled physical experiment is carried out.

I argue that *ordinary and normal* human cognition functions like D'Espagnat's "Veiled Reality" interpretation of quantum mechanics as well. Our brains plot a "state vector"[87] so

---

[85] As cited earlier - From Mathew Iredale , TPM, Issue 54

[86] To repeat my universal premise: –as seen through the filter of a relativized, Cassirerian symbolic form –of a relativized –though still (intentionally) "realistic"- and "materialistic" biological symbolic form

[87] which is what W.J. Freeman's conception essentially entails

to speak, and then act on it. That "state vector" is calculated, (via the intentional functions of the brain), using our primitive calculus, prior experience, stereotypes, (Rosch's "prototypes"?), cultural influences, et al to give us a basis for action –the actual results of which may, in fact, *disagree*[88] with those prior objectives, but which then *again* affects the prediction of other *subsequent* outcomes. This is Merleau-Ponty's "intentional arc" as seen through Walter J. Freeman's anti-hierarchical filter!

The collapse of the brain's "wave function" recycles action back into the unknown world and receives re-entrant feedback through our *same* "gears & levers" which are our *stable* evolutionary artifacts as argued in Chapter 4, (through the macro, human scale artifacts of the primitive "calculus" of the intentional brain!)

But this is our normal -everyday, minute to minute life –it happens every time we decide to cross the street, decide whether a questionable morsel of food will taste good, decide what clothing we will wear –or decide to buy a vacation package to Afganistan as a nice sightseeing experience!

Is rigorous science then better –or so different from -our semi-primitive "calculus" that has worked so well throughout our evolutionary history? Consider this –rigorous science itself must consider the results of its predictions/observations on the (macro) observer himself in *his real world* -it predicts what he would be expected to "see"/ "observe"/ "experience" on the macro, (decoherent) scale itself. Quantum physics becomes therefore a predictive calculus of experiential/cognitive results! It is about the mind and cognition itself!

---

[88] in terms of d'Espagnat's "the reality which *refutes*!"

Chapter 13 — I Think I Have Established my Case Re: Modern Physics

I Think I Have Established my Case Vis a Vis Modern Physics

I must now cease to discuss d'Espagnat's book except for a few casual references.[89] If my conclusions are correct, this is a huge and never-ending discussion which must be continued over the future course of history.[90] But I think I have conclusively established my case vis a vis modern physics at this point!

Surprisingly, there is too much to say *about almost all* of d'Espagnat's book! For instance the dialogues between d'Espagnat and Bitbol and Swirn in the much later pages of his book present just one of many instances. I find this particular dialogue very close to my own perspective —especially Bitbol's views.

Bitbol refers d'Espagnat to Damascius's[91] conception of "*pante aporeton*" which he translates as roughly equivalent to "the abolutely inexpressible".

D'espagnat's dialogue with that conception reminds me very much of my own discussion of "bounds & limits" and the raw "axiom of externality" which followed from them in chapters 4, & 6 in my own book. But that particular discussion must be viewed in the light of my prior discussion in section 3.2: "Turning our Perspective Around" of Chapter 4 which embodies a shift from an "information model" to a pragmatic, operative one. Together they yield essentially the same conception. We don't *parallel* externality, (nor do our "objects" *represent* it), but we must postulate such an "Axiom of Externality" nonetheless!

---

[89] I am running into space and time limitations and need to finish for now to finalize this book. I need also to begin a new one which explores the implications of this one for the future of humankind. It has been in my brain for over 20 years and I think is more important than this one. I seek a better world!

[90] NOTE: I will not claim that mine is the deepest interpretation of my overall conception. There might exist even deeper variations at the level of biochemistry, bacteriology, genetics… for instance. But I will claim that the central core of my thesis will be maintained within any such deepening as an epistemological and paradymal invariant. I believe it is the "Euclidean Geometry" of any future epistemology.

[91] "One of the last neo-Platonists". D'Espagnat 2002, P. 395

We must postulate that "it exists" in the *most minimal*, the *rawest* interpretation of that term! D'espagnat's discussion is very lucid in my own terms which I had evolved without any prior exposure to these philosophical concepts –it fits! Think also of d'Espagnat's conception of "the reality which refutes" discussed again here –in terms of the re-afferent feedback into the intentional aspects of the brain." I too agree that "it is the reality *which refutes*!" It is a very direct correlation.

But these viewpoints are simpler and more easily intelligible within my own particular perspective, I think –within a relativized biology.[92] Conceiving the brain as an instrument of process, of strategy, rather than of information enables a much simpler and easier perspective.

On a general note, while philosophers may talk very deeply and profoundly, they never seem to look down to see the very ground from which they are actually talking, ("plain facts"?). I think I have managed to overcome that difficulty in this book by pursuing Cassirer's perspective of a relativization of *epistemology itself*.[93] It is that very perspective that makes my book so hard to understand however! That is, unless we are willing to adopt Kant's original caution into the very bottom of our minds!

> "If in a new science which is wholly isolated and unique in its kind, we started with the prejudice that we can judge of things by means of alleged knowledge previously acquired –though this is precisely what has first to be called in question –we should only fancy we saw everywhere what we had already known, because the expressions have a similar sound.
>
> But everything would appear utterly metamorphosed, senseless, and unintelligible, because we should have as a foundation our own thoughts, made by long habit a

---

[92] As proposed by Cassirer within his "Symbolic Forms".

[93] Refer to Appendix D, for a comparison of Cassirer's and Niels Bohr's epistemological relativisms.

second nature, instead of the author's." Kant, Prolegomena.

It is precisely my point that the science of cognition is a *new* science, never treated precisely before except by Kant who, I think, went subtly wrong. I believe my thesis of the "strategic brain" supplies the first viable and constructive possibility of such a conception.

An Important and Very Deep Implication for My Thesis:

The above passages inspire a "logical leap", but, I think, a plausible and exciting one! Given the strong correlation I have just argued, is it possible that the very quantum mechanics *itself* might supply the beginnings of the "schematic calculus" I have postulated *for the human brain*? Viewing d'Espagnat's overall picture which I am slowly comprehending, it certainly seems to fit at least the intentional functions of the brain–*and maybe more*!

The point I wish to make here is that there *might already exist* the mathematical beginnings of such a deep "calculus"[94] applicable to the brain paralleling the already existing and developed mathematics within quantum formalism itself! Could it supply the beginnings of the actual mathematics we seek for the brain *explicitly* as well? (Their perspectives are *so similar*!) If this were true, then it would *considerably shorten* the time frame for its actual creation and implementation!

We *must* evolve such a "calculus" to finish my original problem! This has always been my dream and my goal. With it, we may finally begin to treat the mental aberrations of individuals

---

[94] involving very deep, extensively developed and applied varied mathematical disciplines!

specifically rather than by "shotgunning"[95], -and the societal aberrations of humankind itself as well.[96]

I wish to start a "Foundation"![97] As I stated very early in this book, I am specifically asking for help from mathematicians, and, if that is not forthcoming, from the mathematicians who are *also* physicists! I suggest an expansion of physics into a new and deeper realm –*that of the human brain*! This, I think, will be the new ground of science.

So where do we go from here?

The biggest problem still remaining for the science of man is the physical brain itself. Physical science thinks it has solved the essential problem of everything else, (almost), but how large is the scope of its knowledge? A few billion pieces of knowledge, I think. Minsky thinks it is just a few pieces.[98]

But, conversely, how big is the physical brain in itself? It is 100 billion cells alone, and its synapses are of the order of 10 trillion. Without even considering the neurochemical aspects, think of the combinations and the complexity of our original and foundational mechanism which is, furthermore, *self-referential by definition*.

---

[95] The Free Online dictionary: "Covering a wide range in a haphazard or ineffective manner" or to put it another way -by using "weapons of mass destruction"!
[96] This is the subject of my new book-in-progress. Hopefully it will not take as long as this one to fulfillment!
[97] In reference to the title of the famous science fiction book by Isaac Asimov which had a similar import. My dream is to actually do it!
[98] Dreyfus cites Minsky's attempt to specify the magnitude of the mass of knowledge necessary for humanoid intelligence. Minsky estimates the number of facts required as on the order of one hundred thousand for reasonable behavior in ordinary situations, a million for a very great intelligence. If this doesn't satisfy us, we are to multiply this figure by ten! Dreyfus 1992. Minsky apparently thinks that ten million is a huge number! I don't think it is.

Which is the larger, more difficult problem? I think the answer is pretty clear. The focus on the brain will become the primary focus of any future science. I propose that is *already* the hidden purpose of Quantum Physics itself as I think reading d'Espagnat exposes very clearly. I think he is already half way there!

Back to Einsteinian Realism:

In the closing pages of Kumar's "Quantum"[99], he discusses Einstein's final thoughts in his later life on scientific realism versus quantum mechanics.

He cites Wolfgang Pauli's letter to Max Born in 1954: "Einstein's point of departure is 'realistic' rather than 'deterministic'".[100]

> "What really troubled Einstein was not 'dice-playing', but the Copenhagen interpretation's 'renunciation of a reality thought of as independent of observation'[101]"

Kumar states Einstein's final conclusions, (after a lifelong struggle with the problem), this way:

> "The introduction of hidden variables[102] to 'complete' quantum mechanics seemed to be in accordance with

---

[99] Kumar 2008

[100] Letter from Pauli to Born cited in Kumar 2008

[101] Letter from Einstein to Georg Jaffe circa 1954 as cited in Kumar 2008

[102] Repeating a prior footnote for clarification: Here is a citation from Kumar which clarifies the concept: "Einstein's explanation of Brownian motion in 1905 is an example where the 'hidden variable' is the molecules of the fluid in which the pollen grains are suspended. The reason behind the erratic movement of the grains that had so perplexed everyone was suddenly clear after Einstein pointed out that it was due to the bombardment by invisible, but very real, molecules." Kumar, 2008

Einstein's view that the theory is 'incomplete', *but by the beginning of the 1950's he was no longer sympathetic to any such attempt to complete it.* [my emphasis]

By 1954 he was adamant that 'it is not possible to get rid of the statistical character of the present quantum theory by merely adding something to the latter, *without changing the fundamental concepts about the whole structure'*. He was convinced that something more radical was required than a return to the concepts of classical physics at the sub-quantum level. If quantum mechanics is incomplete, only a part of the whole truth, *then there must be a complete theory waiting to be discovered."*[103], [my emphases]

Granted that Einstein envisioned that such a theory would be of the form of a "unified field theory", but I will posit the innate grace of his mind to have been such as to have at least entertained the consideration of my thesis as an alternative —the regrounding of physics into biology in a modified Cassirerian context![104] My thesis is proposed as an exact science of "observables" themselves —of the "observed facts in a (relativized) realistic/Naturalistic context"!

Re-iterating Einstein's prior citation:

> "It is existence and reality that one wishes to comprehend. ... When we strip the [this] statement of its mystical elements we mean that we are seeking for the simplest possible system of thought which will bind together the observed facts." (Einstein 1934, Pps. 112-113)

I think my thesis constitutes exactly such a system.

Is this claim a proof of megalomania? Is this "delusions of grandeur"? Perhaps, but there is just a small possibility —and a

---

[103] Kumar, 2008

[104] i.e. one that is within the context of Cassirer's relativized Biology, itself taken once again within a reduced Maturanian perspective!

desperately needed one for "realism" and "materialism", that it is not!

I hereby modestly and humbly propose a profound "Copernican Revolution" in the very fundamentals of science itself –which is scary to *both* you and me myself as well! The only truly objective criterion which is applicable here however is whether or not it works. As just about every significant scientist admits, science is *all* about observables![105]

That it involves "Copernican Revolutions" across *the whole* of our fundamentals I fully realize, but I think it is worth an investment in time and energy as it finally allows an exact science of "observables" themselves –of the "observed facts in a realistic context" –with the promise of brand new worlds to come. I propose an extended, enlarged and relativized, (though still *realist*), Cassirerian "materialism" as that realistic context!

But I propose that all the invariant relationaliy of each of the SUPERB theories, (to include QM and the Relativities), must be preserved within it, even QM's logical rationalizations of its ad hoc fulfillments as well, but they will be far out from the absolute center of our organization.[106] It is this new center which produces a theory which meets Einstein's requirement that "we are seeking for the *simplest possible system* of thought which will bind together the observed facts." It is this new center which I believe meets Penrose's and Einstein's aesthetic requirements.

D'Espagnat replies:

I had sent Dr. d'Espagnat a copy of a slightly earlier version of my book, and he was gracious enough to reply to me. Though he is now 92 years old, he was gracious enough to try to read the work of an uncredentialed author.

I must confess that I sent it before I had conceived more than a brief understanding of his work. I was exhuberant in

---

[105] This is derivative from Leibniz's "Principle of Observability".
[106] See my "rubber band" model of Chapter 3.

finding a "kindred spirit", and, enthusiast that I am, I acted before I was ready. To quote my prior footnote on the subject: it led, I think, to d'Espagnat's bewilderment when he perused an earler draft of this book I had sent him prematurely in my initial enthusiasm. Though he agreed with my fundamental presuppositions, I believe he was disoriented and confused with my specific section regarding the Aspect experiments. I had misunderstood the import of the Aspect experiments which I, in my too hasty response, had totally mischaracterized -and had tried to force them to fit. I have since reconsidered this particular problem and realized the error of my earler, incorrect perspective.

There is one confirming paragraph in his response to me that I wish to quote however:

> *" ... I can only say is that, while almost all biologists and researchers on the nature of consciousness hold firmly to ontological naturalism (to the idea that science is on the way of revealing to us reality as it is per se) you quite definitely rejected this idea, and that I very much approve of your doing so. Clearly this is one basic point on which we agree.* Unfortunately, I regret to have to say that, for the reason just stated, it is quite clear to me that we cannot proceed together beyond this step.
>
> Bernard d'Espagnat".[107]

In d'Espagnat's case I should probably have urged him to begin with Chapter 4 which begins my actual thesis. I could not presume upon his grace and age by responding further however. This is a very kind and generous man.

---

[107] private correspondence, my emphasis

Conclusions and a serious Proposal for Matrimony of our Offspring:

On Scientific Realism Again

I will conclude this chapter with a claim that my own theory stands as a better answer to the question of "scientific realism" *per se* than does modern physics itself. This is obviously not to say that their answers are wrong, (these are obviously SUPERB answers!), but that they are not *cogent* -specifically in regards to this particular problem –i.e. that of the possibility of maintaining a consistent and believable perspective on "scientific realism" itself! This possibility is clearly compatible with Cassirer's opening position in his "Theory of Symbolic Forms"![108]

Physics in itself does not *require* an assumption of Scientific Realism as the ongoing debate and lack of consensus within QM clearly demonstrates. But Biology contrarily, (considered as a viable symbolic form), *definitely does*! Maturana clearly and brilliantly framed this problem. He divided the world between "autopoietic entities" and "ambience", joined through "structural coupling". This, in its most abstract interpretation, became my first axiom of realist intentionality, i.e. "The Axiom of Externality", and it is fundamental to any position which calls itself "realist"! So too are my other two intentional realist axioms: i.e. the Axiom of "Experience"[109]/"Observation" and the "Axiom of the Interface" which directly derived from Maturana's beginning –i.e. from "structural coupling"! My thesis then provides a viable and cogent perspective, perhaps the only consistent one possible on this deep problem.

Repeating: is my claim above just megalomania? Is it "delusions of grandeur? Perhaps, but, just possibly, perhaps not! To the physicists I address, I humbly apologize for my ignorance as I do not understand your world at any level of depth, but I think I *do* understand your conclusions about this *particular*

---

[108] i.e. Cassirer's "a mere X"
[109] Kant's "intuition"

problem, and I stand with Einstein, (and Kant and Cassirer), in their basic positions on realism itself! I am not the first in attempting to reduce physics to psychology or brain science, but I think I am the first to have done so in a rigorous and ultimately testable manner.

D'Espagnat's answer, for instance, and even Bohr's grounds itself in "what we would expect to observe", but nowhere does it give any definition of "cognition" or "observation" themselves. My thesis actually grounds d'Espagnat's own from what I argue is the truly *equipotent biological* perspective, (i.e. equipotent from the perspective of Cassirer's "Symbolic Forms"), and supplies the requisite specific definitions of "cognition" and observation" as well! As such, it supplies the *realist* perspective, -the ontic reality "which refutes" that d'Espagnat requires, and thus stands as the actual foundation of his own discipline of physics![110]

His book, "On Physics and Philosophy", makes strong and precisely interpreted conclusions for philosophy itself. His book is hard, detailed and long -almost *too* long. He thinks he has arrived at a finished and final philosophy. I don't think he has, but has, rather, helped *profoundly* towards that end.[111]

---

[110] Such a reversal is not unique to my own ideas. Marchal, (2004), pursued a parallel line.

[111] Note: when you finish this chapter, please skip first to Appendix D, (Bohr) and then return to Chapter 14 which concludes my argument.

In conclusion, let me repeat my opening comment: I think my thesis fits d'Espagnat's vision as a hand fits a glove! *I propose that we seriously discuss "a betrothal" of our progeny!*

Structural Divider: The Remainder of this Book is meant to Establish my Claim for a New and Fundamental Re-orientation of our Perspective on Reality Itself –and to Establish a new Basis for Scientific Realism!

Chapter 14: Overall Conclusions & Opinions

It has been a long and difficult journey to this point. If you have followed my admittedly complex argument, then I must indeed thank you for your diligence and patience.

I have always held the very strange and probably naïve conviction that it is the quality of the *ideas themselves* that are important. As I stated very early in this book, I consider most of academia largely incompetent in this specific field of the mind-brain relationship for reasons which should be clear to you by now.[1] I also think that by now you will understand my other

---

[1] I have extreme objections to academia's philosophical classification schemas - which delimit the subsequent range of possibility for the expansion of their own ideas.

By including Kant in their category of [Berkelian] "idealists", for instance, I think that they miss their target absolutely. Kant was, by his own definitions, a *realist* in just the sense that d'Espagnat demands, (I think that his ideas fit, at bottom, Damascius' conception of "*pante aporeton*"). and, in my terminology, was an "ontic indeterminist". He was misled, I think, in trying to account for the extreme success of his contemporary science! This is where and why I disclaim Kant's "categories". "Ding an Sich" is unnecessary and extraneous to his fundamental conception.

That Kant made mistakes, I fully admit. But, as I said earlier, and in reference to his deep commitment to then-current science, -the fact that he was 200 years ahead of his time is a poor argument against his core theses! Move Kant 200 years later, and I think he could have written much of d'Espagnat's book or my own!

I cannot believe that most current philosophers have actually *read* Cassirer's "Theory of Symbolic Forms", as it provides an explicit rationale, (taken under the re-categorization of Kant and Neo-Kantians expressed above), for Bohr's "complementarity" which I will deal with again in my analysis of Neil Bohr's philosophical and epistemological underlayment in Appendix D.

But if we *were* to assume that Kant were "an idealist" then Cassirer *as a neo-Kantian* would be included in that same academic classification schema which I strongly contest. This I regard as an explicit instance of the "balkanization" of academia in general. It is grounded in an *Aristotelian* interpretation of "the concept" itself!

very early comment that I thought that I could never make progress on this problem within academia's narrow confines. I believe this finished conception demonstrates the truth of that early conviction. I believe, now as before, that this is the most complex problem that has ever been addressed by the human intellect –i.e. that of *cognition itself.*

My thesis is ultimately one of optimism however, and of hope for the success of our cooperative human enterprise –i.e. the survival and the betterment of mankind! If my thesis is indeed found viable, (and I do not claim dogmatically that it is), then it will have been worth the effort.[2] Here are the conclusions I have reached:

It is an example and a consequence of academia's own acceptance of the classical concept of the category, (which *produces* the said "balkanization" in the first place), as opposed to Cassirer's redefinition of the same word, ("concept"), as the "Functional Concept of Mathematics"). The former is based in simplistic abstraction, (however complicated and "sophisticated"), as examined early in this book and Cassirer's challenge to it occurs very early in his book "Substance and Function". But the latter is based in *functional rules* – which may extend *across* the classical categories *as a "rule of series"*! See Smart, 1949 for the claim that Cassirer had violated Kant's categories in "Symbolic Forms".

[2] Non-altruist minds might consider the rapacious possibility of pirating Swedish gold! These minds are not possible friends as I freely share my ideas with the whole of the world. I think Man is better than that!

Conclusions:

Given the extreme level, (and it is in fact an *extreme* level), of success and detail of the naïve realistic model[3], (which is what our contemporary science *still remains* I contend), *in all its aspects* —mathematical, logical, neurophysiological, biological, physical, (at a certain level), ..., what are the odds of a theory such as mine ever supplanting it, -of *ever* being accepted, confirmed or utilized?

I think its ultimate odds for acceptance are about the same as were the orginal expectations for Quantum Mechanics itself. It was only the ability of the latter to bring us new things, new insights, new and useful devices which have entailed even its *tentative* acceptance in the form in which Bohr and d'Espagnat conceived it! Its conceptions and worldview are so clearly at odds with our naïve calculus as to make it an extreme stretch for any mind!

And yet "Physical Realists" *still* continually challenge Quantum Formalism in its standard form. Even Einstein resisted it almost to the end of his life. In the end however, he sought some *other* home for his "realism"/"Naturalism" than the standard one.

If my thesis were, (hypothetically), to produce *equivalent* things, comparable useful implementations, then I think it would stand in a like case. But first must come the theory —and *then* the empirical verifications and utilizations. If these two sciences are as close as I have now come to believe they are, then I feel that much of my needed mathematics and mechanics *already exists* in QM and the implementation of my own completed theory may be very close!

Is this wishful thinking? I don't think so —their very similar conclusions and perspectives, *derived totally*

---

[3] at some level of sophistication, i.e. "Naturalism, -but which always postulates "emergence" and "hierarchy"! But this is the exact point I addressed in Chapter 4: "Turning our Perspective Around – a Model of Process"!

independently, are so *very* close! I think we may be much nearer to the end than I had thought.

Consider the possibility of my conception's supplying a specific and practicable cure for Paranoia, for instance?[4] Or of it supplying the ability to profoundly expand the mathematical abilities of a normal brain? How about a cure for "the murdering syndrome" or the "mental" syndrome of extreme pain accompanying an uncurable illness? ... These would be equivalent and powerful utilizations meeting the stated demand.

One of the problems I face is the sheer size of humanity. Manifold beings –7 billion plus- with the tools of communication and duplication multiplying and further deepening the problem of challenging Naturalism more every day! The complexity of generating viable alternatives to this Leviathon is profoundly difficult. And yet I do not surrender as I believe that ultimately I will be confirmed in the "real world"!

Nowhere, for instance, does Naturalism address the problem of meaning itself except in terms of reference –"pointing to"! Somehow the mechanism of the brain would point to an "object" in absolute reality,[5] and it is the properties and interrelationships of the latter, (to include its relationship to ourselves), which would constitute its meaning!

But how could we ever know this "God's Eye Reality" in the first place? How could we ever bridge the gap between merely *appropriate* action, (whatever works!)[6], and reality *per se*? It seems possible only in an unwarranted, *a priori* and anthropocentric assumption that ontic reality itself is "trimmed down" to match the limits of our own particular, evolutionarily

---

[4] As I footnoted early on, I first came to this problem in view of my mother's Paranoid Schizophrenia.

[5] D'Espagnat's arguments against "multitudinism" in the previous chapter are obviously pertinent to this discussion.

[6] See "Bounds and Limits" figure in Chapter 4. You might also consider this question from the viewpoint of Edelman's theory of non-specific antibody production as discussed in Appendix B..

derived cognitive abilities![7] .[8] This is an assumption of a miracle and, I think, profoundly naïve! Contrary to P.S. Churchland, I choose *not* to accept it! But is this what we *meant* by "meaning" in the first place?

How would we address Kant's initial biological problem – the problem of the organization of primitive biological process under this conception? How very much simpler is Hilbert's "implicit definition" under the prior premise of mechanism? How very much simpler is it not to take the *mathematical* perspective of perceptual artifacts as "positions in a structure" for such a mechanism? "Meaning" is Hilbert's unique gift to our problem. It allows "knowing" itself in our deepest conception of that word!

I have maintained that our concepts and our percepts are *mutually* defined in order to implement the simplest and most efficient *operational* "calculus" between them –i.e. that which is most useful for performance.[9] But it is the very *simplicity* of the "calculus"which is my focal point. To reiterate an earlier question: why then are our scientific laws so simple? My answer: because that was *the very purpose* of our naïve realistic (schematic) model in the first place! But Naturalism's unstructured "mechanism", (that is, unstructured *before the fact*),[10] necessarily involves a dream, a hope. It is perhaps best summarized in P.S. Churchland's assertion, (paraphrasing): "and then nature performed a 'good trick'". I assert that it was a *very* good trick indeed! But like stage magic, it involved smoke, mirrors and misdirection as well as an appeal to mass prejudice.[11]

But "mechanism" is itself Naturalism's *own* most basic premise! But how much simpler is it not, even under that very

---

[7] see Chapter 4: "Turning our Perspective Around –a Model of Process"
[8] Would a hypothetical super-monkey Naturalist scientist reach a like conclusion? How about a super-Paramecium from the same "school"?
[9] See Chapter 4
[10] See Durant's paraphrase of Kant's "card catalogue" in Chapter 12
[11] i.e. an argument ad populum!

premise itself, the idea of a *strategic* rather than an *informational* brain? There is no way that a mechanism can *know* its input/output domains, but there *is* a way that it can base *strategies* on it! It is based, in d'Espagnat's terms, in an unknowable reality which can, (*and often does*), refute!

    I have concluded after 50 years of research in many, many fields that the only logical, -that is to say *logically concise* – not *just* "logical" but "logically *concise*" , "well-formed" theory of "Scientific Realism" *per se* comes from the perspective of biology.

The Perspective of Biology and Einsteinian Realism

    As you have seen from the preceding chapters this is not my overall "global" conception which latter deals with a *symmetry* of symbolic forms across the sciences as discussed in chapters 7 through 9.[12] But it is rather to say that the perspective of biology gives the most cohesive, simplest statement of the specific problem of "Scientific Realism" taken in Einstein's sense.

> "It is existence and reality that one wishes to comprehend. ... When we strip the [this] statement of its mystical elements we mean that *we are seeking for the simplest possible system of thought which will bind together the observed facts.*" (Einstein 1934, Pps. 112-113, my emphasis)

That other theories, other perspectives are logically consistent and *profoundly* useful, I fully acknowledge. D'Espagnat, I think, presents a workable rendition of the whole

---

[12] This global picture has a strong affinity with Niels Bohr's early and initial conception of the possible utilization of multivalued functions, (AIP), -see Appendix D: Niels Bohr- and, of course, with Cassirer's Symbolic Forms from which my overall conception is derived.

of the problem¹³ from the perspective of modern physics. It becomes so entangled in its complications however that it really doesn't stand as a viable theory of scientific realism in Einstein's sense, (even as a "veiled" one). It is *not* the "simplest possible system of thought which will bind together the observed facts"! In fact, it is not *simple* at all!

D'Espagnat's "Veiled Reality"entails a retreat from the consequences of his unexamined and unelaborated conception of cognition itself. It centers "on what we would expect to observe", but *nowhere* are we given any clue as to what "observation" itself might be! His protracted treatments of "decoherence", "entanglement", and Von Neumann's "cuts" for example, illustrate this point very clearly. "Complementarity", even though it is taken as a basic Quantum Mechanical principle is *nowhere clearly defined*¹⁴ –except as an *ad hoc* and *gratuitous* addition. Within my own thesis –and specifically in my *foundational* incorporation of Cassirer's "Symbolic Forms" within it-as foundational to "the interface", I have provided a clear and simple rationale!

My theory *begins* with an exact explanation, a specific proposal of exactly what cognition is, (and, notably, it generates a specific theory of meaning as well)!¹⁵ And that theory then generates –easily and centrally to itself¹⁶ –into its center core – ideas that are profoundly complicated and twisted under the perspective of quantum mechanics or under physical theory in the larger sense of the term. Einstein dealt with this exact problem as

---

[13] Which qualifies as SUPERB for its accuracy, but rates far below that for its organization. Maturana is surely very close to this rating, if only for his brilliant conception of "structural coupling"!

[14] See, for instance, Bohr's online transcript from his later years: http://www.aip.org/history/ohilist/4517_1.html , especially Session V. See Appendix D: On Niels Bohr and Cassirer.

[15] QM's "Complementarity", for instance, is clearly just an aspect of Cassirer's "Symbolic Forms" which I incorporate at the very core of my conception. Again see Appendix D.

[16] See heading: Aesthetics in Chapter 3

did Bohr as well, and after 50 years of dialogue, they both still had a definite problem with it.[17]

Einstein finally concluded that "localization", ("multitudinism" in d'Espagnat's terminology), had to be given up to be able to retain *any* legitimate sense of scientific realism: "Does the Moon actually exist even when I'm not looking at it?", (Einstein paraphrase) Answer: Something, or rather *Somewhat* exists, but we as mechanistic organisms are not equipped to *localize* it! But if you give up multitudinism, you are giving up "materialism" *itself* in the ontological sense of the word.[18]

Thus we arrive at d'Espagnat's and my own highly similar joint conclusion of the ontological "something", (d'Espagnat), or my own "somewhat"[19] –i.e. the "thing"/"what", the externality *which can refute* and which guides our *subsequent* intentional and strategic functioning!

I am willing to believe that d'Espagnat's vision of reality is consistent, (though not exhaustive), as he reaches the same conclusion that I do. but his *physical*[20] conception of the overall picture is far less concise and organized than my own biological one which *began* from Maturana's brilliant vision of the actual origins of the problem.

Cassirer's Symbolic Forms

To repeat a section of Chapter 3, early on, in my early 20's when I first found Cassirer's "Theory of Symbolic Forms", (see Chapters 7 & 8), I visualized his profound conception of the equipotence of varied *specifically scientific* but *different* perspectives[21] as a network of "rubber bands" -as representative of

---

[17] See Kumar, 2008
[18] though not in the relativized sense I have suggested under "Symbolic Forms"
[19] My first intentional axiom of realist reasoning, the "Axiom of Externality"!
[20] i.e. from the standpoint of Mathematical Physics
[21] ignoring for the present, the larger panorama of Cassirer's perspective

the innate and invariant relationality between them![22] Given such a network, each such perspective, each beginning was like grabbing a given nexus in the network and pulling it towards my eye[23] and making it the focus and starting point for any viable process of theorization.

But *any* nexus is a candidate –as long as we limit Cassirer's perspective to the specifically scientific, and experiencially viable theories. Each theory *must preserve* the invariants of the relationality of actual experience, (the phenomena)! These viable theories I identify with Penrose's "SUPERB" theories! Cassirer's scientific epistemological relativity was for me the natural extension of Einstein's Special and General Relativity and the relativity of Galileo's laws of motion.

---

[22] but this "relationality" must be taken in a context-free, *invariant* sense!

[23] Or rather, the absolute and beginning center of my (intentional) theorizing! This is clearly a Quinean perspective.

A Claim for a Fundamental Re-orientation in our Conceptions of Reality:

I will now make a claim for a *fundamental* reorientation of our overall picture, and compare the gnarled complications of naturalism's and of physics', (as elaborated by d'Espagnat), current pictures of reality with that of the fictional King of Petrolia we met in chapter 7. The King, you will remember, was forced to make continual *ad hoc* repairs to the structure of his worldview as contrasted with the case of the nuclear technician he was arguing with. In particular, it so complicated the *aesthetics* of that structure, (but clearly not its Quinean possibility!), as to allow us to actually *dismiss* his vision as being a fundamental theory of reality. This is not a totally negative judgement on d'Espagnat's conception, (or on that of the "Physical Realists"), but it definitely relates to the problem of scientific realism as I proposed in that chapter.

It seems that Quantum Formalism must itself likewise keep supplying *ad hoc* (cruder) answers and "work-arounds" as the questions deepen –decoherence, the dividing line between the macro and the quantum world, the Aspect experiments, the possibility of a scientific realism, and the results of any possible future experimentation. But Bohr raises a further complication:

> "...that is something which is typical of quantum mechanics that one can not know all the paradoxes, *but when they come up, one can solve them*" [24]

But how could he know that? The above is more in the nature of a religious conviction than a scientific one! Will we ever find the "ultimate particles, -the Higgs Boson, Dark Matter, … But what new underlayment, what new conceptual complications to the overall structural integrity of the theory will those complications subsequently require?

My own thesis, coming from the perspective of a relativized biology, answers these questions *from the very*

---

[24] See Bohr, AIF Session V:

beginning and from its very conceptual core!²⁵ The question and the choice becomes, then, one of "theoretical beauty"²⁶, (i.e. "musicality"/ simplicity of organization), as seen from Penrose's and Einstein's aesthetic criteria for theories.

That each (SUPERB) theory must be "data-viable", ("phenomenologically viable" might be a better phraseology), and that each must be able to account for the other, (or at least not lead to contradictory phenomenal conclusions²⁷), is pretty much a given.

An afterthought:

Each of the SUPERB theories of science must, at its bottom it seems, include some correlate of the central, beginning perspective and broadest vision of Cassirer's "Theory of Symbolic Forms" —i.e. "each asks different questions", (Cassirer, paraphrase). Alternatively, using Niels Bohr's conception of "multi-valued functions", each must include a linkage to the "mathematical origin" —i.e. to the "zero".²⁸ This translates to be the ability to "walk around in a continuous manner" and to pass from one perspective to another and to discover whether or not "we have changed the meaning of the words", (concepts).²⁹ But

---

²⁵ It is an explicit requirement that it must always preserve the invariant relationality of the *other* SUPERB theories however but possibly in a "distributed" sense as discussed previously.

²⁶ See Penrose, 1989, p. 421

²⁷ See the Bohr quote in Appendix D: "Does [Einstein] think that, if he could prove they were particles, he could induce the German police to enforce a law to make it illegal to use diffraction gratings or, opposite, if he could maintain the wave picture, would he simply make it illegal to use photo-cells? That was, of course, in all friendliness, but it was the idea to say that this [is a] problem we cannot get over, and that means that actually *we got something new in the quantum.* That was the point."

²⁸ See Bohr's reference to "passing through the origin" in Appendix D.

²⁹ It seems better to picture it in terms of "Riemannian Sheets", than of the square root function —as this is proposed as a universal linkage of all the forms of science!

this "origin" correlates precisely with Cassirer's profound –but broadened conception!

Cassirer's conception, his broadest view of his "Theory of Symbolic Forms" must be contained within *any* physical or, I believe, any even non-physical theory of reality as the "origin", as the actual *center* of that "origin"! This is why for instance, it is not a contradiction in the physical realm for quantum physics to move from the wave to the particle conception of light. It is because this "origin", this center must be contained *within the domain* of *each* of the alternatives. In fact, under this conception, it must be contained within the domain of any viable theory of reality.

It provides an allowable pathway to transition from one viable SUPERB alternative to another such alternative *without contradiction*! But each still remains as a viable *perspective* within this realistic overview. This conception is made possible, however, only by transitioning through Cassirer's "origin"[30], his "zero" implicit in each as a legitimate part of its *own* domain!

Back to my claim:

But it is the question of "shape", and simplicity of organization which arises here. It is a question of the *structural simplicity* –taken at the highest level -of the resultant "schematic calculus"![31]

To repeat myself: Theories have "shape" in the same sense that great music has "shape" -not only in their individual themes, but as overall compositions. I will now add a conclusion: that "Scientific Realism" must embody the *best* shape! Occam's razor, (least assumptions), is only the very tip of the iceberg.

---

[30] And actually through my own conception as well

[31] See Chapter 4 for my conception of the "schematic model"!

My Ultimate Conclusions:

My overall conception is an *extension* of Cassirer's redefinition of the very meaning of the word "concept" into his "Mathematical Concept of Function", but *expanded still more* by me through the young Hilbert's conception of "implicit definition" which led to my own extension and redefinition again of that word, ("concept"), as the "Concept of Implicit Definition", (C.I.D.). It was this extension and enrichment that allowed the conceptualization of our very biological *percepts themselves* as (structuralist) elements of order –as "positions" in a structure". This, in turn, enabled us to conceive the brain as a *strategic* rather than an *informational* construct!

I then married that final redefinition of the concept back to Cassirer's *other* brilliant and original conception of "Symbolic Forms" *but as seen through Maturana's eyes*, (i.e. through the "eyes" of structural coupling[32]). I therefore reached my final vision: i.e. of the "interface" and the "Somewhat" beyond that! It is the logical *interplay* of these ideas which must be carefully deliberated to fully understand my thesis.

Is my "strategic brain" a truly "outrageous"[33] proposal? It certainly is! However totally "antirealistic" it may sound, I argue that my thesis is nonetheless more compatible with contemporary science –and with Einstein's "Scientific Realism" as well -than any alternative currently proposed. My thesis preserves science and ordinary experience as well.

---

[32] Remember that biologically, structural coupling implies only *adequacy*, not mirroring –as in "congruent structural coupling" and was the basis of my rejection of Maturana's more reduced and limited form of the underlying idea. See my illustration: "Bounds and Limits", Chapter 4, for the conceptual grounding of "Somewhat" as opposed to "Something". From a biological standpoint, there are two *separate* domains: ambience, (externality) and the autopoietic entity. I conclude that they are connected through "the interface" which I argued is itself a Realist *intentional* imperative!

[33] E.g. Crick's usage of the term. I do not consider his thesis in any sense to be other than quite ordinary from a materialist standpoint.

Chapter 14   A Claim for a Fundamental Reorientation of Knowledge

I will insert the whole of the opening quote from W.V.O. Quine again here, as I think it must be considered as a whole.

> "The totality of our so-called knowledge or beliefs, from the most casual matters of geography and history to the profoundest laws of atomic physics or even of pure mathematics and logic, is a man-made fabric which impinges on experience only along the edges. Or, to change the figure, total science is like a field of force whose boundary conditions are experience. A conflict with experience at the periphery occasions readjustments in the interior of the field. Truth values have to be redistributed over some of our statements. Reevaluation of some statements entails reevaluation of others, because of their logical interconnections- the logical laws being in turn simply certain further statements of the system, certain further elements of the field. Having reevaluated one statement we must reevaluate some others, which may be statements logically connected with the first or may be the statements of logical connections themselves. But the total field is so underdetermined by its boundary conditions, experience, that there is much latitude of choice as to what statements to reevaluate in the light of any single contrary experience. No particular experiences are linked with any particular statements in the interior of the field, except indirectly through considerations of equilibrium affecting the field as a whole.......
> Furthermore it becomes folly to see a boundary between synthetic statements... and analytic statements...Any statement can be held true come what may, if we make drastic enough adjustments elsewhere in the system... Conversely... no statement is immune to revision... even the logical law of the excluded middle... and what difference is there in principle between such a shift and the

shift whereby Kepler superseded Ptolemy, or Einstein Newton, or Darwin Aristotle?"[I]

We have *already seen* deep contradictions "at the periphery" -eg Raichle, W.J. Freeman, Edelman, Maturana —even Bohr[34]....which force us to profound changes "in the interior of the field" comparable to Quine's "even the law of the excluded middle". In fact, they force us beyond even "objects".

> "One could even end up, though we ourselves shall not, by finding that the smoothest and most adequate overall account of the world *does not after all accord existence to ordinary physical things......Such* eventual departures from Johnsonian usage", (Samuel Johnson is said to have demonstrated the reality of a rock by kicking it!), "could partake of the spirit of science and even of the evolutionary spirit of ordinary language itself."[II]

My book supplies exactly *such* a perspective[35]. How could "the color phi"[36], "chinese rooms", "cats on mats", ... tenuous purely philosophical arguments at best, be more important than these deep biological and physical facts? This is *such* a theory, but, at the same time, it also allows us "to have a life!"

Consider once again the parallel between the *most* SUPERB[37], (according to Roger Penrose), of modern physical theories,[38] and my own conclusions:

---

[34] See Appendix D: Niels Bohr

[35] i.e. "*such an eventual departure* from Johnsonian usage"

[36] "the color phi" actually provides a powerful counter-argument against Dennet when science is considered as the organization of observables per se —as it must ultimately be!

[37] His "CAPS"

[38] "Quantum mechanics: Explains "hitherto inexplicable phenomena...The laws of chemistry, the stability of atoms, the sharpness of spectral lines...the curious phenomenon of superconductivity... and the behavior of lasers are just a few amongst these." (P.153) "No observational discrepancies" (*at all*) "with that theory are known."

"There is a very precise equation, the *Schroedinger equation,* which provides a completely deterministic time-evolution for this [quantum] state. But there is something very odd about the relation between the time-evolved quantum state and the actual behavior of the physical world that is observed to take place. From time to time - whenever we consider that a 'measurement' has occurred - we must discard the quantum state that we have been laboriously evolving, and use it only to compute various probabilities that the state will 'jump' to one or another of a set of *new* possible states." (ibid, P.226, his emphases)

In this "more optimistic" view, it is only "*in relation to the results of 'measurements'*" that concrete reality emerges –i.e. that a specific rendition of space-time is enabled.

Now compare this one last time to the re-afferent model I presented in the Freeman Appendix of Chapter 4. Each evolves a "state equation" and then performs a "measurement", (action into the world), which then causes a *new* state equation, (Schrödinger/W.J. Freeman/Merleau-Ponty) to be formed until the next "measurement" is performed.

Quoting Penrose once again:

"What kind of a picture of 'physical reality' does" [quantum physics] "provide us with? ...Many physicists find themselves despairing of *ever* finding such a picture. They claim instead to be happy with the view that quantum theory provides merely a calculational procedure", [an algorithm], "for computing probabilities and not an objective picture of the real world. *Some, indeed, assert that quantum theory proclaims no objective picture to be possible –at least none which is consistent with physical facts.*"[III]

Niels Bohr, the recognized "father" of quantum theory said that a realistic picture of ontology was unattainable! He characterized his new science as a pure algorithm, (i.e.: a rote, purely pragmatic but *profoundly* and *overwhelmingly* useful procedure), instead. What the actual reality beneath it is, he said, we cannot know and cannot picture. His theoretical world could not, (cannot), fit any normal sense of the real world. And yet it works and leads to the production of new things –transistors, nuclear power plants, etc. –and worse!

How close these conceptions are! But quantum physics, according to Penrose is considered the *most* "SUPERB" theory in our current repertoire. It gives you something to think about, doesn't it?

I invite comments and would welcome constructive help in my lifelong quest for a pragmatic answer to this, our deepest and most urgent problem. I sincerely believe it will determine the future of our species as it lies at the bottom of our deepest and most destructive human dilemmas.[39]

Note: I will respond to any decently proposed questions at: jiglowitz@rcsis.com

Please put some verbiage corresponding to "In Response to your Theories" in the subject line as, else, I will probably delete it as "spam" unread.)

Specifically Biological Conclusions:

I consider my most important result, (though you may think this strange), the *Naturalist* one: i.e. that "mind" is the

---

[39] Note: I will not expose my humanistic or ethical views, (and they are quite broad and fiercely important to me), as it is vitally important that this problem be solved scientifically and soon, and advancing my personal beliefs would only hinder the process. I will only say that I think this is the most urgent and the most important problem that humankind has ever faced. Without its input, I think we will exterminate ourselves very shortly –I agree with Stephen Hawking on this matter though I think that he himself is part of the problem.

(logically reduced[40]) "concept" of the brain![41] I hold that it is both legitimate and important within the (reinterpreted and relativized) Naturalist framework and leads to definite and practical empiric lines of research.

That Naturalism is itself thereby relativized detracts neither from its utility *nor* from its importance -no more than did the introduction of relativity or indeterminacy into modern physics lessen *its* viability or importance. Rather, it produced profound and immediate practical results.

Naïve realism is a biological and behavioral algorithm superb for normal life, and Naturalism, its natural extrapolation, is valuable beyond measure -as well it should be under my hypotheses. It is to the ultimate empirical results, (or not), of my thesis, however and finally, that I will equate its ultimate value.

So where do we go from here?

The biggest problem still remaining for the science of man is the physical brain itself. Physical science thinks it has solved the essential problem of everything else, (almost), but how large is the scope of its knowledge? A few billion pieces of knowledge, I think. Minsky thinks it is just a few pieces.[42]

But, conversely, (repeating myself), how big is the physical brain in itself? It is 100 billion cells alone, and its synapses are of the order of 10 trillion. Think of the combinations and the

---

[40] In the sense of reduction to the rule of Cassirer's "Mathematical Concept of Function"

[41] Alternatively, it is the brain's rule of ontogenic coupling

[42] Dreyfus cites Minsky's attempt to specify the magnitude of the mass of knowledge necessary for humanoid intelligence. Minsky estimates the number of facts required as on the order of one hundred thousand for reasonable behavior in ordinary situations, a million for a very great intelligence. If this doesn't satisfy us, we are to multiply this figure by ten! Dreyfus 1992. Minsky apparently thinks that ten million is a huge number! I don't think it is.

complexity of our original and foundational mechanism which is, furthermore, *self-referential by definition.*

Which is the larger, more difficult problem? I think the answer is pretty clear. The focus on the brain will become the primary focus of any future science.

Devil's Advocate:

Though I have argued against our knowledge of externality, and for a schematic organization of process, *could* not our external, metaphysical world *still* be like the objects of our cognition. Of course it could! The possibility is suggested in my conception of "interface".

Since implicit definition defines *our* objects within, conceivably it might, as well, define the "objects" of external reality without! But this is a profession of extreme faith, and not of science.[43]

> "If anyone adopts such a belief, he or she does it as a leap of faith. To make such a leap does not make us *ipso facto* irrational; but we should be able to live in the light of day, where our decisions are acknowledged and avowed as our own, and not disguised as the compulsion of reason."[IV]

I, however, do not choose to, (nor do I have to), make such a leap of faith. I propose that what we have is a viable, (and truly real!), working model that simply "does the job", i.e. it is at least compatible, and probably beneficial[V] vis a vis absolute externality.

---

[43] It is a question of bounds and limits again. Or, more simply, of the distinction between an upper bound and a *least* upper bound. Reality clearly sets definite upper bounds to (evolutionary) development, but does it convey to the organism a *least* upper bound, (which would be defining)? The former encompasses (raw) "structural coupling", but the latter would be necessary for "*congruent* structural coupling". It is an assumption equivalent to the "parallel postulate", you see!

Come, isn't it the height of arrogance to presume, (under the Naturalist presumption), that this race of apes, barely able to scribble for a mere few thousand years, has been able to divine the nature of absolute reality? How much more probable is it not, (changing the metaphor), that we are merely constructing "*a hive*"?

Why do we think we know even the *boundaries* of all the possible solutions to all of the problems of reality? Whence comes our arrogance that we feel we have solved the ultimate problems of the universe and of our existence in it?

Is it not more believable, (under the very Naturalist assumption), that we have merely expressed our own particular mode of existence, -that human civilization, (incorporating human minds), like a swarm of bees, has simply built a hive?

What is this logic we are so sure of? Ultimately, biologically, it is an expression of the "structural coupling" of the race with its environment. But the invariants of that coupling are derived from the structure of the uniquely human brain.

Other brains, other modes of coupling almost certainly would embody another protologic. Ordinary logic, (i.e. "associationist" logic -after Dreyfus' term), denies its biological roots. It believes it has touched eternity and verity. How? Why? What teleological mystery does it hide?

When we thought that man was created by God in his image and that God gave us this open channel to truth, then there was a meaningful rationale for such a view. But when man became purely and simply a *material* animal, derived mechanistically and randomly by material combination, then this mechanistic process lost all justification as correlating with anything other than its own mechanical necessities.

But it works! How and why? Perhaps that is itself the answer. It is an operative process that works well in the world in which it lives! This provides no guarantee of its ontological posits at all however –it is an operative process that works –and that's all!⁴⁴

So Why Bother?

But if this is the ultimate answer, if this "ontic indeterminism" is the conclusion we must reach, what is the point of it all?

Throughout this book I have admitted the extreme intuitive difficulties of my thesis. But modern physics has much the same difficulty –its picture of reality, though intensely beautiful and exotic, offends those same normal sensibilities.

The (why bother) answer for physics is that that very picture produces desirable, powerful, and practical results right at the human, (naïve), scale, and which we cannot deny. The transistor, nuclear power, working telephones and radios, ... are necessary and practical consequences of that *very* theory –and they would be impossible without it.

I propose that this will be very much the case for my conception. Though admittedly offensive to our (naïve) realist sensibilities, *if it is correct,*⁴⁵ it will lay the scientific and mathematical theoretical ground necessary for the "quantum advances" in neuroscience, for instance, which will finally and specifically, (rather than non-specifically and destructively), cure the terrible aberrations of mental illness.

But the mind-brain puzzle has far larger implications than that. It deals with the *problem of man* in all its aspects. It deals

---

⁴⁴ But conversely, if Naturalism claims to hold the whole of rational thought, it must own up to the whole of its social and neuro-scientific consequences as well. If its course were to lead to the destruction of the human race –then that too must be considered a measure of its success or failure. I think the odds for the latter are higher than they have ever been. As I said elsewhere, crazies have always been loose in the world, but now they have the means!

⁴⁵ and I do not *dogmatically* assert that it is. The future of science must answer this question.

with all his social, ethical and artistic parts.[46] The final implications must not be underestimated.

This is the "why bother". Even offensive theories can yield useful and powerful results, necessary to man! The final test, the final judgment therefore, must be made on results. But, before results can be obtained, it is necessary, first, to entertain *the possibility*!

My reconception of fundamentals, though radical, is absolutely consistent with the historical progress of science -of physics, biology, mathematics and logic. It solves the biological and the philosophical problems inherent in the mind-body problem, and exorcises the "homunculus" once and for all.

It provides an Archimedean fulcrum to overturn our naïve realistic presuppositions, (inherited by "Naturalism"), and lets us get on to the serious business of creating a science of mind and brain. It provides a viable context in which I believe workable and testable theories are now, finally, possible.

No substantial progress will ever be made in dealing with "mind", or in the treatment of its terrible, destructive aberrations, (both individual and societal), -until the mind-body problem itself is solved and *workable tools* are developed.

To deal with the mind, we must deal with its "objects" and the relations between them. To deal with the brain, we must deal with its process. To constructively and *specifically*[VI] affect the processes of mind[VII] *via* the brain, the relationship between the two *must* be understood!

The simplistic orientations of naïve realism, ("though grown up and sporting a beard" -to coin a phrase), just will not stand any longer. Great issues, to include the most profound social, ethical and spiritual aspirations of the race, depend upon the resolution of this problem -and upon its consequent: the establishment of a mature and viable neuroscience.

---

[46] I think it would be a real mistake to discount the possibility of real, purely physical implications from my thesis. In the transition beyond "objects", wholly new degrees of freedom may be possible for physics itself. Note [2011] d'Espagnat's conception exposes just such a possibility!

There is too much pain in our world, and too much *need*, -dependant upon real solutions to these problems, to cling to the playgrounds of our intellectual youth.

How do we live?

So, (given my thesis), *what is the point?* Do we exist, therefore merely contemplating our navels, lost in the "ontic indeterminism" of metaphysics? No.

I, for one, rarely even *think* about metaphysics, but love and feel pain, pay attention to passing cars, and generally live my life as you, (or any dogmatic Naturalist), would. I practice Descartes' interim life strategy of normalcy, (by necessity), and pretty much live my life as I always have. I speak the language of Naturalism because it is good and fecund language and because it is, well ..."natural"!

When I *choose* to consider the connection however, I know that by following my inbuilt model, (and extending it through the discovery of new science, let's say), I am in harmony with that nameless externality. I do not *use* my model, you see, I *live in* it!

My *Own* "Act of Faith":

But what do *I*, personally and as *my* act of faith, believe? (I, after all, get to have beliefs as well!) Though I do not believe in the necessity of spatially and temporally separate *metaphysical* objects, (consistent, certainly, with the views of modern physics), nor in the metaphysical "aether" in which they are *still* (incredibly!) conceived[47], I, (personally), believe in the metaphysical existence *of other minds*![48] (That there is *still* more,

---

[47] What is "physical realism" itself but a belief in exactly such?

[48] I also believe in a *continuity of sentiency*, at least with the higher animals -for reasons which should be perfectly obvious by now. Just where the "cutoff point" may be, I would not be presumptuous enough to speculate. Might not *these* be

–an absolute externality, "phenomena substantia"– I also believe as should be totally clear by now.)

But regardings those other minds, specifically *as minds*, (as per my second thesis), I believe they are all precisely products of implicit definition, variations on, (values of), a single universal function. They are, I believe therefore, *continuous variations of me*. We are all, I believe consequently, *more than brothers*, but "states" of the same being.

"You" are "me" in a different "place", (state) –there is no necessary spatial or temporal separation between us, i.e. there is no necessary *metaphysical "aether"* between us! But somebody already *said* all that, didn't they?

> "'I tell you the truth, whatever you did for one of the least of these brothers of mine, you did for me. ... whatever you did not do for one of the least of these, you did not do for me.'" (Mat. 25:40-45)

---

the "extra-terrestrial"/ alien intelligences we have so long desired to meet and welcome? But, if so, *how have we greeted them?*

Chapter 15: Epilogue

How do you convince a bird, living in a dying tree, to leave its accustomed perch, its familiar nest, and go to inhabit another? You may praise the new view, and describe fantastic horizons invisible to the old. You may catalogue the prospects of juicy worms, temperate climes, and soaring flights through inestimable thermals.

But the bird, clutching stubbornly to its tattered branch, may only envision the loss of its well-defined routines. The path to an easy patch of straw for its nest or a worm-rich meadow might become convoluted or even impossible because of distance or predators! It cannot even envision the possibilities of the new place unless it is willing to chance an exploratory flight.

Its world is simple and uncomplicated –or at least the complications are well known. This has been my problem here. I believe the mind-body problem is the most difficult in the history of the human intellect. It hinges on the problem of cognition –and *that* is the problem of everything! Its solution, I feel, involves a brand new "roost" –a new intellectual perspective with horizons different but incomparably broader than before.

Admittedly however, though it proffers "sunsets of unmatched vividness", and "new and fertile meadows", it involves a definite risk as well. It may turn out, after all, that the "nest" I propose lies over *fallow* fields and iron-hard soil where no "worms" might survive!

You are right, therefore, to be conservative and cautious in the selection of your ultimate habitat, but you are *wrong* if you are timid in your *survey* -your future may depend on it. I invite you to conquer your fear of vertigo and try your wings in an exploratory flight to this very different tree of knowledge.

> "Safe (that is, probable) hypotheses are a dime a dozen, and the safest are logical truths. If what science is seeking is primarily a body of certain truths, it should stick to spinning out logical theorems. The trouble with such safety, however, is that it doesn't get us anywhere." (P.S. Churchland)[1]

There are really just two schools of thought on the mind-body problem. One holds that the relationship between the mind and the brain is inherently unsolvable. It holds that the natures of mind and brain are (1) either absolutely incommensurate, (are of different kinds), or (2) the problem is beyond intrinsic limitations on human understanding.

The other school holds that the relationship is perfectly direct and unproblematic, albeit totally one-sided and exceedingly complex. The first offers no practical hope whatsoever for the dysfunctions of the human mind, but the latter destroys the reason for caring in the first place.

Its solution is that we are all automatons, (functionalist) "zombies"! Mind, in its ordinary sense, is a fantasy, a "figment" of the imagination! What, then, does it matter whether *another* automaton makes "pain" noises rather than "happy" noises?

Less delicately, what possible objection could there be to the Dachau "fetus series" or to the atrocities in Bosnia and the rest of the world?

The solutions offered by both schools, moreover, are counterintuitive, limit the scope of empirical investigation and involve significant logical difficulties. I have offered a new alternative capable of resolving the whole of the problem and commensurate with the whole of the human spirit.

My thesis opens the further and distinct possibility of an actual "physics", i.e. a mathematical and scientific mechanics of mind and brain, as it defines, for the first time, an appropriate context in which it could be formulated.[1] Just as the SUPERB[II] theories of Newton, Maxwell, Einstein and Bohr were literally *unthinkable* in the cosmological context of Ptolemy or in the physical (and gravitational) context of Aristotle, neither can the

---

[1] See Chapter 13 where I propose Quantum Mechanics as possibly the most <u>direct</u> route to that explicit mechanics!

Chapter 15   Epilogue

SUPERB theories which must eventually encompass the mind and the brain arise without the context -and the continuum - which will make them possible.

I believe the mind-body problem is the most important problem in the history of our (human) species. Subsuming both science and ethics, it will ultimately determine our future as a civilization.

Though this sounds overly dramatic and even downright pompous, reflection shows that it is not. Answers to what we are, and *why* we are will determine what we can do and what we *will* do.[2] Profound belief determines actual practice!

The bounds of future civilization will be set by our ultimate understanding of our own being. This problem demands, therefore, the greatest latitude and the greatest tolerance to radical ideas. It is too important to be treated otherwise.

It has been said of scientists, (and it certainly applies to philosophers of mind as well), that they live, alternately, in two disjoint worlds. They do not take their reality home with them. The reality they believe as professionals is not the reality they believe when they dodge cars on the freeway or make love. *None* will put out a saucer of milk for Schrödinger's cat.

Is Dennett prepared during his self-stimulating monologue, (whilst sitting in his rocker and listening to Vivaldi), to accept himself solely as a "center of narrative gravity", solely as the cumulative product of temporally and spatially separate and discrete processes, (the "Final Edition" published on his "Demonic Press"), lacking "figment" or "qualia"? I, personally, am perhaps willing to accept *him* as such, but I am certainly not willing to accept *me* as such.

Like Dennett, I have been wrestling with this problem for over 50 years. I came to it not from philosophical curiosity or

---
[2] Consider Nazism, as just one recent example.

"epistemic hunger", but as a result of personal tragedy -the loss of a loved one, (my mother), to the maw of mental illness. Frustration -and anger- at the inability of science to help her and a survey of the dismal "mythological",[3] (Freudian and quasi-Freudian), state of then-current thinking on the subject[III] caused me to begin a personal and private search, of necessity based in logical and abstract theoretical criteria -but aimed at an empiric goal.[4]

Emerging from my "cave", (of contemplation), just a few years ago, I was surprised and fascinated by the illuminating and brilliant bonfires which had been lit on the plains of biology and philosophy. Since then, with more than a little trepidation, I have been scouting each of the major encampments so lit.

I have concluded that I have something still new and novel to say. I think that my torch, crafted as much by art as by science, carries a unique Promethean flame. I think I have solved the essence of the problem of mind-brain. Now I, like Benjamin Franklin, Rousseau's "backwoods philosopher", stand before the sophisticates of Paris in my bearskin cap.[IV]

Though my thesis admittedly opens new and fundamental problems -more, perhaps, even than it solves, that very fact unlocks whole new worlds of possibility for scientific advance and in itself constitutes an argument for serious consideration.

If, in fact, we have already "arrived", if you are satisfied that we do, in fact, already possess in rough form a valid picture of

---

[3] echoing Einstein's characterization of Freudianism

[4] Since then, my perspectives have widened. I have come to believe that the tragedies of mental illness are echoed in the tragedies of the human social condition -the wars, the hatred, the arrogance, the exploitation of man by his fellow man, these are other aspects of the same basic problem. Under the perspective of dogmatic Naturalism, these are plausible and normal, and therefore necessary. I do not believe they are.

the whole of our reality, then the very *poverty* of that reality as regards the human condition must make you very sad -and kindle the hope that something more is possible. I think it is!

Science to date has provided the tools for an enlarged and more sophisticated physical life, but taken away the reasons for living it! It is time to deepen science itself.

Appendix A: The Dennett Appendix and the Color Phi, (from Iglowitz 1995)

Perhaps the hardest hurdle for my conception is what I call "the static problem".[1] The axiom systems of current mathematics tend to create uniform, "static" fields of objects: the integers, for instance, or the real numbers.

True, there are special, unique objects within them, pi, or e, or 1 for instance, but these are not promising for the kind of usage we will need to see for viable mental objects.[2]

To this point, the model I have proposed stands more in the sense of a Platonic "form", and lacks the viability of Aristotle's conjunction of "form and matter" for the existence of *actual, special* objects.

Let me try to suggest the beginnings of a solution for the existence of such objects within such a system. Let me try to suggest a rationale for actual perceptual objects!

Daniel Dennett, (though he is a confirmed anti-mentalist), has provided an inspiration. It derives from his treatment of the "color phi" phenomenon, -though his conclusion must be stood on its head. I suggest that the answer to the "static problem" and the ground of viable perceptual objects lies in recognizing intentionality as a primary component of brain process. It is a necessary and complementary (system of) "axiom(s)".

Towards a Working Model of *Real* Minds: Dennett, Helmholtz and Cassirer

I really *liked* Daniel Dennett's "Consciousness Explained"[1]. It is not because I could agree with his conclusions, (except in a *certain sense*), that I liked it, but because it is a *brutally* candid and

---

[1] See the earlier discussion in Chapter 13 under the heading of "How Then Do We get to the Aspect Results? "

[2] See Chapter 13: d'Espagnat, for another very recent thought on the subject.

forthright exposition of the Naturalist position, proceeding with compelling logic, and without hedging.

I respect that! It is, moreover, a phenomenologically *pure* position. I think it is, (agreeing with his own parenthetical question), really "Consciousness Explained *Away*" however, rather than "Consciousness Explained" because, at the end, "we are all zombies".[3]

There is one crucial argument he makes against the existence of mental states, (i.e. "figment"), however, in which I think he has correctly identified a profound antinomy -and, I believe, a necessary and major modification to our ordinary conception of mind. He has argued it from "the color phi".

"The color phi" names an actual experiment, suggested by Nelson Goodman, wherein two spots of light are projected in succession, (at different locations), on a darkened screen for 150 msec intervals with a 50 msec interval between them. The first spot, however, is of a different color, (red, say), than the second, (green). Just as in the case of motion pictures, (the "phi phenomenon"), subjects report seeing the continuous motion of a

---

[3] I know, I know! I must, in threat of disingenuousness, quote his footnote to this comment: "it would be an act of the utmost intellectual dishonesty to quote this statement out of context."

But the context he demands is 470 pages of careful redefinition and argument against all the normal senses of mental function and existence -qualia, figment, the "substance of mind". The upshot is that it is O.K., (i.e. socially correct), to be a zombie! But the sense in which his statement would normally be understood *out* of context is essentially what it *still* means within it. He attempts to make any objection, (or any comment on its own prima facie unintuitiveness), unraisable. There is *another* cult, (besides the Feenomanists!), in the jungle, you see! :-)

single spot, but interestingly, they report that it changes color, (from red to green), midway *between* the two termini![4]

Dennett bases a very interesting, (and, I feel a very important), argument against the very possibility of a "Cartesian Theatre", against a unity, (and "figment" = substance), of consciousness on this well documented and reproducible experiment. Dennett's argument, in brief, is this:

Mental states, the "Cartesian Theatre", if they exist, are subject to the laws of causality, of time precedence. For one event to affect another, it must occur *before* it. Let me, for discussion's sake, label the events described. Let E1 be the ("heterophenomenological"[5]), perception, (hereinafter to be called by me "h-perception"), of the first, (red), spot. Let E2 be the h-perception of the red-changing-to-green, and let E3 be the h-perception of the final green spot.

Dennett argues, based on the principle of causality that E2 cannot occur until after E3. Since there were only two actual, (physical), events, (the first and second projected spots), he argues that the h-perceived midpoint, (the "mental event", i.e. red-changing-to-green), cannot occur until *after* the reception of the second actual event, (green projection), as it was that which provided the very sensory data *necessary* to the h-perception of change.

Other than a (mystical) hypothesis of "projection backward in time", there remain for Dennett just two possibilities

---

[4] and not, for instance, that it is red all the way till its terminus, with a final and sudden change-to-green.

[5] Dennett introduces the criterion "heterophenomenological" to describe "mental events", which he does *not* believe in, to describe whatever-it-is that is named by them, i.e. to talk about them as they are (linguistically) used by real bodies and brains, (which he does believe in), but with a neutral metaphysical commitment.

for an internal, "Cartesian Theatre" consistent with the experiment: the "Stalinesque" and the "Orwellian" hypotheses.

The first involves the creation of a "show trial" staged by a subterranean "central committee", (after the fact of both real events, of course, and involving a "delay loop"), wherein the complete, (and partially fabricated), sequence, (red ->red-changing-to-green -> green), is "projected", (i.e. achieves sentiency).

Under this hypothesis, the whole of our sentiency, our consciousness, occurs "after the fact". The second possibility, the "Orwellian" hypothesis, is that the actual events are received by our sentient faculty *as is*, but that our memory then rewrites history, (just as the thought police of Orwell's "1984" did), so that we *remember* not two disjoint and separate events, but the connected, and pragmatically more probable sequence red -> red-changing-to-green -> green.

Dennett argues that ultimately *neither* theory is decidable -that either is consistent with *whatever* level and kind of experimental detail science may ultimately supply, and that, therefore, the only pragmatic distinction between them is purely linguistic, and therefore trivial.

He argues that there *is* no "great divide", no actual moment, (nor existence), of sentiency, but only the underlying brain process, (which *all* theories must countenance), itself. Based on the "spatial and temporal smearing of the observer's point of view", he expounds his thesis of "multiple drafts" wherein there *is* no "theatre" -only brain process- and its various "speakings", (drafts).[6]

And yet the observer *himself* has absolutely no problem with these events! *His* perspective is very clear: E1 -> E2 -> E3.

---

[6] Though I do not accept the idea, you might also consider Maturana's perspective on this point –in his "third order coupling", ("languaging").

It is our interpretation, (and rationale), for this sequence that causes the problem.

I think Dennett has a very strong argument, but I want to refocus it. Nondecidability is all very well and good, but it is a much weaker line than the one he started out with- on the possibility of *synchronization*! In a very real sense, I feel it is very similar in intent and consequence to Einstein's famous "train" argument against simultaneity.

Consider, (with Einstein), an imaginary train moving (very fast)[7] down a track, with an observer standing midway on top of the moving train and observing two (hypothetically instantaneous) flashbulbs going off at either end of the train.

The train goes by another (stationary) observer standing (hypothetically infinitely close) to the track as the bulbs go off. Suppose that the moving observer, (OT), reports both flashes as simultaneous. He argues that since both photon pulses reach him simultaneously, (granted for all frames on the local, infinitesimal scale, and thus supposedly agreeable by (?) *both* observers who are assumed infinitely close –i.e. side by side), that therefore the pulse from the rear of the train, having to "catch" him, must have left its source sooner than the pulse from the front which added his velocity to its own and so must have left later.

Relative to OS, (stationary observer), however, the two sources travel the same distance to a *stationary* target, (himself). Since OT and OS are momentarily adjacent to each other, (i.e. within a local, infinitesimal time frame), they should be able to agree that the two pulses *arrive* there simultaneously.

What they *cannot* agree on, however, (in that instance), is whether the events, (the flashes), *occurred* simultaneously –*nor that the other could have thought, (i.e. could have observed),*

---

[7] nearing the speed of light

them so! Time, in Dennett's words, is "smeared"![8] We could, of course and significantly[II], *vary the parameters* of the stated problem to make *either* event "earlier" and the other "later".[9]

The argument[10] is that from the standpoint of one observer, he must maintain that the *other* cannot see them as simultaneous, and vice versa! Thus from OS's standpoint, if he sees them as simultaneous, then, since he is stationary, they *occurred* simultaneously. But if they occurred simultaneously, and since OT is moving, then *OT cannot*, (OS argues), see or conceive of them as simultaneous, (and conversely). And yet both observers pass through an infinitesimal local frame of reference, (side-by-side). Time is "smeared"!

Just as Einstein's two observers, near the limits of physical possibility, cannot agree whether the two lights were *simultaneously* flashed at the ends of the train or not, (i.e. cannot establish a common temporal frame of reference), nor that the other could observe them locally as such, neither, given Dennett's pointed argument, can we establish a common temporal frame of reference for "the world" and "the mind" at the limits of cognition.[11]

I agree with Dennett that "the color phi" identifies a legitimate and critical aspect of the mind-body problem. The

---

[8] Are the observers, (and the experimental apparatus), then "heterophenomenological"?

[9] i.e. if the front pulse arrived at the correct interval before the rear pulse, OT could argue that they were, in fact, simultaneous, but OS would obviously argue to the contrary. This would be a better match to Dennett's specific problem.

[10] assuming the legitimacy of "simultaneity" itself

[11] For macroscopic science, these limits are at the scale of the speed of light. For atomic physics, they are at the scale of Planck's constant. And for the brain, I suggest, they are at the scale of minimal biological response times, i.e. in the 100 msec. range.

spatial and temporal "smearing" of the percept and the non-explicit reference of qualia that he demonstrates forces a profound extension to our traditional conception of the "theatre".

But his dimensional "smearing" actually fits very well[12] with the model I am proposing. I submit that it is *more plausible* in terms of the "focus" and "function" of an operative "schematic object" than in terms of his "multiple drafts", "demons" and "memes" in the "real world".

Cassirer on the Color Phi:

Dennett's objections to the ordinary "Cartesian theatre" are admittedly valid, but so were those of Cassirer and Helmholtz long before him:

> "For example, if we conceive the different perceptual images, which we receive from one and the same 'object' according to our distance from it and according to changing illumination, as comprehended in a series of perceptual images, then from the standpoint of immediate psychological experience, no property can be indicated at first by which any of these varying images should have preeminence over any other. Only the *totality* of these data of perception constitutes what we call empirical knowledge of the object; and in this totality no single element is absolutely superfluous. No one of the successive perspective aspects can claim to be the only valid, absolute expression of the 'object itself;' rather all the cognitive value of any particular perception belongs to it

---

[12] when taken "heterophenomenologically" -i.e. with a neutral ontic commitment. Heterophenomenology works both ways!

only in connection with other contents, with which it combines into an empirical whole."

"...In this sense, the presentation of the stereometric form *plays 'the role of a concept*", (my emphasis), "'compounded from a great series of sense perceptions, which, however, could not necessarily be construed in verbally expressible definitions, such as the geometrician uses, but only through the living presentation of the law, according to which the perspective images follow each other.' This ordering by a concept means, however, that the various elements do not lie alongside of each other like the parts of an aggregate, but that we estimate each of them according to its *systematic* significance...." (Cassirer, 1923, pp. 288-289, citing Helmholtz)

But Cassirer's own drastic reformulation of the formal [technical] "concept" itself must be considered for an understanding of his meaning here. The "concept", for Cassirer as we have spent a lot of time understanding, *is a function*! It is like "the form of a series", independent and distinct from what it orders.[III] This is the "systematic significance" which he purports.

I urge, extending Cassirer's insight and in the sense of my conclusions of Chapters 3 and 5, (re: C.I.D.), that the stereometric form itself, the actual percept,[13] then plays the role of, (is), a function.

From the standpoint of (relativized) Naturalism,[14] if we take the mind to be a schematic, but specifically a *"predictive"* and

---

[13] This, *the percept as concept*, is clearly at odds with, but, (I have argued), a legitimate extension of, Cassirer's ideas. He did not have the perspective of "the schematic object".

[14] cf. Chapter 5

"intentional" schematic model, (which extension I will suggest shortly), rather than a static and "representative" one[IV], then the temporal and spatial "smearing" of the percept do not have the implications against the "theatre" *per se* that Dennett attributes to them.

I have argued that the percept itself is conceptual, (albeit specialized, invariant and constitutive), and therefore, following Cassirer, functional. It is an entity of order and process –and it *is* "smeared". That is the normal nature of *functions* –functions *are* smeared! [Note June, 2010: Reconsider the continuum itself!]

What Dennett explains by "multiple drafts", (and the "demonic" process he envisions beneath them), I explain by "focus". We focus the percept, (via implicit definition) according to operational need.[15]

An Extension of the Schematic Model: A Brief Sketch

Let me frame the following in the language of ordinary Naturalism, (this *will* be a short appendix). I want to sketch a very large canvas very quickly.[16] In "the color phi", I think that Dennett has identified a very important difficulty in our ordinary conception of mind. It suggests an enlargement and a more sophisticated perspective on the schematism I have argued

---

[15] See Appendix D for a discussion of Cassirer's "Symbolic Forms" vis a vis Bohr's "multi-valued functions".

[16] I could, of course, try to footnote every misconception and every possible claim of inconsistency, but I have already done that. I think I have paid my dues. "Predictivity", "intentionality", et al are, under my thesis, perfectly valid conceptions *within* the Naturalist "form" - and I may consistently use them *as such* without self-contradiction! Within the context of my larger perspective, they are model-model correlations, synthetic a priori "slices" across the phenomena.

heretofore. Though I think I have successfully laid the solid foundation, let me now briefly sketch the design of the cathedral itself, i.e. the design of *real minds*!

I have dealt, previously, with the schematic object. I argued that the object of perception is a schematic artifact of reactive brain process, specifically "designed" to optimize a simple and efficient "calculus" of response.

But the converse side to that argument is that a calculus was actually enabled! What are the (Naturalistic) implications of that calculus, and of the schematic model?[v]

A Thought Experiment

Follow me in a thought experiment! Keeping your eyes fixed to the front, you perceive, (in your perceptual model), this paper in front of you, the wall behind it, and, perhaps, the pictures of your family. There may be pens and pencils, books. You may hear music from the stereo next to you, (and perhaps still in peripheral vision). There may be a window, and the lights of the neighbor's house beyond it.

But there is no wall behind you! There is no car in the driveway outside of your house -indeed, there is no "house" at all. There is no city, no taxes, no friends. The sun does not exist in this model. There is no government, no "universe", -no tomorrow! The (purely) perceptual model is incomplete as a model of "reality" and it is, (Naturally!), inadequate even to keep you alive!

There is something else necessary for completeness of the model detailed in this book, i.e. a new perspective on it. It is an *intentional* aspect. It is necessary to supply the object behind your back and the reality "over the hill"! It supplies the connection to "tomorrow" and "yesterday". It supplies "causality". It is necessary for the completeness of a model of "*the world*". It is necessary, (specifically following Dennett!), even for the individual "objects" of perception itself, (E1 and E3 for instance).

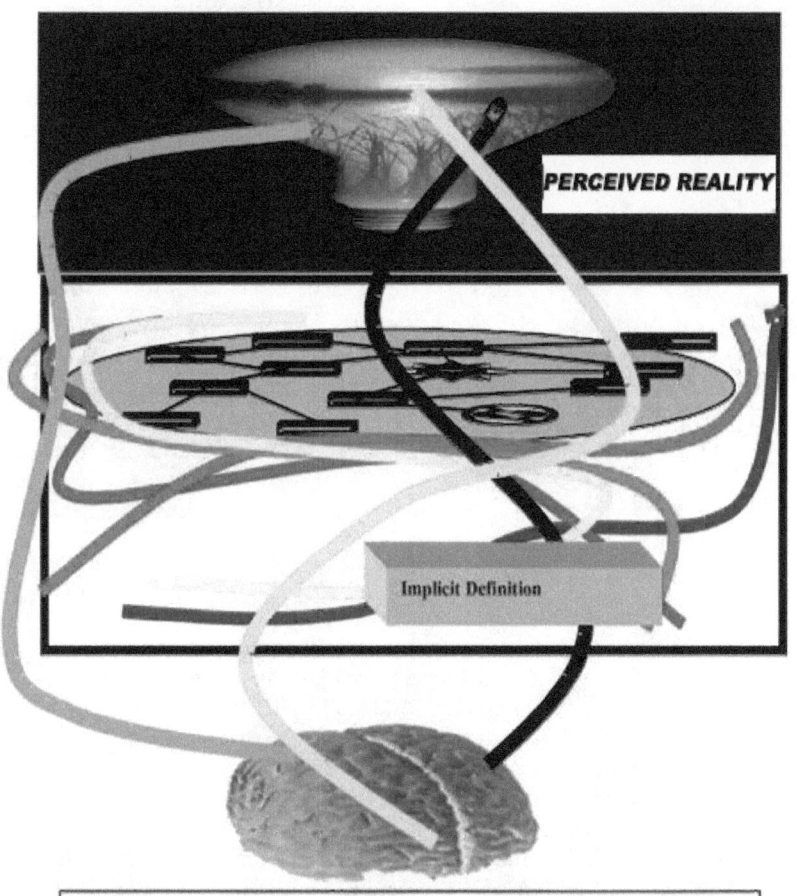

Figure 31. A Model of Mind — Consciousness: A phenomenon of the brain's intentionality!

This model, I suggest, is where E2, (the object of Dennett's perplexity), lives. It cohabits there very comfortably with E1 and E3 which, I argue, are *also* predictive and schematic objects. There is a seamless integration, (above the scale of 100 ms, let us say), of what we normally think of as our pure percepts and the intentional fabric within which they are woven.[17]

This model, I believe, is the actual "home" of mind, and the legitimate purview of a truly scientific psychiatry.[VI]

> "Now what is a phenomenal space? Is it a physical space inside the brain? Is it the onstage space in a theater of consciousness located in the brain? Not literally. But metaphorically? In the previous chapter we saw a way of making sense of such metaphorical spaces, in the example of the 'mental images' that Shakey, [a robot], manipulated.

---

[17] Let us turn Dennett's argument around. Dennett argues strongly and convincingly that "figment", (mental states), are logically inconsistent with our, (his), ordinary (naïve) views of cognition and reality. If, instead of accepting his conclusion however, we choose to accept the *reality* of that figment –E1, E3, *and E2*, –if we believe that E2 is *actually* perceived, (whatever it may be), then his argument takes on a different import and works against the very ground in which it was framed: i.e. his ordinary view of cognition and the Naturalism, ("objectivism"), in which he embedded it. The "color phi", he says himself, embodies a precise and *reproducible* experiment –both you and I would expect to *see* E2!

I consider the "phi phenomenon" *itself* more interesting than the "color phi", however. The credibility and intentional depth of a series of oversized, rapidly sequenced still pictures, (a movie), is quite suggestive. Its potential for an uncanny parallelism with our ordinary experience suggests that the latter, (i.e. ordinary experience), is *itself* a predictive and integrative phenomenon grounded in a schematic, intentional model in precisely the same manner as I propose the "color phi" to be.

In a strict but metaphorical sense, Shakey drew shapes in space, paid attention to particular points in that space, based conclusions on what he found at those points in space. But the space was only a *logical* space.

It was like the space of Sherlock Holmes's London, a space of a fictional world, but a fictional world systematically anchored to actual physical events going on in the ordinary space in Shakey's 'brain'. If we took Shakey's utterances as expressions of his 'beliefs', then we could say that it was a space Shakey *believed in*, but that did not make it real, any more than someone's belief in Feenoman would make Feenoman real. Both are merely intentional objects.... *So we do have a way of making sense of the ideas of phenomenal space -as a logical space.*"[18]

But this is my *exact conclusion* of Chapters 3 and 5. Dennett and I are not so very far apart after all -save in our metaphysics, (wherein we are *very* different). Mind is a *logical* entity -i.e. its "space" is a *logical* space.

But Dennett's "mind" is based in abstractive, associationist logic (after Dreyfus' usage[VII]), and *dead*, and mine is based in a functional logic, (the constitutive logic of Kant and of biology), and *live*. We are *not* zombies!

On the issue of metaphysics, on the other hand, surprisingly Dennett specifically argues that "nature does not build epistemic engines."[VIII] Why, then, does he think that *he*, either as a physical engine of process, (and the "demons" of process), *or* as a

---

[18] Dennett, 1991, pps.130-131, my emphasis.

linguistic engine of "memes", –*is epistemic*, (i.e. metaphysically so)? Or that *his book* is so?

I don't think that he, or it or I, are. This was my exact conclusion of Chapter 5.

Appendix B: Lakoff, Edelman, and "Hierarchy"[1]

(Drawn largely from an appendix to an earlier rendition of my ideas –Iglowitz, 1998)

I had not seen George Lakoff's "Women, Fire, and Dangerous Things" nor Gerald Edelman's "Bright Air, Brilliant Fire" until about 1995. It was remarkable to me, therefore, to see how closely Lakoff's logical and epistemological conclusions resembled those of Cassirer[2], (considered as the combination of Cassirer's dual theses: his logical thesis of "the functional Concept of mathematics" and his epistemological thesis of "Symbolic Forms"), -and how *closely* Edelman's biological and philosophical answers, based in Lakoff's and his own original work, resembled my own conclusions.

There is an uncanny parallelism of structure, (though not of consequence), between the paths we have followed to arrive at our conclusions. Our structural differences are differences of degree –but important differences. I believe that Lakoff, (and Edelman), have gone too far in the case of logic, and not far enough in the case of epistemology.

They fail[3], crucially thereby, to provide the grounds for an answer to the ultimate problem: i.e. how can "mind" or "consciousness", (normally taken) coexist with the existence of the brain?

---

[1] [Note: This is the original Lakoff/Edelman appendix from Iglowitz 1998 sans the discussion of mathematical "ideals" which latter is presented elsewhere in this book. Pretty much everything else is reproduced.]

[2] Of which Lakoff, apparently, was unaware

[3] -innocently for Lakoff who never promised such an answer, but more pointedly for Edelman who did

George Lakoff:

Lakoff grounds his work in logical reflections of Wittgenstein[4] which questioned the adequacy of the classical logical Concept and in the work of Rosch and a host of modern empirical researchers which further challenged that classical Concept by demonstrating exceptions in actual human usage of language and concepts across cultures -and even within our own legitimate contemporary usage. From these grounds and his own original work, Lakoff drew strong conclusions about the nature of logic[5] –and the human mind- itself.

The Classical Concept

The classical concept[6] is defined "by necessary and sufficient conditions" -that is, by set theoretic definitions on properties. It is an elementary theorem of logic that the whole of the operations of sentential logic, for instance, may be grounded solely in the primitive operations of intersection and complement.[7] More generally, logical sets and categories,

---

[4] E.g. Wittgenstein's "family resemblances"

[5] compare Cassirer: "... Every attempt to transform logic must concentrate above all upon this one point: *all criticism of formal logic is comprised in criticism of the general doctrine of the construction of concepts.*" –cited at the beginning of my Chapter 3.

[6] Lakoff is concerned with primarily with categories, but the distinction is technical and not necessary to this discussion. Cassirer dealt specifically with concepts, but he covered essentially the same ground.

[7] Or, equivalently, on other subsets of set operations.

(concepts[8]), are defined on presumed "atomic properties" and are commensurable wholly based on the set-theoretic possibilities of those sets –i.e. union, intersection, complement, etc.

Concept-sets, (within this classical perspective), express a hierarchical "container schema" moreover, (using Lakoff's language). Though Lakoff frames his discussion to the same end slightly differently, by this I mean that whenever we classically specify a genus, we do so by eliminating one or more of these atomic properties, (by intersection of the properties of species), at the same time thereby specifying an expanded extension, (union) –i.e. the set of "objects" which the genus concept encompasses.

The delimitation, (by property containment), of the genus category is contained within, (is a subset - an intersection of), that of the species category while the *extension* of the species category, conversely, is contained within, (is a subset of), the extension of the genus category.

In specifying a species on the other hand, we do so by adding one or more properties –ultimately "atomic properties" to the properties of the genus and this species concept encompasses a diminished, (intersectional), extension of the extension of the original genus.[9] This classical categorization therefore expresses an absolute, rigid and *nested* hierarchy of levels and containment. In Lakoff's terms it expresses a hierarchical "container schema".[10]

Ultimately, (because they *are* nested), at the limits these processes specify

---

[8] See prior footnote: categories vs. concepts

[9] "Cross categorization", the "other . . . classical ... principle of organization for categories" refers to the various possibilities at any stage of genus or species categorization – on the particular choices of which "atomic properties" are to be eliminated or added. Cf Lakoff pps. 166-167

[10] ibid

(1) a largest concept: "something", (defined by *no* atomic properties), whose extension is "everything", and

(2) a smallest concept: a particular "object" in reality, (or possible reality), defined by *all* its atomic properties[11].

Given the classical paradigm then, reason necessarily begins with "something", (the most general concept), and points, inexorably, to some particular *"thing"*, i.e. a specific object.[12]

But Lakoff plausibly argues that concepts[13] in legitimate human usage are actually determined by *any* rule, (to include the classical rules of set operations on properties as just one *special case* of a rule), or even *by no rule at all*!

Thus metaphorically based categories, such as the Japanese concept of "hon" are generated, (determined by), a metaphoric rule of extension and metonymically based categories are generated by a rule of metonymy. (Metonymy is the case where one particular instance of a category is made to stand for the category.) "Don't let El Salvador" become another Vietnam" is an example Lakoff uses of a metonymically based category.[14] Here "Vietnam" stands for the concept of all hopeless, unending .... wars.

In the case of "radial categories" on the other hand, such as the concept of "mother", (to include birth mother, adoptive mother, foster mother, surrogate mother, etc.), or of "Balam"[15] in

---

[11] to include spatio-temporal properties

[12] or *the exact converse* – i.e. beginning with some specific object or objects in reality or possible reality and ending with everything!

[13] he would say "categories"

[14] P. 77. Actually I like his "ham sandwich" better, but it was pre-empted by Edelman!

[15] The category which is the source of his title and includes, among other things, women, fire, and dangerous things.

the Dyirbal aboriginal language in Australia, they are determined by simple historical accident –they are not generated from the central model by general rules

[but] .. must be learned one by one."[16] (Extensions from the central model are not "random" however, but are "*motivated*', his emphasis, "by the central model plus certain general principles of extension.")[17]

He argues his case rigorously and scientifically by exhibiting myriad examples that are *not* compliant with the classical Concept and analytically by demonstrating the degradation of concepts in actual bi-cultural environments –i.e. where one culture and language is being overrun by another, ("language death"), as is the case with the Dyirbal aboriginal language in modern Australia.[18] The degradation is characterized by the loss of blocks of suborganizations, not of random individual elements.

Lakoff's logic is not trivialized by this "free formation" of concepts however, (as it might seem it *would* be[19]- logic being [paraphrase] "mostly concerned with categories"), as he bases

---

[16] Lakoff, P.91

[17] As I will repeat later, this discussion of Lakoff's thesis is woefully inadequate, but it will have to do for the purposes of this appendix. He states as the "main thesis of [his] book .. that we organize our knowledge by means of structures called *idealized cognitive models*, or ICMs, and that category structures and prototype effects are by-products of that organization." Ibid, p.68

[18] See Lakoff, pps. 96-102

[19] If, according to Lakoff, (1) legitimate concepts may be formed on any principle or no principle, and if, also according to Lakoff, (2), most of the business of logic is concepts, (categories), then it would appear, (at first glance), that (3) logic could prove *any* conclusion. But if logic can prove anything, then it can prove nothing! Thus it would appear, on the face of it, that his purported impossibility of a rigorous, comprehensive structure for categories in general would imply the invalidation of logic in general.

logic and the relevance of concepts ultimately in a *preconceptual context* rather than in the concepts themselves. Concepts, (categories), he argues, are not created in a vacuum, but within preconceptual schemas: "idealized cognitive models", (ICMs). The latter are ultimately determined, (he argues), by the function of the body in the external world–all describable from "body in the world".

> "There are at least two kinds of structure in our preconceptual experiences:
>
> A. Basic-Level structure: Basic-level categories are defined by the convergence of our gestalt perception, our capacity for bodily movement, and our ability to form rich mental images.
>
> B. Kinesthetic image-schematic structure: Image schemas are relatively simple structures that constantly recur in our everyday bodily experience: CONTAINERS, PATHS, LINKS, FORCES, BALANCE, and in various orientations and relations: UP-DOWN, FRONT-BACK, PART-WHOLE, CENTER-PERIPHERY, etc."[20]

These schemas, however, being at the *basis* of our reasoning[21], are necessarily mutually relativistic and equipotent and we utilize them on a "best fit" rationale. The concepts that arise within them need not be commensurate *across* them. Thus he arrives at a relativism of logic and concepts.

Lakoff's Concept/category in many ways resembles Cassirer's[22] and he rejects, (as does Cassirer), the classical

---

[20] Lakoff, p.267, his CAPS.

[21] rather than categories

[22] There is an uncanny parallelism of argument throughout between Lakoff's and Cassirer's treatment of logic. Consider, as an example, the following:

"necessary and sufficient conditions", (as he phrases it), which ground set theoretic abstraction and the Aristotelian generic Concept. His logical and ultimately epistemological relativism, (in his "idealized cognitive models"), is also very similar to, (though it is not as abstract and comprehensive as), Cassirer's "Symbolic Forms" which is described in my Chapters 7 & 8 in the current writing.)

Cassirer and Lakoff's Logic

Cassirer rejected the logical sufficiency of classical categorization as does Lakoff, but he did *not* reject the possibility of the possible existance of an absolute, comprehensive structure for concepts/categories, (which Lakoff does). Instead Cassirer retained *an overall formal structure* for categorization in the notion of a mathematically functional rule or series.

"Category cue validity defined for such psychological (or interactional) attributes might *correlate*", (his emphasis), "with basic-level categorization, but it would not *pick out* basic-level categories; *they would already have to have been picked out* in order to apply the definition of category of category cue validity so that there was such a correlation." (Lakoff: P.54, my emphasis) This is almost an exact parallel to one aspect of Cassirer's argument against the classical concept, and the "theory of attention", (see my Chapter 3), –and for a "new form of consciousness".

Discussing Erdman, Cassirer writes: "...instead of the community of 'marks,' the unification of elements in a concept is decided by their 'connection by implication.' And this criterion, here only introduced by way of supplement and as a secondary aspect, proves on closer analysis to be the real logical *prius*, " (his emphasis), "for we have already seen that 'abstraction' remains aimless and unmeaning if it does not consider the elements from which it takes the concept to be *from the first* arranged and connected by a certain relation." Cassirer, "Substance and Function", p.24

Cassirer did not question the *legitimacy* of the classical schema, but he did question its *necessity* and *sufficiency*. (Which is pretty much where Lakoff and myself stand as well.) He argued that the latter is, in fact, a special and *limit case* of the Concept and of the possibilities of logic. Cassirer maintained that many concepts –and *specifically* the very concepts of mathematical and physical science[23] –demonstrate another mode of concept formation and specification than the classical scheme, (this is the subject of my Chapters 3 and 5. Both concept formation upward, (genera), and downward, (speciation), can obey *another* rule-based law, i.e. the properties of their extensions can embody a series *other than the specific series of identity*!

As a crude example, one member of the extension of a concept, (using an example drawn from numeric sets), might contain the numeral "2", another extension the numeral "4", another "8", "16"… rather than the numeral "2" being in all of them. Thus the concept would express, (and be *the rule* –i.e. be formed on the principle of), the series 2,4,8,16,… *across* its extension rather than being based in the series of identity: 2, 2, 2,…. , (the classical schema). The extension of a category, therefore, may be defined based upon the possession of some property belonging to a series or function *on* properties rather than on the possession of some *identical* property(ies). Concepts can be specified by functions *other than* identity. [24]

---

[23] Cf Cassirer, "Substance and Function", "Einstein's Theory of Relativity". Incidentally, the original title for "Substance and Function" was "Substanzbegriff und Funktionsbegriff", i.e. Substance Concepts and Function Concepts!

[24] Cassirer's "series" could be ordered by *radically variant* principles, however: "according to equality", (which is the special case of the "generic concept"), "or inequality, number and magnitude, spatial and temporal relations, or causal dependence"[24] –so long as the principle is definite and consistent. But please

Cassirer has supplied a clear counterexample and an alternative to the classical schema, (which I explained at length and further extended as the subject of [the old]Chapter 3). Simplistically, (and as crude illustration), we may have three pieces of "metal" in front of us for instance, wherein *none* of their properties are the same!

The first is a one pound piece of gold, (color: yellow, specific gravity: a.aaaa...., conductivity: b.bbbb...., etc.), the second a two pound piece of lead, (color: gray, specific gravity: l.lll..., conductivity: m.mmm...., etc), and the third a three pound piece of tin: (..., ..., ..., etc.) None of these properties need be *identical* however. They are related as "metal", (and are specified as "metal objects"), because the color of each, (for instance), *is a value of the function*: $COL(x) \varepsilon$ {yellow, gray, silver,...), the specific gravity of each is a value of the function $SG(x) \varepsilon$ {lll..., ggg..., ...}, and so on.

These objects, (the objects called "metal objects"), can "cross party lines", so to speak —i.e. they are not the product of strict set-theoretic intersection of atomic properties. In the illustration their intersection across these properties is null! The extension of scientific and mathematical concepts, (*specifically*, Cassirer argues), need have *no* atomic properties in common[25]. Repeating a short citation from my Chapter 3:

> "Lambert pointed out that it was the exclusive merit of mathematical 'general concepts' not to cancel the determinations of the special cases, but in all strictness fully to retain them. When a mathematician makes his formula

---

remember that these are *principles of category construction* rather than *properties* of categories. see my Chapter 3

[25] Compare Wittgenstein's "family resemblances".

more general, this means not only that he is *to retain* all the more special cases, but also be able *to deduce* them from the universal formula."[26]

But this possibility of deduction does not exist in the case of the scholastic, (Aristotelian), concepts, "since these, according to the traditional formula, are formed by *neglecting* the particular, and hence the reproduction of the particular moments of the concept seems excluded."[27]

"The ideal of a *scientific* concept here appears in opposition to the schematic general presentation which is expressed by a mere *word*. The genuine concept does not disregard the peculiarities and particularities which it holds under it, but seeks to show the *necessity* of the occurrence and connection of just these particularities. What it gives is a universal *rule* for the connection of the particulars themselves.... Fixed properties are replaced by universal rules that permit us to survey a total series of possible determinations at a single glance."[28]

Consider "the ellipse as a simple mathematical example of a genus" for instance. Its species are functionally related –and fully recoverable– in the defining equation of ellipses in general.

Conversely in the specification of species and subspecies, ("downward"), the process does not necessarily lie in the addition of (identical) atomic properties either, (the members of the extension of a subspecies, which is *also* a category, need not contain (any) identical atomic properties by the same reasoning),

---

[26] Cassirer, "Substance and Function", P.20-23

[27] ibid P.20-23, my emphasis

[28] ibid P.20-23

but can be accomplished instead in the identification of the value of a function –i.e. a sub-function whose possibility is implicit within the genus.[29]

Ultimately, (and recursively), the question proposes itself: *need there be* a lowest, "bottom" level concept *at all?*[30] Speciation is no longer necessarily intersection or containment,[31] (it is no longer necessarily nested), so there is always the possibility of another, further rule of assembly for a subspecies of *any* species – at *any* level![32] There is thus no longer a *necessary* logical focus on an ultimate "thing".

Cassirer argues that the ultimate "objects", (the "theoretical objects"), of mathematics and physical science are "implicitly defined" by, (and express), the fundamental laws of the science itself. He argues that they are instances of complex speciation based in the general functional rules, (the laws), of the sciences themselves and not objects "in reality".

Some of Lakoff's categories, it is true, are also rule based, (other than the classical rule), but in the case of his "radial categories", they may be formed by historical accident. Lakoff concluded that categories may be formed by classical rules, other

---

[29] Since we can build a genus without commonality, so can we build a super-genus. Turning our perspective around, then, we may speciate downward from that super-genus without the utilization of commonality!

[30] The other pole is clearly impossible. There is clearly no Concept, (category), of all concepts under Cassirer's vision as it would necessarily be defined on "the rule of all rules". But some, (most), rules are obviously inconsistent with other rules –disallowing the concept.

[31] Since there is no longer a necessary presumption of nesting, the implication that there must be a "least member" is no longer justified.

[32] Remember that under Cassirer's Concept, we do not eliminate properties to speciate, but rather functions.

rules or "*no rule at all*"! But this characterization divorces him from the possibility of any universally comprehensive categorical structure.[33]

Cassirer includes this special latter case as an *ad hoc* rule, (series), however, rather than as an example of "no rule". It would correspond to the special case in mathematical set theory wherein a set is defined by the explicit listing of its members.

Cassirer's conception may be likened to a line segment bounded on one end by the classical criterion of identity of properties across members, (a "unity"), with the central section composed of any and all functional rules, (i.e. rules of series/regular functions on those properties), and bounded at the other end by the rule of explicit listing, i.e. no *other* rule, (a "zero"). This view reconciles the two conceptions, I think, and might be acceptable to Lakoff.[34] What it does besides, however,

---

[33] Cf: the discussion of the crucial role of comprehensiveness vis a vis mathematical ideals near the end of this Afterword.

[34] Compare Lakoff, p.146 : "in the classical theory, you have two choices for characterizing set membership: you can predict the members (by precise necessary and sufficient conditions, or by rule), or you can arbitrarily list them, if there is a finite list. The only choices are predictability (using rules or necessary and sufficient conditions) and arbitrariness (giving a list). But in a theory of natural categorization, the concept of *motivation*", (his emphasis), "is available. Cases that are fully motivated are predictable and those that are totally unmotivated are arbitrary. But most cases fall in between —they are partly motivated." [Note 2010: But Cassirer's "Symbolic Forms" are *definitely* motivated —as *intentional perspectives!*]

Cassirer suggested another, (and more classical), "middle ground" wherein the principle of "necessary and sufficient" is not grounded in an identity of properties, but in a functional relationship between them. The relationship between their proposals is more complex than is possible to describe here, but as a thumbnail sketch of my opinion, the deficiencies in the classical category that Cassirer resolves in his "Functional Concept of Mathematics", Lakoff

is reveal a comprehensive structure across the *whole* of categories/concepts.

I have suggested a further extension beyond Cassirer's "Functional Concept" and sets of n-tuples however in the arguments of [my old] Chapter 3. Just why is the color of "gold-metal" *yellow* instead of gray? Why is "gold" a *particular* n-tuple rather than some *other* mix of possible place-values? Physical scientists will never agree with Lakoff, for instance, that it could be just an (accidental) property of a "radial category", nor, possibly even with Cassirer, that it is simply an element in a multi-place series.

They will insist that it must be a *necessary* property determined by physical law. Cassirer apparently glimpsed this connection in his conception of the "ideal objects" of the sciences, but he never fully exploited it. (I have pursued it in my "Concept of Implicit Definition".[35])

Both Lakoff and Cassirer followed the paths of their logical conclusions to see the essential flaw in "naïve realism", (as Cassirer termed it), and "objectivism", in Lakoff's words, (I have used the term "naturalism"). If the classical logical schema of strict hierarchical containment were legitimate, and, more importantly, if it were necessary and sufficient, then the *only* possibility of science, as the resolution of experience and reality with logic, would lie in the absolute objective existence, (however reduced), of our ordinary objects.

---

attributes to his Cognitive Models whereas the deficiencies in classical metaphysics are resolved by both of them very similarly in the epistemological relativity of "Symbolic Forms" by Cassirer and of "ICM's" by Lakoff. Cassirer's is the more general of the two solutions to the latter problem, however, as it is not framed within a specific image of the world, but within the constraints only of abstract epistemology as Kant definitively iterated them.

[35] Cf my Chapter 3

If valid logic and conceptualization is broader than that, however, then the possibility of reality is considerably enriched. Valid conceptual, (or utilitarian cognitive), "objects" need not then express "membranes" around spatio-temporally contiguous properties of *ontological*, (i.e. metaphysical), objects or groups of such objects![36] They can "cross party lines"!

Cassirer had no problems with such an implication. It was implicit, of course, in his neo-Kantian origins. Lakoff did. In his laudable commitment to realism, he was forced to consider the minimal necessary requirements of such a (scientific) realism.[37]

Putnams' Requirements

Lakoff lists Putnam's requirements of "internal realism"[38] as:

(1) "A commitment to the existence of a real world external to human beings

(2) a link between conceptual schemes and the world via real human experience; experience is not purely internal,

---

[36] This discussion constitutes my answer to one of the more difficult objections to my first thesis wherein it is objected that "schematism" is "just a level of abstraction", (Richard Reiner, private communication). The discussion above shows why it need not be!

[37] The criteria of Putnam's, Lakoff's and Edelman's basic realism are, I have argued in my chapters 3 and 4, essentially the same ones definitively identified by Kant. Kant is grossly mischaracterized as an "idealist". He was, in fact, the penultimate modern realist in just the sense demanded by these thinkers. See chapters 3 and 4.

[38] Which he uses as the jumping off point for his own "experiential realism". Edelman, incidentally, has adopted Putnam's definition pretty much "as is".

but is constrained at every instant by the real world of which we are an inextricable part

(3) a concept of truth that is based not only on internal coherence and "rational acceptability", but, most important, on coherence with our constant real experience

(4) a commitment to the possibility of real human knowledge of the world."[39]

He has extended and refined Putnam's position somewhat from this basis, (his "basic realism"), to be able to answer certain further questions that arise, but this is a reasonably concise rendition of his stance vis a vis realism.

I have discussed his position, (as reiterated by Edelman), briefly in the preface to my [old] Chapter 2, [Chapter 6 here], wherein I agreed with (1) – (3), but strongly qualified (4). I had argued the equivalent of his essential conclusions as the subjects of my [old] chapters 3 and 4, [Chapters 4 and 6 in this MS respectively]: i.e. the bare and unstructured "axiom of externality", and the bare and unstructured "axiom of experience" respectively. These are purely intentional postulates, foundational to Scientific Realism.

Because of his conclusions, Lakoff was further forced into a position of epistemological, (as well as logical), relativism – against what has been called a "God-eye view of reality".[40]

Lakoff's relativism, necessary because of his logical conclusions but challenged in his own mind, (admirably, I maintain, as I consider myself a strong realist as well), by his fervent commitment to science and realism. It is ill-defined

---

[39] P.263

[40] cf my Chapter 5 for a discussion of Cassirer's arguments on the same subject and of my extension of them.

however. Though he talks about relativism at length, he never clearly defines it. He begins by noting the anathema which "relativism" is considered by the scientific world, but argues that there are, in fact, many different forms of relativism. (Neither he, nor I, advocate a "relativism of everything".) The most cogent interpretation I can give to his position, (Whorf aside), is that he advocates a cognitive and logical relativism based on bodily function, (in the world), which leads to a relativism of contexts, (ICM's), which employ different categorical, (conceptual), schemas. Within each of these ICM's, however, there *does* exist a structure consistent with rigor,[41] -but ultimately the ICM's themselves are relativistic.

I like what Lakoff has done, (hugely!), but his ICMs, the relativism in which he has based them, and his epistemology are deficient insofar as they are *all* derived from, (grounded in the concept of), the human body and the functions of that body in the world. This is his overview, and this is the context within which they are framed. That very body in the world is conceived *in the primary set theoretic sense*, (he would call it the "container schema" ICM), however!

But if they all may be *described* within the container schema, (the body in the world), then ultimately all of his ICMs and his epistemology are *theoretically reducible* to a container schema! [2010 Similarly to my critique of Maturana's ultimate thesis, I maintain that] this is a contradiction of his own position

---

[41] "The main thesis of this book is that we organize our knowledge by means of structures called idealized cognitive models, or ICM's, and that category structures and prototype effects are by-products of that organization.." Lakoff, 1987, p.68, his emphasis.

against a "God's eye" picture of the world.[42] It is the *generality* of Cassirer's solutions[43] and of my extensions of them, (founded ultimately in a neo-Kantian perspective), which allows the solution of the general logical and ultimately of the epistemological problems.

Though Lakoff rejects the view that "anything goes" –that any conceptual system is as good as any other, nowhere does he approach the possibility of a *scientific, mathematical* relativism which would give rigor to his conceptions –save within a tacit objectivist context. It is the possibility of a general and comprehensive structure of the Concept which allows the true relativity of the essential forms/ICMs.

I will argue shortly, (in the sense of mathematically conceived "ideals" –[2010 see Chapter 9]), that the various "generators" of such an ideal must each be capable of generating the *whole* of the "space" of that ideal –to include all possible alternative generators as well. Thus each (legitimate) structure must be *comprehensive* to be translatable, (i.e. capable of itself being generated by *another* set of generators). But its concepts/categories/objects may be *distributed* in the translation.[44]

This is intelligible only outside of the classical conception of logic, and is the essence of my conclusion of Chapter 5. Lakoff's "Concept" is certainly broader than the classical concept, but he takes his arguments too far –against *any* rule of concept formation.

---

[42] I.e. all his arguments against it are reducible within it. I will have more to say on this subject shortly and will suggest a way out of his dilemma.

[43] and their origins in science and mathematics

[44] cf my Chapter 5

Please do not misunderstand me. I loved Lakoff's book. It is brilliant, far reaching, and, I believe, essentially valid. He develops and documents his arguments solidly, but I think his strongest point is in his clear and cogent examples from our own normal usage[45], (as well as from extensive empirical studies), which makes his essential case almost unanswerable.

His conception is considerably richer than it is possible to describe within the confines of an appendix, nor is it as simplistic as I have characterized it. We have huge areas of agreement and possible interaction, (his and Rosch's "basic level categories" have a natural correlate in my "schematic perceptual objects", for instance.)

Lakoff's ICM's

Lakoff's ICMs are biologically based —on the human organism. Human cognition and human reason consists, for Lakoff, in the application of the best fit of these inbuilt ICM's, (and their respective categories), to a given problem or situation. They constitute an "embodied logic" deriving from the nature of the human organism itself. There is an obvious parallel between Lakoff's "embodied logic" and the more general case I have argued. I have argued that logic is indeed embodied, *but at the primitive level of cellular process!* This more general characterization allows the crucial epistemological move,[46] (which Lakoff's does not), *beyond* the "God's eye view" he disclaims.

---

[45] Cassirer's case was grounded primarily in scientific examples.

[46] Through what Maturana and Varela call "structural coupling"

An Important Distinction: Biology as a Symbolic Form

The distinction is important because at the cellular level of phenomenology biology becomes a *pure form*, (in Cassirer's sense within his "Symbolic Forms" and compatible with Cassirer's Hertzian premise). This is especially transparent in Maturana and Varela's book, for instance, (see Chapter 6), i.e. in its explicit constructiveness and the subsequent purity of their phenomenology.

Citing a few pertinent examples quoted earlier in Chapter 6:

Maturana:

"Our intention, therefore, is to proceed scientifically: if we cannot provide a list that characterizes a living being, why not propose a system[47] that generates all the phenomena proper to a living being? The evidence that an *autopoietic unity* has exactly all these features becomes evident in the light of what we know about the interdependence between metabolism and cellular structure."

"*Autopoietic unities specify biological phenomenology as the phenomenology proper of those unities*", (my emphasis), "with features distinct from physical phenomenology... because the phenomena they generate in functioning as autopoietic unities depend on their organization and the way this organization comes about, and not on the physical nature of their components."

---

[47] i.e.: an "axiomatic system"!

"Ontogeny is the history of structural changes in a particular living being. In this history each living being begins with an initial structure. This structure conditions the course of its interactions and restricts the structural changes that the interactions may trigger in it", (my emphasis). "At the same time, it is born in a particular place, in a medium that constitutes the ambience in which it emerges and in which it interacts.

This ambience appears to have a structural dynamics of its own, operationally distinct from the living being. This is a crucial point. As observers, we have distinguished the living system as a unity from its background and have characterized it as a definite organization. We have thus distinguished two structures that are going to be considered operationally independent of each other, (my emphasis), "living being and environment."

These are purely *constructive* and *operational* definitions, (or capable of being made so within "structural coupling"), in the precise sense of Hertz and Cassirer and clearly mesh with the substance of my Chapter 5. They are Hertzian "images" with a definite, predictive logical structure. They are clear examples of Cassirer's "each asks questions, each from its particular standpoint"!

At the level of cellular biology therefore, biology becomes a *pure form*, and, as such, it, (and the logic I posit within it), is capable of legitimate embodiment[48] within the now viable scientific epistemological relativism espoused by Cassirer and myself. It is this deeper placement, (and not as reductive physics), which allows an escape from the inconsistent "God's eye view"

---

[48] i.e. as a legitimate, fundamental "symbolic form"

implicit in Lakoff's and Edelman's theses, and enables a truly consistent and viable epistemological relativism.

It is because of Lakoff's Wittgensteinian origins, I think, that he has gone too far, (-and not far enough). Had he started from Cassirer instead, the case might have been different. I will return to Lakoff presently to suggest a "cleaner" solution to his problem consistent with his apparent needs —in the mathematical notion of "ideals". [again see Chapter 9] There is a way to save it, but I think it is too limited and inconsistent with the dictates of modern biology as espoused, for instance, by Edelman.

Edelman:

Gerald Edelman has adopted Lakoff's, (and Putnam's), logical and epistemological conclusions as the philosophical underpinning to his own theories of "Neuronal Group Selection", (TNGS), and "re-entrant topobiological maps". He proposed the combined result as an actual answer to the problem of mind-brain. Though Edelman's is a very plausible theory of brain development and function, it is limited to dealing with "mind" only *reductively* -i.e. as strictly biological and therefore physical *process* and falls to the same objections that I, (and the preponderant Naturalist camp as well), have raised.

"Mind", normally taken, is therefore *superfluous* therein! Edelman explicitly denies the "homunculus", (as do I), but his "Cartesian theatre" is specifically a physical and spatial one. It is spatially and temporally distributed. Though he does not explicitly deny the existence of "mind" as ordinarily taken, he tacitly reinterprets it and reduces it to a description of process. He fits very comfortably, I feel therefore, within the naturalism, (and "objectivism"), which Dennett, Churchland, et al espouse.

I do not question the insightfulness or the importance of Edelman's work —it is profoundly important and very solid —but, because of its limitations, (derived from Lakoff), it falls short of an answer to the problem of consciousness, retains internal inconsistencies, and does not resolve the mind-body dilemma.

Starting with the nature and limitations of embryology, Edelman makes a case for a very different concept of "recognition systems". His exemplar "recognition system" is the immune system for whose investigation he won the Nobel Prize. The immune system, he argues, does not depend on information about the world –i.e. we do not create new antibodies from informational templates resident in newly arrived antigens.

Rather, science finds that the body randomly generates a huge diversity of antibodies *before the fact* and reactively selects from this pre-existing diversity "ex post facto" as he phrases it. This, the immune system, is a system of process, not of information.

> "A recognition system … exists in one physical domain", (for the immune system it is within an individual's body), " and responds to novelty arising independently in another domain, (for the immune system it is a foreign molecule among the millions upon millions of possible chemically different molecules) by a specific binding event and an adaptive cellular response. It does this *without* requiring that information about the shape that needs to be recognized be transferred to the recognizing system *at the time when it makes the recognizer molecules or antibodies.* Instead, the recognizing system *first* generates a diverse population of antibody molecules and then selects *ex post facto* those that fit or match. It does this continually and, for the most part, adaptively." Edelman, P.78

Cognition, our *ultimate* "recognition system", he argues, is a parallel case and must be reconceived accordingly. Because of the sheer size, and the place and time sensitivity of embryological neural development, the neural system, (he argues), is progressively "pruned" ex post facto from random preexisting variety over the stages of its development in like manner to the immune system.

"given the stochastic (or statistically varying) nature of the developmental driving forces provided by cellular processes such as cell division, movement, and death, in some regions of the developing nervous system up to 70 percent of the neurons die before the structure of that region is completed! In general, therefore, uniquely specified connections cannot exist."

"the principles governing these changes are epigenetic – meaning that key events occur only if certain previous events have taken place. An important consequence is that the connections among the cells are therefore not precisely prespecified in the genes of the animal." Edelman, pps. 23- 25

Of the great diversity of (preexisting) neural connections generated at any stage, particular connections are reinforced and kept, or pruned and deleted, in tune with place and time dependent events the scenario of which is too complex "by several orders of magnitude" to be embodied in the human genome. This pruning is achieved operationally, not informationally. Embryological development is too complex, too dependent on place and time to be prespecified. His argument in some ways parallels my own of [old] appendix A wherein I argued that there simply hasn't been enough time in evolutionary history, (nor ever will be), to create such an information engine.

In his "ex post facto" adaptive "TNGS", Edelman argues a criterion of *competence* , (as, indeed, did Darwin –and as did I in my first chapter), rather than one of information in the evolution and development of organisms –and specifically of the *human* organism.

"The immune selective system has some intriguing properties. First, *there is more than one way to recognize*

509

successfully any particular shape. (my emphasis) [49] Second, no two individuals do it exactly the same way; that is, no two individuals have identical antibodies. Third, the system has a kind of cellular memory." Edelman, P.78 (These comments are directly relevant to my discussion of bounds and limits and the "parallel postulate" of cognitive science.)

---

[49] You might want to look at my "Bounds and Limits" diagram here –Chapter 4, Figure 18

God's and Edelman's Eye

He too disclaims the possibility of a "God's eye view" of reality by an organism.[50] But competence, as I have argued, does not imply *parallelism*. It is the question of bounds and limits that I have argued previously,[51] and Edelman falls into the same epistemological trap as does Lakoff, (and Maturana and Varela as well). Other than this failing, however, I believe his overall position and arguments are very strong.

On "Presentation"

Edelman challenges ordinary logic and ordinary epistemology, (the classical, "objectivist"/"naturalist" views), for some of the same reasons that I do. In his TNGS, he has framed the same problem, and reached largely the same conclusion that I did under the issue of "presentation".

> "some of the reasons for considering brain science a science of recognition", [under his special definition of "recognition systems" cited above]. " The first reason is almost too obvious: brain science and the study of

---

[50] cf: my "Axiom of Externality" and "Axiom of Experience", (Chapters 3 and 4).

[51] Let me repeat a footnote of my Chapter 1: The question, of course, is whether "information" is necessary to competence. I will argue, (in Chapter 4), that it involves a distinction between "bounds" and "greatest lower bounds" of biologic survival. A given organism, (to include human beings), must reflect a *lower* bound of competence in the world. But "information" requires that it reflect a *greatest* lower bound, and this is inconsistent with the fundamental premises of evolution. It is the "parallel postulate" of cognitive science.

behavior are concerned with the adaptive matching of animals to their environments. In considering brain science as a science of recognition I am implying that recognition is not an instructive process. No direct information transfer occurs, just as none occurs in evolutionary or immune processes. Instead recognition is selective."

"a potent additional reason for adopting a selective rather than an instructive viewpoint has to do with the homunculus. ...the little man that one must postulate 'at the top of the mind', acting as an interpreter of signals and symbols in any instructive theory of mind.... But then another homunculus required in *his* head and so on, in an infinite regress... selectional systems, in which matching *occurs ex post facto* on an *already existing* diverse repertoire, need no special creations, no homunculi, and no such regress." Edelman pps. 81-82

Presentation in any sense other than an eliminative one *requires* a homunculus, and this is the problem that Edelman believes he has solved- in essentially the same way that I did. But, in doing so, he believed he had solved the whole of the mind-body problem.

Re-entrant Maps

To this point, (his theory of "TNGS"), his argument is very plausible and compatible with my own conclusions. His rationale from that point onward, however, bears examination.

His theory of re-entrant topobiological maps, (reactively linked cortical surfaces), is quite plausible and highly interesting, but, ultimately, it is tied to a truly topological correspondence of those maps with the "real" world, (contrary to his conclusions of the first part of his thesis –see Chapter 4, Figure 12 "Edelman's Epistemological Error"). "Maps… correlate happenings at one spatial location in *the world* without a higher-order supervisor…"[52]

These maps themselves *do*, therefore, embody a "God's eye view", (contrary to the implications of TNGS). I have suggested a different orientation of Edelman's schema in the discussion of my Chapter 4, wherein I suggested we step back from our human (animal) cognitive prejudice and consider the larger "global mapping" also described by Edelman, (which relates "*non-mapped*" areas of the brain to the topobiological maps), as the primary focus of biological process. (See illustration in Chapter 4: Figure 13 "A Metacellular Perspective). Under this perspective, the "objects" of our topobiological maps may be reconceived, not as God's-eye renditions of ontology, but rather as *organizational foci*, (efficacious artifacts), of process.[53]

---

[52] Edelman, p.87, my emphasis

[53] An aside: While I hope it should be clear by now that I have no affinity for traditional idealism, I think it is worth quoting a short passage from Edelman as it talks about levels of "strangeness" in theories:

"and Berkeley's monistic idealism –suggesting that inasmuch as all knowledge is gained through the senses, the whole world *is* a mental matter –falters before the facts of evolution. It would be very strange indeed if we mentally created an environment that then subjected us (mentally) to natural selection." Edelman, p. 35 [Note:2011] My answer: Why not? It is a SUPERB organizational principle, and, as such, is easily incorporated as a component of a schematic model!
Berkeley aside, Edelman seems very put out with the very strangeness of the (recursive, re-entrant?) complication of such an idea. The complication, he implies, boggles the mind! But much of modern science is even more mind-boggling. My thesis proposes an even greater "boggle", but results in an integration of epistemology and an actual solution to the mind-body problem.

Edelman rationalizes his biological solution to the problem of the brain and the mind upon Lakoff's, (and Putnam's), answer. To him that answer is important because it allows a rationale for the brain which is *not* based in information as, in fact, he has concluded that it *is* not, (inconsistently with his theory of re-entrant maps, I maintain). He therefore reaches a conclusion very similar to my own. But again, like Lakoff's, his conception is too limited and incorporates an inherent contradiction. His concept of the world, like Lakoff's is based in a container schema. We, you and I and Lakoff and Edelman, *are organisms too after all.* But then "TNGS" requires that even *our* brains are not informational![54] It is the *generality* of Cassirer's "Symbolic Forms" –and of my extension of it –the generality of the Concept and the generality of the *scientific* relativism which allows a

---

Modern epistemology is radical at both the extremely small and at the extremely large (and fast) scales. It is *only as algorithms* they are comprehensible. And yet everyone, (read this as "most realists"), seems to accept that at the *middle scale* epistemology must be simple. Consider instead the *truly* mind boggling possibility I propose that the middle scale is algorithmic *as well!* Does this not explain "the prototype" which Rosch demonstrated and which ground Lakoff's and Edelman's very logical theses. Prototypes and the logical relations between them would, under this view, represent the "objects" and the "calculus" of algorithmic biology. If this thesis be accepted, then continuity, temporarily removed from epistemology by modern science, is restored across the board. This is a major epistemological and scientific result and worth the price we must pay for it. So was quantum mechanics!

[54] I think that Edelman would comment here, as he did on another occasion, that this conclusion would "boggle the mind"! Maybe so, but I think we'd better get used to such a state. Modern physics? Edelman's own conclusions?

consistent and meaningful solution[55] to the problems of the brain, mind and epistemology.

What Edelman has *not* solved: the problem of the Cartesian Theatre!

What Edelman has not solved is the *other* problem, the problem of the "Cartesian theatre"[56], (i.e. "mind", ordinarily taken), and this is the most important problem. It is that which we normally *mean* when we use the terms "consciousness", "sentiency", etc.

Its comprehensive solution is the subject of Chapters 1 through 5: the Concept of Implicit Definition and its integration with biology as the unified rule of Maturana's ontogenic coupling. Edelman's solution remains an essentially naturalist, (objectivist), one itself however and is, I argue moreover, epistemologically inconsistent. It is compatible with the rest of the eliminativist camp in that ultimately all his correspondences, (his stated epistemology to the contrary), are from topobiological maps, *themselves topologically corresponding to "the (real) world"!* (See figure 12, Chapter 4 –Edelman's Epistemological Error!)

His "mind" is purely process, spatially and temporally localized –and known! He accounts for "conscious" *behavior* but not consciousness. His *is* "a God's eye view".

Edelman is very derisive of Penrose's "Emperor's New Mind",[57] but I think he has missed a major aspect of it. Penrose,

---

[55] by allowing a reorientation of the problem to a consideration of forms rather than of information

[56] after Dennett

[57] "Penrose's account is a bit like that of a schoolboy who, not knowing the formula of sulfuric acid asked for on an exam, gives instead a beautiful account of his dog Spot." Edelman, P.217

(though he doesn't say so explicitly), and the rest of the "quantum people" are trying, (Gödel aside), I think, to supply a "non-localization" –i.e. a spatial universality to the brain's perceptual and cognitive objects- to make headway on the problem of knowing. They are trying to conceive an answer to Leibniz' problem of the "one and the many" within a *physical* space.

The "chaos theory people" stand in a similar motivation I think, but attacking the logical problem of the object from a perspective of localized process, conceiving our objects as "attractors". But even were such solutions meaningful, (and they *are* interesting), they would miss the requirement of a (structuralist) *self-standing logical space in depth* which the Concept of Implicit Definition, as combined with the schematic model of biology, supplies and which furnishes the foundation of "meaning" and "knowing". Dennett glimpsed such a possibility[58] for a Cartesian theatre based in logic in Shakey the Robot's program, (as I cited previously[59]), but his naturalist/objectivist metaphysical prejudice enervated the concept before it could bear fruit.

But ordinary logic,[60] (Shakey's program for instance), is inadequate to the problem. It is essentially dimensional: linear, planar, multi-dimensional, missing the integration in depth – missing the autonomy and (logical) self-sufficiency which is necessary to knowing and to meaning. [61] [62]

---

[58] but using an inadequate logic

[59] cf the "Dennett Appendix" - "the color phi"

[60] "associationist logic" in Dreyfus' term

[61] Wittgenstein's objection is clearly pertinent here. He raised the question of the necessity for one to have *another* rule: i.e. another rule to apply any given rule. C.I.D./biology, however, supplies a consistent rationale. "One" *is* a

That aspect of ordinary mind we call the "Cartesian Theatre" does not work as a linear, a planar, or even as a multidimensional space[1] –even as a *logical space*. As I argued in chapters 3 and 5 each requires "presentation", either physical or logical. Nor do such conceptions supply "knowing", "meaning" or "motivation", except as unnatural and gratuitous appendages.

C.I.D. and the schematic model focus logic and cognition in biology. Biology has *innate* depth and structure –derived from the single principle of efficacy as coupled with Darwinian survival –of ontogenic coupling, and these necessarily pass to the logic and the cognition which are embedded in it! The Concept of Implicit Definition as coupled with the schematic model[63] supplies an integration and a rationale in depth –*and an autonomy*- implicit in its biological roots.[64] Edelman got very close to this answer, but his efforts were frustrated by his epistemological beginnings.

Cassirer, ("symbolic forms"), Rosch, ("prototypes" and "basic levels"), and Lakoff, (ICM's), demonstrate that dimensional logic is not adequate to the realities of the human mind. Nor, even putting aside the problem of "information", (Maturana and

---

rule, "one" doesn't *apply* the rule. "One" is the single, "ex post facto" and unified rule of ontogenic coupling!

[62] and which could provide the enrichment necessary to the possibility of future scientific development moreover. All the other proposals yet presented are essentially *just* explanatory –i.e. logically reductive- and hold little promise for further exploitation.

[63] i.e. the "*concordance*" mentioned in the introduction

[64] It supplies "the rule which we need to apply the rule which we need to apply the rule ..." demanded by Wittgenstein. Ultimately it is a *constitutive* rule. But one doesn't "apply" this rule. Rather, "one" *is* a rule –namely the constitutive rule of ontogenic coupling as the term is used by Maturana and Varela.

Varela, Freeman, Edelman), can such a logic supply meaning or motivation except in a very unnatural and perverted sense. It is biology itself which supplies this aspect –in the concept of a schematic model and an enlarged logic. This is my argument of as culminated in Chapter 11.

On Epistemology:

But let me be more generous to Lakoff and Edelman. In basing their conceptions on our ordinary world, or, to call a spade a spade, on our ordinary naïve realistic conception of the world, (people, baseballs, cars and all the things they do), they are *trying to preserve experience!* This they identify with realism. They seek to preserve their logical and biological conclusions with the *objects* of that ordinary realism,[65] and their relativism is a laudable and understandable attempt at a reconciliation.

I have explained my answer to the same problem in terms of the multiple possible axiomatic foundations of mathematical systems, but another line of understanding is possible. Consider the notion of "mathematical "ideals" as presented in Chapter 9.

Those mathematical "ideals" described there open a door to a better conclusion to Lakoff's and Edelman's arguments, and a simpler understanding of my own. *None* of the "generators" implicit in these mathematical ideals stands prior to any other, nor

---

[65] cf Lakoff's discussion, (p.262) of the "objects" of our experience –his chair, for instance. "It is important not to read Putnam out of context here, especially when he talks about objects. *An 'object' is a single bounded entity....* Putnam, being a realist, does not deny that objects exist. Take, for example, the chair I am sitting on. It exists. If it didn't, I would have fallen on the floor." (my emphasis). Compare this reference with my modification of Kant's position on "objects" which I advocated in the footnote in Chapter 5.

does it "create" the figure comprehended. Each stands, rather, as an equipotent and relativistic "logical", (i.e. explanatory), basis fully exhausting the actuality of the figure.

But we must consider this example in the larger context of mathematics. Not only can such descriptions be relativized in relation to a fixed coordinate system, but the very coordinate systems themselves stand in like case. Axes need not be orthogonal, nor need they be rectilinear, (e.g. polar coordinates are possible). Nor need they be fixed. They may be in translation —e.g. relative motion, (which translates to special relativity), and they need not be Euclidean, (nor Hyperbolic nor Spherical). Russell, for instance, further argued[66] that our descriptions of phenomena might even be based in projective geometry.

But need they be *even spatial?* Can we not conceive of such explanations being framed as abstract transformations, which latter are *not* defined on spaces, but on abstract sets! Abstract sets, however, fall naturally within the scope of axiomatics wherein I grounded C.I.D.

Such a relativism of descriptions, combined with a scientific relativism of logic and epistemology themselves as argued by Cassirer, Lakoff, and myself, (superseding the traditional "container schema" and broadening the very ideas of "set" and "object" themselves), points to the further possibility for such an "idealistic", (in the mathematical sense), foundation of logic *itself.*

Need mathematics, or logic, be *necessarily* grounded in objectivist sets, (ultimate "atomic" –i.e. *least* objects -and a *fixed "Universe"* of such objects), or could it not pick itself up by its own bootstraps, (following the cue of mathematical "ideals"[67] and

---

[66] Russell, "Foundations of Geometry", 1956

[67] though presently itself conceived in set-theoretic terms

the findings of Cassirer and Lakoff), and stand without them?[68] This is a question –not an easy one to be sure- for abstract mathematics and the future of logic.

If we think of "experience" in the abstract –i.e. as the "axiom" without interpretation, (i.e. "impartially" in the sense of "basic realism"), – then I think an "ideal" in this sense is a very reasonable way of understanding it – beyond any particular "generator", beyond any particular interpretation.[69] But it is not necessarily a *spatial* interpretation either. Ideals are broader than this.

On a narrower focus, the possible generators of a mathematical ideal rigorously parallel the explanatory possibilities which can absolutely preserve the objects of ordinary experience and naïve realism, (conserving shapes, boundaries, etc.). As such, the ideal they ground is entirely commensurate with Lakoff's and Edelman's conceptions and logically validates their (limited) relativism.

Within the perspective of that same "basic realism", the "experience" we deal with need not be taken as ultimately *informational* however,[70] but can be taken as specifically organizational and operative instead[71] as I have argued in my Chapter 4 and consistently with Edelman's "TNGS".

Though *connected* with externality, (as representative of successful- .i.e. *adequate* process[72]), it need not be further taken as

---

[68] This would be the truly transcendental logic after which Kant sought.

[69] "context-free" in Van Fraassen's term

[70] This my qualification on Putnam's 4th requirement of basic realism

[71] contrary to Putnam's 4th requirement

[72] "ex post facto", in Edelman's words

conveying information about that externality. It need not be taken as *paralleling* externality. The latter presumption, I have argued, goes far beyond the needs and the implications of Darwinian biology.

The deeper issue is that of an adequate definition of "experience" itself. Need we identify it with the absolute and necessary preservation of ordinary objects? Or, might we not, consistent with the foundations of their own conceptions and the work of Rosch upon which it is grounded, consider even our ordinary perceptual objects *as "prototypes" of a larger experience*? Prototypes are objects of utility, of efficacy, after all, they are not foundational objects.[73] Could not our ordinary objects be considered, (as I have argued), as prototypes, ("schematic perceptual objects"), of a *biological* calculus?

"Experience" in a *modern* sense must be broadened to include the experience of the results of scientific experiment, and *that* experience, at least insofar as modern physics is concerned, is not commensurate with the preservation of objects either,[74] nor is it commensurate with ordinary spatiality.

Without even considering the deeper implications of QM or of Relativity, one need only consider results of the "twin slit" experiment or the implications of its multiple execution to see the point. Not even *cardinality* is preserved![75] Similarly, consider

---

[73] see Lakoff for a discussion of Rosch, prototypes, and the logical significance of the latter. It is a very illuminating discussion.

[74] Note 2011 –see d'Espagnat on "multitudinism" for instance which totally rejects ordinary objects on grounds of experimental physics.

[75] In answer to a question I asked on this point, a physicist correspondent of mine replied that "Yes, you can have many slits one after another, (it is better with Mach-Zehnder interferometers than slits, with the same result that one doesn't know if the photon went through or was reflected by a mirror…. We can say that one photon may be in an arbitrary number of places at once."

Penrose's "most optimistic" view of quantum mechanics, (most optimistic for objectivism/naturalism, that is):[76]

> "I shall follow the more positive line which attributes *objective physical reality* to the quantum description: the *quantum state*.
>
> "I have been taking the view that the 'objectively real' state of an individual particle is indeed described by its wavefunction psi. It seems that many people find this a difficult position to adhere to in a serious way. One reason for this appears to be that it involves our regarding individual particles being spread out spatially, rather than always being concentrated at single points. For a momentum state, this spread is at its most extreme, *since psi is distributed equally all over the whole of space*, (my emphasis),...It would seem that we must indeed come to terms with this picture of a particle which can be spread out over large regions of space, and which is likely to remain spread out until the next position measurement is carried out...."

The particle -this *smallest part of our "object"*- is *not* included, (spatially, reductively, nested), *within* the spatiality of the atom or within the molecule -or even within the *human scale* object of which it is the theoretical (and supposed material)

(Wlodek Duch, private correspondence) My point was that even the *cardinality* of this basic object, (the photon), was *purely arbitrary* —it could be 1 or 2 or 3 or 1,000,001 or ..., depending on the branching structure of successive slits and the design of the experiment. But innate cardinality is perhaps the *most basic* "property" we ascribe to ordinary objects, so I think the conclusion is significant.

[76] Repeating a section of a prior appendix

foundation. Naturalism/objectivism can no longer support, therefore, even a consistent hierarchy of spatial scale![77]

At the human level, of course, it is a very useful tool, and that is just what I propose it is -constructed by evolution! Science and logic suggest *other*, non-scaled and non-hierarchical organizations -i.e. they support *any* other efficacious organization. It is a simple matter of utility.

Appendix B Conclusion

To conclude this appendix, let me repeat that I truly admire Lakoff's and Edelman's work. It is both profound and crucial to the resolution of the ultimate problem. But then I really like the work of *all* the authors I have cited —even those most contrary to my own conclusions. (I would not cite or spend much time on anything of lesser quality —the problem is too huge and too difficult to be distracted.)

Dennett's work, for example, is very beautiful to me in his honorable and perceptive pursuit of the hard implications of naturalism. P.S. Churchland, as another example, has a "clean" mind and frames the problem wonderfully from the perspectives of biology and philosophy. None of them has resolved the

---

[77] Compare Lakoff, p.195: "In the case of biological categories, science is not on its [objectivist philosophy's] side. Classical categories and natural kinds are remnants of pre-Darwinian philosophy. They fit the biology of the ancient Greeks very well....but they do not accord with phenomena that are central to evolution. ... Objectivist semantics and cognition and, to a large extent, even objectivist metaphysics are in conflict with post-Darwinian biology. I'd put my money on biology."

fundamental problem, however, though all have come very close in different aspects of it.

This is a *hard problem*, the hardest one, I maintain, that the human mind has ever dealt with. To solve it requires an intellectual ruthlessness, and specifically, a *ruthless realism*!

Appendix C: Some Further Thoughts on Cassirer and Some Further Thoughts about the Model

It is too bad that Cassirer did not possess some of the qualities of Bertrand Russell. Russell's writing was completely lucid while Cassirer's was "oblique" in the extreme. Brilliant but oblique!

Cassirer evolved his revolutionary new meaning of the word "concept" as" The Mathematical Concept of Function", but never developed it in the explicit way that Russell did. Cassirer never supplied a logical calculus for his idea in the way that Russell did for set theory.

Cassirer accepted, with Kant, that the fundamental organizing precept was the series —I think that was a mistake. And yet I think that his real, *unexamined* fundamental was actually the *real number continuum* —whose "series", as we can clearly see from Cantor's work, will never generate the continuum! I think his actual thought process was in terms of (continuous) "curves".

Cassirer never developed a "calculus" even for his "mathematical concept of function" in any meaningfull way. Apparently Cohen, his teacher, had suggested an expansion into "The Calculus", but I am unable to find a good reference, other than a mere mention of the fact.

How, in fact, could Cassirer's "mathematical concept(s) of function(s)" interact. Where is the supporting calculus? The interaction of multi-dimensional surfaces seems to have a certain relevancy, but Cassirer's "concept" always remains bound in "series" —even within a given dimension of *any* of his concepts— and it is *explicitly framed* as "dimensional", (f(w,x,y,z,...), i.e. a "series"! This is not to fault him —his work is profoundly brilliant as I have acknowledged —but to say that it is unfinished!

Expanding from the series to the continuum may supply the keyway. But how? I think his fundamental insight embodies *continuous* funtions rather than "function" in its minimalist meaning, (yes, I've heard of the Koch function[1]). I think continuous functions are fundamental, and an expansion of the Calculus is a plausible beginning. But again, how?

Another, probably much more efficient approach to this problem which appeals more and more to me at this point is mentioned in Chapter 13, (D'Espagnat), in the "Conclusions" sections.

Another, Related Thought:

Thinking again about the problem of the creation of a brain's "calculus" from another perspective! If logic is *actually* "bio-logic", (as I argued in Chapter 4), then we gain yet another perspective on the problem. I have read that someone, (Henry Markram,EPFL), has actually started a comprehensive project to map the total connectivity of the human brain, (extremely difficult, but not impossible)! Such a map, (whomsoever actually completes it), oriented in physical space, and overlaid with neurochemical activities, could conceivably lead to at least the beginnings of a rough model for such a fundamental "calculus" for the bio-logic! "Chaos theory" with its "wells of attaction" from fundamental generators, etc. is not out of the question here!

Do these reflections violate my fundamental perspective? Absolutely not, as they lie within the physicalist (intentional form) whose relativism I have already dealt with. This is *epistemologically relativized* materialism! Though this is extremely difficult to always keep in mind,[2] *it is the foundational*

---

[1] Wirzcup introduced me to it at 18.

[2] Because of our inbuilt "realist imperative"! Quoting a short extract from Chapter 12 I argued[2] that from a biological perspective it is not important that the "operator" of such a complicated process knows what it is, (specifically), that he is doing. It is important *only that he does it well*. It *is* crucially important that he does it *diligently*, however. It is imperative that he be locked

principle throughout this whole book! This is a problem reflecting "meaning", semantics, not syntax!

The Limitations of Cassirer's Perspective:

Cassirer never expanded his "Mathematical Concept of Function" to what I have called "The Concept of Implicit Definition". Therefore it was impossible for him to move past "the phenomena" – or the homunculus! "Representation" embeds "presentation" within the very word itself!" (Presented to what *homunculus*?) He was never able to conceive of *our very objects themselves* as "rules", as "orderings" of the mind itself –of his "new form of consciousness". He was never able therefore to even conceive them as products *themselves* of the constitutive logic of the brain. *I have*, and it has led to a theory of meaning as well! (This latter, incidentally, may be the most important defect in all the current conceptions of mind and brain.)

Repeating just a couple of paragraphs:

My overall conception is an *extension* of Cassirer's redefinition of the very meaning of the word "concept", of his "Mathematical Concept of Function", (*expanded still more* by me through the young Hilbert's conception of "implicit definition"), to my own extension and redefinition of that word, ("concept"), as the "Concept of Implicit Definition", (C.I.D.). It is this extension that allows our conceptualization of our very percepts

---

into the loop of his virtual reality -that he "pay attention". This introduces the necessity of an *inbuilt realistic imperative* -i.e. a mechanical guarantee of his dedication, (see P.S. Churchland / Hume).

themselves as (structuralist) elements of order —as "positions" in a structure!

But then I *married* that redefinition of the concept back to Cassirer's own "Symbolic Forms" *but as seen through Maturana's eyes*, (i.e. through the "eyes" of structural coupling[3]), to reach my final vision: i.e. of the "interface" and the "Somewhat" beyond that!

Some Random Further Thoughts about my Proposed Model Itself:

In some of my very early writings I noted that it is not really clear whether the primitive model and the intentional faculties are two truly independent entities or "dimensions" of the *same* entity! At this point in my development I would say this: if this model is, *in fact* evolutionarily derived as I claim, then it would make absolutely no sense to have just the "primitive model" with its translatory, ("A/D converter"/ "hierarchical/non-hierarchical converter"), "objects"/percepts *by themselves!* From the very beginning, intentionality must have been incorporated, so that *action* on those "objects" would have been possible and useful–at whatever early stage we assume. The primitive (converter) model and the intentional model therefore must have been co-evolved simultaneously! But this directly implies that the "objects" of the primitive model must always

---

[3] Biologically, structural coupling implies only *adequacy*, not mirroring —as in "congruent structural coupling" and is the basis of my rejection of Maturana's more reduced and limited form of the underlying idea. See my illustration: "Bounds and Limits", Chapter 4, P.212 for the conceptual grounding of "Somewhat" as opposed to "Something". From a biological standpoint, there are two *separate* domains: ambience, (externality) and the autopoietic entity. I argue they are connected through "the interface"!

Appendix C   Further Thoughts about Cassirer & about the Fundamental Model

have been what I will, (metaphorically only), call "fuzzy objects". I am not talking about truth values on a zero to one scale, but something that is defined, but *loosely defined*, (contiguously, I believe), from its process!

This begins to illustrate the interplay between our intentional functions, (and theorizing is one of them), and our primitive "objects". It makes a case for the "fuzziness" in the primitive contiguity of the latter. Evolution must have constructed them so in complement to the also evolving primitive intentional faculties of the brain![4]

---

[4] As I suggested in Chapter 1, perhaps as the objects of a "higher level language" utilized by the intentional faculties.

Appendix D: On Niels Bohr and Cassirer's "Symbolic Forms"

Let me begin this appendix with excerpts from the transcript of a taped interview with Niels Bohr conducted by Thomas Kuhn, and Erik Rudinger in Bohr's later life.[5] I believe I can demonstrate a definite and close parallelism between Bohr's viewpoint on epistemology and realism -and Cassirer's![6]

It is too bad that the interview was conducted in English rather than Bohr's native Danish, as his precision in English -and this subject demands a *very* high level of linguistic precision -is barely minimal.

That Bohr was truly a great mind we both recognize, but that does not make his mastery of English better than it was —as the quoted sections reveal. Nor does it make his self-admitted minimal philosophical breadth adequate to the broader problem. So we must try to interpret his remarks from the flow and the context -like trying to re-interpret a bad "Google® machine translation" of a paper in another language[7], or like a reading of Shakespeare in its original language and spelling. This involves some work[8] admittedly, but it is worth it because it is a unique source and explains a general dilemma for those attempting to explain and validate his profound achievements —specifically of interpreting his foundational principle of "complementarity".

---

[5] On November 17, 1962

[6] Note: I recently incorporated and expanded this discussion as an explanatory route towards understanding Cassirer's thesis of "Symbolic Forms" in Chapter 8. You might want to review the relevant sections of that chapter.

[7] I've had to do this on several occasions

[8] and absolute transparency —to show the sources of inferences

Appendix D — Bohr's Epistemological Roots

I think there is a striking parallelism of perspective between Bohr's very early and personally original usage of multivalued, purely mathematical functions[9], (as he utilized them in his own final epistemological perspective), and Cassirer's broader and philosphically more deeply based theory of epistemological relativity[10], i.e. his "Theory of Symbolic Forms".

From the interview, we find that Bohr was trying to approach the deepest problems of our mental world even as a very young man.[11] He built his original mathematical perspective of multi-valued functions into his eventual perspective on the deepest problems of the mind —to include cognition and epistemology.[12] He was able to conceive the possibility of simultaneous and equipotent legitimate, *but alternative*, foundational, (i.e. epistemological), perspectives from this viewpoint.

In his later application of this conception of "multivalued functions" to the specifically physical world as "complementarity", he was able to comprehend the fundamental wave/particle duality of light. Therein Bohr incorporated that same "multi-valued" and "branching" epistemological perspective within the specific context of physics and Naturalism. But Cassirer, long before him, had gone one step deeper to incorporate such a "multi-valued" viewpoint into a deeper

---

[9] like the multivalued square root function: $F(x) = x^{-2}$ over the non-negative Real line

[10] using Swabey's characterization

[11] just after he had finished his high school examinations

[12] Granted that he began with things like "free will", but the course of the interview shows how he broadened his beginnings all the way up to alternative explanations of "real things" themselves —e.g. to biology and to the wave/particle duality of light!

perspective –across the *whole* of the various sciences' differing perspectives on reality itself!

The sciences were Cassirer's beginning for his conception of "Symbolic Forms" –based on Hertz's reflections.[13]
Heinrich Hertz:

> "The [scientific] images of which we are speaking are our *ideas* of things; they have with things the one essential agreement which lies in the fulfillment of the stated requirement, [of successful consequences], but further agreement with things is not necessary to their purpose. Actually we do not know and have no means of finding out whether our ideas of things accord with them in any other respect than in this one fundamental relation."[I]

Cassirer argues that it is the *method* by which science derives the future from the past which is significant, however. We make "inner fictions or symbols" of outward objects, and these symbols are "so constituted that the necessary *logical consequences* of the images are always images of the necessary natural consequences of the imaged objects". [14] [II] But this analysis –and "image"– must be interpreted carefully:

> "... [though] still couched in the *language* of the copy theory of knowledge –... the concept of the 'image' [itself] had undergone an inner change. In place of the *vague demand* for a similarity of *content* between image and thing, we now find expressed a highly complex *logical*

---

[13] See Chapter 8!
[14] my emphasis

relation, [my emphases], a general intellectual *condition*, which the basic concepts of physical knowledge must satisfy."[III]

Its value lies "*not in the reflection of a given existence, but in what it accomplishes as an instrument of knowledge,*"[IV] [my emphasis], "in a unity of phenomena, which the phenomena must produce out of themselves."

Cassirer's continues:

"… Even in 'nature' the physical object will not coincide absolutely with the chemical object, nor the chemical with the biological –because physical, chemical, biological knowledge frame their questions each from its own particular standpoint and, in accordance with this standpoint, subject the phenomena to a special interpretation and formation."

Specifically he asserts that each scientific physical perspective adopts differing fundamental Hertzian "images", differing "objects" because each "frames [its] questions from its own particular standpoint [and therefore] subjects the phenomena to a *special* interpretation and formation." He concludes:

"…But instead, a *new* task arises: to gather the various branches of science with their diverse methodologies – with all their recognized specificity and independence – into one system, whose separate parts precisely through their necessary diversity *will complement and further one another*. This postulate of a purely functional unity replaces the postulate of a unity of substance and origin, which lay at the core of the ancient concept of being. "[15]

This is *Cassirer's* "principle of complementarity" and is almost a precise restatement, (though an expansion), and an exact parallel of Bohr's beginning "multivalued" perspective![16]

[Note: the origins of this quote derive ultimately from his earlier redefinition of "the concept"[17], instead, as "The Functional Concept of Mathematics"![18] The latter is "the rule of a series" – but here that "series" itself, that "multivalued function" -is specifically, (inside his later "Theory of Symbolic Forms"[19]), that of possible *alternative* epistemologies themselves!]

A Remarkable Parallelism

This parallelism between Bohr's and Cassirer's conceptions, and the former's crucial –and successful- role in modern physics I believe, specifically validates my incorporation of an epistemologically *relativized biology* into the "real world! Based in Maturana's profound ideas, it becomes is a legitimate and primal beginning perspective on the *whole* of the problem of

---

[15] My emphases. See Chapter 7: "Cassirer's Theory of Symbolic Forms"

[16] Incidentally, I think that reading Bohr suggests a converse extension of Cassirer's perspective to include multiple versions of each of the sciences! There would, therefore, be multiple, equipotent versions of biology, of physics, (wave/particle?), of psychology, ... depending on the initial perspectives!

[17] In his "Substance and Function". Cassirer, 1923

[18] See my Chapters 3 & 5: Cassirer's "Functional Concept of Mathematics". Afterthought: It is this derivation which breaks the "balkanization" of the "schools" of academia I referred to in Chapter 13 and which Bohr specifically mentions, (as "the schools"), later in this interview.

[19] Cassirer 1925 -1929

cognition and epistemology. I believe it is the *best* perspective, (utilizing Penrose's and Einstein's aesthetic criteria for the evaluation of theories), on the *whole* of the problem of "scientific realism" itself![20]

If you can picture Bohr's early perspective in this matter, (which he maintained continually to the end of his life), his principle of "complementarity" then makes perfect sense as the explanation of quantum physics' fundamental wave/particle duality of light! But this perspective, in a very real sense, is just another, (though a much more delimited), way of looking at Cassirer's "Symbolic Forms" —as Cassirer had initially limited it to the physical sciences.[21]

Begin excerpts from Niels Bohr interview

> "[Immediately before this discussion, Professor Bohr had talked informally to Thomas Kuhn and Rüdinger about the philosophical conceptions with which the early parts of the interview deal. Some of the questions addressed to him derive from that earlier discussion.]" [22] (Opening note of the Session V transcription from which all citations herein derive.)

---

[20] See Penrose, 1989, p. 421 Also see my argument in Chapter 14.

[21] See my headings "Cassirer's Theory of Symbolic Forms" in Chapter 7, and, more pointedly, "Contra Cassirer" in Chapter 8.

[22] Interview of Niels Bohr by Thomas Kuhn and Eric Rüdinger on November 17, 1962. Niels Bohr Library & Archives, American Institute of Physics, College Park, MD USA.
www.aip.org/history/ohilist/LINK

Appendix D — Bohr's Epistemological Roots

Bohr: "I took a great interest in philosophy in the years after my [high school] student examination. ... At that time I really thought to write something about philosophy, and that was about this analogy with multivalued functions. I felt that the various problems in psychology — which were called big philosophical problems, of the free will *and such things*[23] ... that one could really reduce them when one considered how one really went about them, and that was done on the analogy to multivalued functions.

If you have square root of x, then you have two values. If you have a logarithm, you have even more. And the point is that if you try to say you have now two values, let us say of square root, then you can walk around in the plane, because, if you are in one point, you take one value, and there will be at the next point a value which is very far from it and one which is very close to it[24]."

COMMENTARY -to paraphrase and clarify: consider the function $f(x) = x^{-2}$ over the non-negative real line and more specifically the related functions:

$$g(x) = +|f(x)|\ [25]\text{ and } h(x) = -|f(x)|.$$

Bohr visualizes tracing *continuous* curves within, alternatively, either the range of $g(x)$ or of $h(x)$ which lie, respectively. above and below the X axis. Each of these alternatives serves as an equipotent expression of the logical consequences, ("walking around in the plane" and the "next

---

[23] my emphasis   As mentioned earlier, he subsequently expanded this range to include cognition, epistemology, biology and the "real world" of physics.
[24] Ibid, my emphasis
[25] i.e. the absolute, (positive), value!

point very close to it" –roughly equivalent to continuity), within its particular theoretical perspective.

> **Bohr**: "If you, therefore, work in a continuous way, then you — I'm saying this a little badly, but it doesn't matter — then you can connect the value of such a function in a continuous way. But then it depends what you do. If in these functions, as the logarithm or the square root, they have a *singular value at the origin*, then if you go round from one point and go in a closed orbit" [i.e. continuously] "and [if] it doesn't go round the origin, you come back to the same [value]. That is, of course, the discovery of Cauchy.
>
> But when you go round" [through] "the origin, then you come over to the *other* [value of the] function, and that is then a very nice way to do it, as Dirichlet [Riemann], of having a surface in several sheets and connect them in such a way that you just have the different values of the function on the different sheets. And the nice thing about it is that you use one word for the function, f(z).
>
> Now, the point is, what's the analogy? The analogy is this, that you say that the idea of yourself is singular in our consciousness then you find [26]— now it is really a formal

---

[26] Free will, and consciousness are taken as his example and are the only referent in this specific interview of an application of his multivalued perspective. His elaboration immediately after it explains and expands his perspective more fully to the whole of the sciences!

way — that if you bring this idea in, then you leave a definite level of objectivity or subjectivity. For instance, when you have to do with the logarithm, then you can go around; you can change the function as much as you like; you can change it by 2 & pi; when you go one time round a singular point. But then you surely, in order to have it properly *and be able to draw conclusions from it,* will have to go all the way back again *in order to be sure that the point is what you started on.*[27] —

Now I'm saying it a little badly, but I will go on. -That is then the general scheme, and I felt so strongly that it was illuminating for the question of the free will[28], because if you go round [through the origin], you speak about something else, unless you go really back again [the way you came]. That was the general scheme, you see", (ibid, my emphases)

Bohr: (expanding his perspective to the physical sciences): "… If you have such a thing like this, and you go around here, then you certainly are treating things in an orderly manner, *but you gradually get over into some other meaning of the words.* Now, I say it very badly, but that was the kind of interest [I had]. We were later on very interested in the particle-wave problem. I felt also — but not to do anything with it — that it was more so that if one created a photon, then one had made a knot in

---

[27] My emphasis

[28] Note: "Free Will" is not my focus here, it was his expansion of the core idea —of "multivalued" functions as viable epistemological tools within the core of physics itself -that gained my attention!

existence, a knot which was of a very difficult kind to say, and only when that photon was absorbed, annihilated, that knot was untied. ... But now we know that these are solved by the non-commutation rules, and therefore, the non-commutation rules are certainly something great. *But in order to understand what they mean — You cannot get over that problem of the particle and the wave. And, therefore, it is also so nice that this lies in the complementary description.*"

COMMENTARY:

Consider the meaning of the paragraph above in light of the one preceding it. "... If you have such a thing like this, and you go around here," [i.e. continuously], "then you certainly are treating things in an orderly manner, but you gradually get over into *some other* meaning of the words."

> "But then you surely, in order to have it properly *and be able to draw conclusions from it*, will have to go all the way back again *in order to be sure that the point is what you started on.*[29] ... That is then the general scheme, ... because if you go round, [through the origin] *you speak about something else*, unless you go really back again [the way you came]. That was the general scheme, you see", (ibid, my emphases)

---

[29] My emphasis

Conclusions:

Consider Bohr's schema in its most basic sense. His are multivalued functions which take a shared, unique, (single) value only at the origin, but which branch as equipotent instances beyond it as in the square-root, or Riemannian sheet analogies. But even restricting them to the origin, they do not even begin to constitute the basis of –a fecund analogy for- a viable epistemology, only a methodological route *towards* such a conception. But it, ("complementarity"), works as a fundamental explanatory principle in the field of modern physics! It lies at the basis of the Copenhagen Interpretation! But there is no explanatory basis other than successful consequence.

His conception, however, is very much equivalent to and provides a rationale, now found to be workable in the "real world" for Cassirer's earlier and more deeply based and philosophically legitimized epistemological center, ("origin"), of his "Symbolic Forms".[30] Cassirer provides the basis of his largely equivalent conception *for the whole of the sciences* in his considerations of Hertz's beginnings, (as cited earlier), and in his philosophical conclusions which derived from it. I think this provides an actual scientific basis for an acceptance of Cassirer's thesis. "It is applicable in the real world". But that thesis, *Cassirer's* "Principle of Complementarity", as just stated, is much broader than Bohr's –it is applicable, not just within one particular discipline, (i.e. Quantum Physics), *but across each and all of the sciences*. Repeating:

> "...But instead, a *new* task arises: to gather the various branches of science with their diverse methodologies -

---

[30] i.e. Cassirer's multivalued "principle of Complementarity" as cited above

with all their recognized specificity and independence – into one system, whose separate parts precisely through their necessary diversity *will complement and further one another.* This postulate of a purely functional unity replaces the postulate of a unity of substance and origin, which lay at the core of the ancient concept of being."

I specifically used the latter as the underlayment for my (realist intentional) assertion of the ontic existence of the "interface". It is this "interface" that corresponds to Bohr's "origin" and to the philosophical center, this "gathering into one system" of Cassirer. This congruence of perspectives between Cassirer, Bohr, and myself I believe confirms my *relativized,* (conditional), materialist perspective of ontic indeterminacy for the functionalist/materialist brain as a *viable* epistemological principle!

Back to the Interview:

> **Bohr**, [Perhaps his most emphatic point made in this session]: "... Does [Einstein] think that, if he could prove they were particles, he could induce the German police to enforce a law to make it illegal to use diffraction gratings or, opposite, if he could maintain the wave picture, would he simply make it illegal to use photo-cells? That was, of course, in all friendliness, but it was the idea to say that this [is a] problem *we cannot get over*, and that means that actually *we got something new in the quantum.* That was the point."

> **Bohr**: (expanding again): "... But in between I was just interested also as regards the problems of biology, just what the problems of teleology meant, and so on. Therefore, I meant only that it was a natural thing to me to get into a problem *where one really could not say*

anything from the classical point of view, but where it was clear that one had to make a very large change and that one got hold of something which one really believed in."[31]

Bohr and Kuhn on Philosophers of Science:

Bohr: "...But that is my error, you see. It is not an error now, but it was an error in those days. — I felt ... that philosophers were very odd people who really were lost, because they have not the instinct that it is important to learn something and that we must be prepared really to learn something of very great importance.... There are all kinds of people, but I think it would be reasonable to say that no man who is called a philosopher really understands what one means by the complementary description.[32] I don't know if it is true, you see, because one can tell [there are] all kinds of people, and time goes.... — I think, at any rate here, the thing is preposterous. I do not also know how the thing is here. —

But if you take it on the whole, or a few years ago, they did not see that it was an *objective* description, and that it was *the only possible objective description*."

"So, therefore, the relationship between scientists and, philosophers was of a very curious kind. First of all I would say — and that is the difficulty — that it is hopeless to have any kind of understanding between scientists and

---

[31] ibid

[32] This is where I consider his lack of acquaintance or his own "balkanized" inability to evaluate the prior work of Cassirer is particularly tragic.

philosophers directly. It has to go *over the school.*" [i.e. *go above* – span the walls between the rigid schools/categories of academia]   "I don't know exactly how it is, but let us say, if you go back to ... the Copernican system, then some scientists they thought that it also was beautiful. But they were killed. Bruno was absolutely killed, and Gallilei was forced to recant. But in the next generation, the school-children did not think it was so bad, and thereby a situation was created where it belonged to common knowledge or common preparation that one had to take that into account. I think it will be exactly the same with the complementary description. It may be it's already, but I do not know. "[33]

COMMENT: Please note his usage of "over the school" in the citation above.

Kuhn:

"I think not yet."

Bohr: "How are the philosophers in Berkeley?[34] Do they take it for obvious that these things are right, or do they not? I think they do not, as far as I know them, but perhaps they do."

---

[33] my emphases
[34] I believe Kuhn was teaching at the University of California at Berkeley at the time.

Kuhn:

"Almost none of them has the technical competence necessary to follow these ideas into the problems from which they have been developed, and, therefore, none of them is really in a position to deal with them in any depth."[35]

OVERALL COMMENT: Bohr's utilization of "complementarity", though brilliant, is however at bottom, I think, essentially "ad hoc". It is an epistemological principle that is founded on no good logical or philosophical rationale. It is what he believed in and I think it was a very good "prejudice" that he had, (in contrast to Cassirer's superb rationale –and the one I have supplied in this current book as a key facet of cognition itself), but he had no basis for applying it other than its fecundity.[36]

Similarly, William James'[37] assertion, (as discussed in Chapter 12), that "the relations are given right along with the

---

[35] ibid. Note: Please read the transcript itself, especially Session V, wherein Bohr talks about his philosophical origins.

[36] Fecundity is a powerful, though very dangerous tool in theorizing as it could allow extreme power from trivial assumptions –e.g. "because I say that it is!" Somewhere it must be deepened and fit into an appropriate logical and theoretical context. Within my conception, Cassirer's "multivalued function", i.e. his "Symbolic Forms" validates and grounds my conception of the "interface" as a fundamental of cognition! And it actually *defines* my second intentional "Axiom of Experience" as well. Bohr's "complementarity" allows the actual existence of quantum physics as it grounds the wave/particle duality of light.

[37] Bohr mentions William James as one of his philosophical preferences in this interview. See Chapter 12 for my brief discussion of James' philosophy.

objects"[38] is again essentially ad hoc. It doesn't really do anything at the fundamental level of epistemology.[39] Why then, for instance, does James' presupposition work? It works because we have things like telescopes and microscopes and we can expand our vision or compress our vision until we get down to certain levels. But when we get to the size of the hydrogen atom, we run directly into the "brick wall" of "the quantum"[40] at which point *it no longer works!*[41] "The quantum" *does not work under a compression principle* as the Bohr citation shortly above and Kumar's treatment of this whole era confirm. Kumar's book[42] is especially lucid in this regard. It explains the deep experimental and theoretical basis of this dilemma.

In contrast, my own thesis provides an actual rationale for the *whole* of Bohr's "multivalued" perspective on epistemology. It comes from the perspective of Maturana's biology, from Cassirer's profound epistemology and specifically from my conception of "the interface".[43] It provides, moreover, the very

---

[38] Which lies at the basis of pragmatism

[39] As I noted earlier, I consider James' perspective as a very real candidate for a viable symbolic from –as a psychological perspective. See Chapter 12 for a brief evaluation of James' pragmatism.

[40] See prior Bohr citation: "but it was the idea to say that this [is a] problem we cannot get over, and that means that actually *we got something new in the quantum.*"

[41] An interesting reflection occurs here: compare Penrose's: "over a meter's range, deviations from Euclidean flatness are tiny indeed, errors in treating the geometry as Euclidean amounting to less than the diameter of an atom of hydrogen!" Penrose 1989 (p. 152) –another "flicker" in the swamp gas of metaphysics!

[42] Kumar, 2008

[43] See Chapters 8,9, and 10.

foundations for my second hypothesis, the "Axiom of Experience".

On the subject of "the interface", (viewing it from the epistemologically relativized functionalist perspective of mechanism), Cassirer's "multi-valued" conception of "Symbolic Forms", (his own "complementarity principle"), is implicit. It is a very different and very deep conception but it actually provides an explicit rationale for why and how these ideas work.

I think it is a pity that Bohr had such a limited and narrow exposure to philosophy, being taught initially by a proclaimed "idealist". He himself fell victim, therefore, to the limitations of the same malady, ("the schools'" categorizations), that he discovered in his own listeners thereafter, (see above on "schools"). He classified Kant and the neo-Kantians into what I have argued is the "wrong box", (i.e. "school"). These were not, and *never were* "idealists" of the Berkelian mold, but were rather, as I have argued throughout this book, "realists", but more specifically realist "ontic indeterminists". He missed, therefore, the possibility of communicating with the one mind that would have nurtured and expanded this fundamental idea –i.e. the mind of Ernst Cassirer. What I think he was struggling for is something very much like Cassirer's Theory of Symbolic Forms! Bohr had very similar conceptions of an equipotent set of beginnings. And I think that's exactly where modern science, as demonstrated in Quantum Mechanics, is going. D'Espagnat's conception entails something very much like it. (See Chapter 13: d'Espagnat....)

Bohr talks later in the interview about discovering William Jame's philosophy[44] which gave him some justification for his

---

[44] You will remember my comment in Chapter 12 where I commented that I had trouble with William James' philosophy as I think it is capable of standing as an independent Symbolic Form in just Cassirer's sense. See Chapter 12 for my brief evaluation of William James' ideas.

"realism", ("pragmatism"). But there's a note in that same interview where he is corrected by one of the interviewers that he actually began studying James in 1932. Significantly it occurred *after* the crucial Solvay Conference of 1927, wherein his own "Copenhagen Interpretation" of Quantum Mechanics was validated by the majority of the physicists, (but not by Einstein)! In this interpretation, "complementarity" became a crucial cornerstone to explain the wave/particle "ambivalence" of light. Bohr had basically *already* locked down most of the Copenhagen interpretation of quantum physics.

But, even in this interview given in his old age, wherein Bohr talks about "complementarity" he never really enables or validates it. He claims, contrarily therein however, that he has *already* done so! I've read numerous comments on it and Einstein said he was never satisfied with Bohr's explanations –and I'm not either. But it makes a great deal of sense if you consider his beginnings as described in this interview above and then consider it from the standpoint of Cassirer's Symbolic Forms, especially as regards the scientific forms which was Cassirer's original starting point.

Bohr's personal interpretation of the multiple equipotent beginnings possible for reasoning derived from his own specifically mathematical, and very early perspective. It constitutes an entirely original, non-imitative perspective on the problem of reality.[45] I think it substantiates my understanding of the Copenhagen

---

[45] It's incorporation into the 1927 Solvay Conference was, of course, predated by Cassirer's "Theory of Symbolic Forms" which was published in 1925-1929. It is too bad that Bohr was so ill-read in philosophy, as Cassirer entirely vindicates his conceptions in this regard. I think this is a consequence of Academia's Balkanization! (See Bohr's comments above on "the schools"). As I have repeatedly argued in this book, neither Kant, nor, by inheritance, was Cassirer an "Idealist" – not even a "critical" one –it was a horrible choice of words! Academia –in its "schools" - has "thrown out the baby with the bath"!

interpretation and ties it to the fecundity of modern Quantum Physics in real life. And more specifically it ties it to Cassirer's conception of Symbolic Forms as originally delimited to the scientific forms which grounds my own ideas.[46]

---

[46] I really don't know where in this book to insert this note. I think it is appropriate everywhere, but let me state it here for the record.

When we categorize a thinker as being contained within a given school, or accept or reject him or her as a source of ideas, we are most definitely *not* required to accept or reject each and every one of his or her specific ideas. I personally definitely believe, for instance, in Marx's specific vision of a compassionate and non-parasitory possibility for economics and societies, but reject what I consider his " throw-away" and "smart-ass" comment that "religion is (just) the opium of the masses!" And why should I not have this freedom? Similarly, I accept Kant's essential visions of "substantia phenomena" and "intuition", (experience), while rejecting his "categories". Again, why should I not have this freedom to cross academic categories?

Cassirer's new and reformulated definition of "the Concept" specifically allows it! But this is the rationale I just argued for Bohr's and Cassirer's "complementarity principles". I have the freedom to pick and choose across the whole range of the "schools" of academia based on the quality of the *individual ideas*. I am not required to accept this "balkanization" within it.

I think the latter is inherent within the very narrow and *still* classical and set-theoretically based vision of "the concept" that academia has accepted and which has been treated much earlier in this book. But the acquisition of this freedom was exactly the fortunate consequence of my very early discovery of Cassirer's "Substance and Function".

The working principle is set forth *on page 4* of this revolutionary work. (I doubt that you can conceive of just *how* revolutionary it is for that in itself would involve your prior possession of a "concept" larger than the classical one *—however* "sophisticated" it had become.)

"Every attempt to transform logic must concentrate above all upon this one point: all criticism of formal logic is comprised in criticism of the general doctrine of the construction of concepts." I think he should have had it stamped on the very cover of the book! This was his opening line for the development of his "Mathematical Concept of Function" which is a specific redefinition of the very word "concept" itself and is a primitive for his

Conclusions: A Confirmation of my Epistemological Foundations

Bohr and Cassirer were basically "on the same wavelength", but sadly they didn't speak the same language. I believe this appendix validates my biological conclusions from the words of one of the greatest physicists of all time, himself the founder of the *most* modern of modern physical science!. I think it validates my own claim for a scientific realism –but one grounded in biology!

"Symbolic Forms". It is an *entirely new* and different way to look at reality itself! See Chapters 3 & 5.

Appendix E: (Cassirer speaks to Shapiro & Mac Lane) –a Continuation of Chapter 2

Both Mac Lane and Shapiro are embedded in the Anglo/American traditions of philosophy which is essentially one of pragmatism, (with the possible exception of their *own* mathematics which they treat as a special case)! Their conclusions as well as their conceptions of what is real, or *of what could even exist* must be considered within that overall, broadest context however.

But there are other "schools" of philosophy –the "Continental school", for instance, which is quite different and broader, wherein Kant, as just one example, is *not* trivialized as he is in theirs. I think it would be highly productive for mathematicians working "at the leading edge" to broaden their philosophical reading to go "across the schools"[1] -and across the traditions to expose them to new and different, -and perhaps more fertile possibilities for their own subsequent ideas.

At the conclusion of chapter 2, there remained an open question: *is* there, *does there exist* a "general group"? *Is* there, does there exist a "general integral domain", "a general field",...

I broke off that discussion at that point as I felt it would adversely affect the overall flow and presentation of the overall book which is already complex admittedly.[2] This appendix is intended to take it up again at exactly that same point and to elaborate it.

To repeat, *is* there, *does there exist* a "general group"? *Is* there, does there exist a "general integral domain", 'a general field",...

---

[1] See Bohr's reference, coming from the perspective of Quantum Mechanics, to "the schools" in Appendix D and my comments on it.
[2] Please remember my early comment that it was not me, but the problem itself that made it so!

Shapiro and Mac Lane say no, but then what is it that they *are* referring to when they say that "the group" has such and such properties...", or "the field is such and so...."? Are they referring simultaneously to each and every instance of such specifically –*and to nothing else?*

But what is their *rule* of reference? It is essentially the rule of identity, the rule of the classical, Aristotelian generic concept against which Cassirer has argued so effectively, and against which he argues that his "mathematical concept of function" has (tacitly) replaced. He argues convincingly that this replacement has enabled the development of the whole of modern science!

Do the concepts, *per se*, of "the [general] group", "the [general] field, et al actually exist in themselves then? Cassirer argues that they do.[3] He says that they, (in their rule of formation), define a "new form of consciousness"! They consist in the rules, rather than the contents, of a series, they are "a new expression of the characteristic contrast *between the member of the series and the form of the series*"[4]. But the rule itself, this "new form of consciousness" is now internal to the mind(/brain) itself and, (as I have argued in Chapter 5), that as such it *does* exist! It is a relational and *operative* concept of the strategic brain itself, (in Cassirer's new and deeper meaning of "concept").

Note: See my heading "Contra Cassirer" in Chapter 5 wherein I dispute his ultimate refusal to hypostasize[5] his new concept, his new rule. Biology itself and a demand for a possibility of scientific realism[6] argue otherwise –it is an actual theme and argument of this current book generally.

---

[3] See Chapters 3 and 5 of this book
[4] Cassirer, 1923, p.26
[5] "hypostasize": to make real, to assert as actually existing
[6] See chapters 13, 14 and Appendix D, (Niels Bohr)

Let me now go back, relevant to this specific issue at the conclusion of Chapter 2, to an expansion of the ideas of one of my deepest and profoundest sources: Ernst Cassirer[7], (a "Neo-Kantian" philosopher), once again. Cassirer is perhaps the deepest and most widely read philosopher I have ever been acquainted with —as evidenced by my central placement of several of his deepest ideas *as foundational* throughout this very book, (see especially Chapters 1,3,5,7…). He was probably the most proficient in his knowledge of modern science of any philosopher I have ever read —and in his knowledge of the actual historical development of the same.[8] To say that he was acquainted with modern mathematics is a gross understatement —he was *intimately* familiar with it!

Because of space limitations and the specific limitations of my own particular background, these citations reveal only his conclusions, but not his lengthy but extremely detailed and convincing arguments which support them. I hope I can convince you to read at least the relevant chapters in the original.

Central to his book, "Substance and Function", (in the course of his development of a redefinition of the very word "concept" itself[9] into his "mathematical concept of function"), is a historical analysis of number, algebra, and geometry as used in modern mathematics. His starting point, (or at least the place I will start citing from), will be his remarks concerning Gottfried

---

[7] I think Cassirer will ultimately emerge as the most brilliant and relevant philosopher of the 20th century.
[8] It's a personal prejudice, but I think he has read and understood everything ever written about the subject —and about much more besides!
[9] see chapters 3.5,7 and appendix D of my book for references.

Leibniz, the co-inventor/discoverer with Isaac Newton of the (formal) calculus[10]. [11]

In our "popular culture", -that "sound-byte" culture such as it is, Newton is regarded as a scientific genius who liked apples, but Leibniz is regarded as a fool, thanks to Moliere and Leonard Bernstein.[12] The truth is quite otherwise.

Our actual heritage from Newton resides in the practical science of classical physics he developed from that same "calculus" and which validates his intellectual brilliance, and most of the scientific advances, (which incorporated it), until the end of the 19th century -which validate the conception itself in the "real world". Our actual heritage from Leibniz resides in his *independent and simultaneous* development of that very same calculus. But Leibniz, primarily, was a philosopher, (but a profoundly deep one especially in the origins of mathematic), and so Newton's success with the same idea "in the real world" therefore conversely validates the deep underlying mathematical and philosophical ideas resident in the mind of Leibniz which led him to it.

Let me start a series of excerpts from Cassirer's "Substance and Function" -specifically from his chapter 3 –"The Concept of Space and Geometry". [13] In the beginning of that chapter,

---

[10] i.e. differential and integral calculus

[11] Note: Mathematicians and logicians should also carefully examine his preceding Chapter 2: "The Concept of Number" as well as Chapter 3: "The Concept of Space and Geometry" which latter will be the source of my citations for this brief appendix. These two chapters of his book are highly interdependent.

[12] i.e. "The best of all possible worlds!", the song!

[13] It would be worthwhile, I think, for mathematicians and logicians to study the whole of the book as it will expose entirely new possibilities, beyond their limited, specifically philosophical presumptions inherited from the "schools" - which they think every "sane" mind *must* accept. They may perhaps

Cassirer began with a detailed analysis of many of the geometrical developments of modern mathematics, but then returned to re-examine the ideas of one of his own foundational philosophers, Gottfried Leibniz:

> "*Characteristic (Kombinatorik) as pure "doctrine of forms" (Leibniz)",* (his emphasis).

> " ... again we are led to the Leibnizian conception of mathematics.[14] According to this conception, *mathematics is not the general science of magnitude but of form,*" not the science of quantity, but of quality. Characteristic, (*Kombinatorik*) thereby becomes the fundamental science; we do not comprehend under it the doctrine of the number of combinations of given elements, but the universal exposition *of possible forms of connection in general* [15] and their mutual dependency."[16]

Note: you might want to refer back to the very similar perspective of Goldblatt earlier in Chapter 2 of my book which reflects a very similar perspective on current developments in the very foundations of mathematics as grounded in relations -i.e. functions, mappings:

---

ultimately conclude that pragmatism is indeed the most rational, but this conclusion is totally unwarranted without first truly examining the alternatives! But it is a fact that leading-edge science is actually forcing us to do so.

[14] Very early in this chapter Cassirer made an argument for geometry, and not number, as the basis for mathematics and for his own "mathematical concept of function".

[15] My emphases.

[16] Note the connection with my initial argument in Chapter 4 regarding the "schematic calculus".

Goldblatt: "One of the primary perspectives offered by category theory is that the concept of *arrow*, abstracted from that of *function* or *mapping*,[17] may be used instead of the set membership relation as the basic building block for developing mathematical constructions, and expressing properties of mathematical entities. Instead of defining properties of a collection by reference to its members, i.e. *internal* structure, one can proceed by reference to its *external* relationships, with the other collections. The links between collections are provided by functions, and the axioms for a category derive from the properties of functions under composition."[1] [18]

Back to Cassirer on Leibniz:

"Wherever a definite form of connection is given, which we can express in certain rules and axioms, there an identical 'object' is defined in the mathematical sense. The relational structure as such, not the absolute property of the elements, constitutes the real object of mathematical investigation."[19]

Note: This is precisely pertinent to Shapiro's arguments – and to Mac Lane's -i.e. to the "what" of those "objects", and to the "what" of it that actually does exist –and how! It also seems to embody a direct parallel with the modern mathematical perspective of structuralism, (see Chapters 2 and 3).

" ...Two complexes of judgements, of which the one deals with straight lines and planes, the other with the

---

[17] i.e. relation
[18] Note this is an example, but an imperfect one from the standpoint of Leibniz and Cassirer as we shall shortly see. *What* is it that is *in* those "collections"?
[19] This is precisely pertinent to Shapiro's arguments –and to Mac Lane's

> circles and spheres of a certain group of spheres, are regarded as equivalent to each other on this view, in so far as they as they included in themselves the same content of conceptual dependencies along with a mere change of the intuitive 'subjects,' of which the dependencies are predicated."

This is his prelude to his introduction Hilbert's "implicit definition" which is the main topic of my own Chapter 2 and central to my thesis! The above citation also correlates pretty closely with Wilder's "presumptive and permissive".[20]

> "In this sense, the 'points' with which ordinary Euclidian Geometry deals can be changed into spheres and circles, into inverse point-pairs of a hyperbolic or elliptical group of spheres, or into mere number-trios without specific geometrical meaning, without any change being produced in the deductive connection of the individual propositions which we have evolved for these points."[II]

Comment: Hilbert's "tables, chairs and beer-mugs"?

> "The particular elements in ... mathematical construction are not viewed according to what they are in and for themselves, but simply as examples of a certain *universal form of order and connection*; mathematics at least recognizes in them no other "being" than that belonging to them by participation in this form. For it is only this being that enters into proof, into the process of inference and is thus accessible to the full certainty, that mathematics

---

[20] See chapter 2

gives its objects." (Note again that this is a modern structuralist perspective.)

Cassirer on Hilbert via Leibniz:

> *"Geometry as pure "doctrine of relations", (Hilbert)"*, (his emphasis)
>
> "… In contrast to the Euclidian definitions, which take the concepts of the point or the straight line as immediate data of intuition, from which fixed content they proceed, the nature of the original geometrical objects is exclusively defined by the conditions to which they are subordinated. The beginning consists of a certain group of axioms, which we assume, and their compatibility has to be proved'

This "proof of compatibility, of course, involves the question of "how"? I assert that Hilbert's *relative* proofs as addressed by Shapiro in chapter 2 make the most sense within Leibniz's overall context which grounds itself in relation rather than objects -and within the context of the immediately following citation.

> "From these rules of connection that we have taken as a basis follow all the properties of the elements. The point and the straight line signify nothing but structures which stand in certain relations with others of their kind, as these relations are defined by certain groups of axioms. Only this systematic 'complexion' of the elements, and not their particular characters is taken here as the expression of their essence. In this sense, Hilbert's geometry has been correctly called a pure theory of relations. In this, however, it forms the conclusion to a tendency of thought, which we can trace in its purely logical aspects from the first beginnings of mathematics."

"...At first, it might seem a circle", [a logically circular argument], "to define the content of the geometrical concepts exclusively by their axioms: for do not the axioms themselves presuppose certain concepts in their formulation? This difficulty is disposed of when we clearly distinguish the psychological beginning from the logical ground ... but we look for all closer determination from their successive insertion into various relational complexes. It is by this intellectual process that the provisional content first becomes a fixed logical object, while the elements in their apparent absoluteness signify only an "π ρδτερον π ρὸς ημας" [21], (an untranslatable –by me, Greek phrase which I assume means something like "just a tentative beginning". I would appreciate some feedback on this assumption).

"Intuition seems to gasp the content as an isolated self-confined existence; but as soon as we go on to characterize this existence in judgement, it resolves into a web of related structures which reciprocally support each other. Concept and judgement know the individual only as a member, as a point in a systematic manifold; here as in arithmetic, the manifold, as opposed to all particular structures, appears *as the real logical prius*. ... The determination of the individuality of the elements is not the beginning but the end of the conceptual development;

---

[21] There is an oddity in scholars of Cassirer's era. For him it was normal to state a term in its original Greek which they assumed that any educated reader was familiar with. Since I do not currently have access to a translator, I will simply mark it as above. It is on p. 94 of Cassirer, 1923.

> it is the logical goal which we approach by the progressive connection of universal relations!"

Please refer to chapter 2 of Substance and Function", ("The Concept of Number"), which developes this argument in detail as specifically applied to number, and which he subsequently extends and deepens to geometry as the *real* basis of mathematical thought. Note: the above citation, I feel again, is an explicit argument for the "structuralist" perspective of modern mathematics!

> "The procedure of mathematics here points to the analogous procedure of theoretical natural science for which it contains the key and the justification."[III]

Note: I used the immediately prior citation as a footnote in Chapter 1 in justification of Hilbert's methodology. It is supportive of Niels Bohr's concept of "complementarity" in modern Quantum Physics, which latter I argued in Appendix E, is essentially comparable to the epistemological consequences of Cassirer's "Theory of Symbolic Forms" -which I have utilized throughout my own thesis.

Now I'm going to leave this specific topic, (of the connection to Shapiro's and Mac Lane's objections -I think I have made my case), but, since I am in Cassirer's fertile backyard, I will cite just a single further passage which relates to several of my other theses.

> "...The role which we can still ascribe to experience, does not lie in *founding the[22] particular systems*, but in the

---

[22] i.e. any *particular system* —see my chapters 3 and 5 on Naturalism's epistemological mistake!

selection that we have to make among them. It is reasoned that, as all systems are equally valid in logical structure, we need a principle that guides us in their application. *This principle can be sought only in reality,* since we are not here concerned with mere possibilities, but with the concept and the problem *of the real itself,* in short, it can be sought only in observation and scientific experiment.^{IV}

How close all this is to Quine's perspective, to d'Espagnat's "reality which refutes", and to my own thesis of the strategic brain. It is also precisely relevant to the beginnings and the very definition of my "axiom of experience", -and to my ultimate perspective of ontic indeterminism!

Note: a final "aside" about Ernst Cassirer: Cassirer's "Substance and Function" is certainly one of the most brilliant and profound books ever written about philosophy, logic and mathematics —it is relatively easy to read -*and short*! (about 456 pages).

His "Theory of Symbolic Forms", I think is in reality just a (brilliant) *appendix* to it, but that "appendix" is over a *thousand pages long*, -and "oblique" in the extreme. It is necessary to cull through his interesting and extensive supportive material[23] to try to extract his ultimate purpose and conclusions.[24] D'Espagnat seemed to really like Cassirer's "Substance and Function", but

---

[23] Anyone who has ever attempted Cassirer will be aware that "extensive", as applied to Cassirer, means something like having examined the entire history of science in detail ever since it began!

[24] Chapter 7 is my attempt at a summary and analysis of that book.

could never really accept it because of his prior classification of Cassirer as, (in the "school" of), "an idealist", a "Neo-Kantian".

But there is another book to be considered here. It is "The Philosophy of Ernst Cassirer".[25] It was in preparation at the time of Cassirer's death, and so lacked the inclusion of the particular philosopher's response to his critics normal in that series of books addressing the work of the major philosphers by Tudor publishing. I have read this book, and the one part which grabbed my attention was Smart's[26] critique of Cassirer's "Symbolic Forms". He remarked that Cassirer had actually violated Kant's "categories' in his "Substance and Function" and "Theory of Symbolic Forms". I think Cassirer was open enough to have entertained Smart's objection seriously and brilliant enough to envision the new possibilities it opened.

I agree with Smart in this, and *that*, I think it is the crux of the issue. Kant's "categories' were never innate in his own fundamental position, nor in Cassirer's "Neo-Kantian" position behind him. They were an attempt to justify the enormous success of then-contemporary Newtonian physics in the world, and so were an attempt to rationalize philosophy with science.

Today, however, science gives us a quite different picture –and those categories no longer work! "The phenomena" need no longer be tied directly to them[27] but are open to new, and very different interpretations.

---

[25] Smart/Tudor 1949

[26] Smart, H. "Cassirer's Theory of Mathematical Concepts" in The Philosophy of Ernst Cassirer. Tudor Publishing 1949

[27] This goes back to Penrose's "billiard ball" universe which is *still* the vision of most neurophilosophers. You see, even in rejecting Kant, they carry his worst mistakes forward into their own conclusions!

Agreeing with Smart, I think it was this tacit enlargement of Cassirer's fundamental idea —in that "appendix" to "Substance and Function" that allowed the brilliance of his final conception of a fundamental *symmetry of perspectives* —but, in the same act, made Cassirer's own treatment of it so convoluted. I think he still retained a subconscious tacit, (but contradictory I argue), familial loyalty to Kant's categories" —and to "the schools"![28]

If d'Espagnat were to consider Cassirer's work in this larger perspective, -I don't think he would have any problem with it —nor do I think, would Kant himself. He always said that his ideas were a beginning, not an end. But you see, I began reading Cassirer[29] as a very young man, Cassirer was my "teacher" about Kant – and I *never* interpreted either of them in this limited way. Very young I was impressed with Cassirer's injuncture that

---

[28] I argued a similar line as applied to Niels Bohr in Appendix D.

[29] Cassirer actually constituted a "father-figure" for me for reasons I will not go into here but which caused me to take everything he taught me very seriously. He was a great and gentle soul.

Kant considered his work as a beginning, albeit a *profound* beginning on the problem of cognition and not an end.

If we interpret Cassirer's "mere X" within Maturana's essential context therefore, then we arrive at d'Espagnat's "something", and at my "somewhat". It is the reality *which refutes*!

Appendix F: An Outline of my Overall Argument

That my book is hard, (perhaps the hardest you have ever attempted), I fully acknowledge. It is hard because it begins not from just a mere physical relativism, but from the perspective of an even deeper and specifically a *scientific epistemological* relativism. The latter is based in the preservation of mathematical invariants[1] (equations) *across* the various scientific disciplines in our SUPERB[2] theories.[3]

This must be very strange to you. How could this seemingly *purely philosophical* aspect be important to the actual scientific solution to our problem? How could this have significance in our practical world? I believe it is the *only* perspective which makes actual sense out of the mind-brain dilemma!

I grant that almost all of you acknowledge purely physical relativism in a certain sense. It is embodied in your acceptance of Copernicanism and Galilean Relativity, (you utilize the latter every day –*it is why classical physics works!*), and not Ptolomeanism for instance. And yet Kuhn's quote from a medieval source, (cited previously), showed our innate resistance to even that easier and more primitive perspective:

> "Those clerks who think (think how absurd a jest)
> That neither heav'ns nor stars do turn at all,
> Nor dance about this great round earthly ball;
> But th'earth itself, this massy globe of ours,
> Turns round-about once every twice-twelve hours;
> ... So should the fowls that take their nimble flight
> From western marches towards morning's light, ...

---

[1] i.e. its relationality seen in a *context-free* setting
[2] Penrose's term, his CAPS!
[3] See Chapter 8: Contra Cassirer, (What are the Real Parameters?), and Chapter 9 on mathematical "ideals".

> And bullets thundered from the cannon's throat
> (Whose roaring drowns the heav'nly thunder's note)
> Should seem recoil; since the quick career,
> That our round earth should daily gallop here,
> Must needs exceed a hundred-fold, for swift,
> Birds, bullets, winds; their wings, their force, their drift,
> Arm'd with these reasons, 'twere superfluous
> T'assail the reasons of Copernicus;
> Who, to save better of the stars th'appearance,
> Unto the earth a three-fold motion warrants"[4]

(Roughly translated: if the Earth is spinning at such a speed, why doesn't a cannon ball, fired in the opposite direction destroy the breach and go in the opposite direction from which it was fired? Therefore the Ptolomean, rather than the Copernican picture of the universe is the correct one! It seems like a pretty plausible argument, doesn't it?)

The *fundamental and ultimate* perspective embodied in this book, going far beyond simple physical relativism, lies in an extension and delimitation[5] of Cassirer's "Theory of Symbolic Forms" which details a profound *epistemological* relativism in our very conceptions of science and of reality itself –but always grounded in rigid mathematical invariants of the phenomena as described in our "SUPERB theories". It is this origin which allowed me the freedom to begin the dialogue as I did in Chapter 1. It allowed me to present the problem in strictly materialist, (simpler), terms without however an *absolute* commitment to that materialism itself, (and its specifically metaphysical claims). It

---

[4]Kuhn, Thomas "The Copernican Revolution" Harvard Press, 1957
[5] See Chapter 8: Contra Cassirer, (What are the Real Parameters?), and Chapter 9 on mathematical "ideals"

allowed me to treat the problem from the perspective of functionalism, -from the perspective of brain as a machine!

Immediately following, I will present a simplified flowchart of my overall argument, citing only a few of the most relevant points which I hope will serve to clarify and simplify the overall picture.

Appendix E: Argumentative Outline – a Flowchart

In opening a dialogue on the fundamentals of *any* problem, and, specifically here, on the mind-brain problem, we must *necessarily* begin from some philosophical and some specifically *intentional* philosophical perspective -be it the intentionality behind Realism, behind Idealism, behind Dualism, …Solipsism,….

These define what we are willing *to believe* and is part of the actual definition of the term "intentionality" itself! But this is always a *closed* context. The only alternative is to claim direct communication with God and I don't think that any one of us qualifies! My own perspective on this problem is that of a scientific Realist –but *not*, at the final telling, a materialist one.

Cassirer's "Theory of Symbolic Forms", (which Swabey has correctly characterized as a genuine *epistemological* "theory of relativity"), taken in its *realist*[6] sense and beginning from a materialist interpretation of biology[7] is my starting point. Each perspective", (and in my extension of his beginning, *each mind*), begins from a different starting point. It begins with *different* fundamental questions. Each begins with a unique epistemology! My own perspective begins within Cassirer's relativized biology. It is not aimed at academic word games, but at pragmatic, usable and verifiable answers in the context of neurophysiology leading to new results *useful* to humankind. It will not be easy reading, but I believe it actually fulfills the role I proposed for it. I propose it as science!

> ERNST CASSIRER: "... *Even in 'nature'* the physical object will not coincide absolutely with the chemical object, nor the chemical with the biological –because physical, chemical, biological knowledge *frame their questions* each from its own particular standpoint and, in accordance with this standpoint, subject the phenomena to a special interpretation and formation. .... The One Being,", [i.e. the ultimate metaphysical/ontological object], "to which thought holds fast and which it seems unable to relinquish without destroying its own form, eludes *cognition*." [my emphases] It becomes "a mere X"
>
> HEINRICH HERTZ:"The images", [i.e. scientific entities], "of which we are speaking are our *ideas* of things; they have with things the one essential agreement which lies in the fulfillment of the stated

---

[6] See Chapter 8: Contra Cassirer: (What are the real parameters?)
[7] i.e. Maturana's

requirement", [of successful consequences], "but further agreement with things is not necessary to their purpose. *Actually we do not know and have no means of finding out whether our ideas of things accord with them in any other respect than in this one fundamental relation.*"

CASSIRER COMMENTS ON THE ABOVE: "... [though] still couched in the language *of the copy theory of knowledge* –... the [Hertz's] concept of the 'image'", [itself], "had undergone an inner change. In place of the *vague* demand for a *similarity of content* between image and thing, we now find expressed a highly complex *logical* relation, a general intellectual *condition*, which the basic concepts of physical knowledge must satisfy. ... Its value lies not in the *reflection* of a given existence, *but in what it accomplishes as an instrument of knowledge*, in a unity of phenomena, which the phenomena must produce out of themselves", [my emphasis].

A system of physical concepts must reflect the relations between objective things and their mutual dependency, but, Cassirer argues, this is only possible "in so far as these concepts pertain *from the very outset* to a definite, homogeneous intellectual orientation", [a unique and specifically *logical* framework]. It is only within a distinct logical framework that these "images" are significant at all! The object cannot be regarded as a "naked thing in itself", *independent* of the essential categories, (and logical frameworks), of natural science: "for only within these categories which are required to constitute its form can it be described at all."

CASSIRER: "The naïve realism of the ordinary view of the world, like the realism of dogmatic metaphysics, falls into this error, ever again. It separates out of the totality of possible concepts of reality a single one", (e.g. mathematical physics), "and sets it up as a norm and pattern for all the others. *Thus certain necessary formal points of view, from which we seek to judge and understand the world of phenomena, are made into things, into absolute beings."* [my emphasis]

Cassirer's broadening of perspective, (and it is a *genuine* "Copernican Revolution" in Kant's sense), necessitates and validates his conclusion of the innate symmetry and *a relativity of interpretations* for phenomena. "With this critical insight ... *science renounces its aspiration and its claim to an 'immediate' grasp and communication of reality.*"

"If the object of knowledge can be defined *only through the medium of a particular* logical and conceptual structure, we are forced to conclude that a *variety of media*", [my emphasis], "will correspond to various structures of the object, to various meanings for 'objective' relations."

This is the assertion of symmetry and the foundation for his thesis of "Symbolic Forms" which becomes, in a slightly delimited form, my entry point into the problem of the mind-brain relationship. It allows multiple, *equally viable* scientific perspectives on any given problem and enables my initial —but provisional- approach to this problem from the standpoint of a biological materialism. It allows us to approach the problem of the brain as a machine!

Berkeleian Idealism	Realism	Dualism, (in whatever form)	Solipsism	...

My own perspective is that of a realist —but it is a very sophisticated "realism" that I urge. I begin within the perspective of biology, (essentially Humberto Maturana's), informed with the conclusions of W.J. Freeman which latter specifically breaks our *hierarchical* commitment to externality, (Kant's "substantia phenomenon"). It allows us to view that externality instead through the filter of Maturana's "structural coupling". I arrive, therefore, at the first of my three specifically *intentional* axioms of realism –i.e. "the [raw] axiom of externality"! Realist Axiom #1: There exists an "externality" –Kant's "substantia phenomenon"- which corresponds to the first of Putnam's prerequisites for realist (intentional) reasoning.

My second intentional axiom of realism is "the axiom of experience". It is derived as a consequence and an extension of David Hilbert's brilliant mathematical insight of "implicit definition", (seen within a provisional context of "materialism" – later relativized) –which I use to extend Cassirer's *other* brilliant logical insight –i.e. his redefinition of the word "concept" itself into his "mathematical concept of function". My own further extension of his result, (based on Hilbert's insight), into "the Concept of Implicit Definition", (C.I.D.), derives from considerations of mathematical structuralism which enable my new, *expanded* concept, (C.I.D.), to encompass within itself even our percepts *themselves* as "positions in a structure", (i.e. purely organizational nexuses of the brain). It obviates the necessity of a "Cartesian Theater" as Dennett envisions it and supplies an actual theory of meaning to my constructs, leading to my definition of "experience" as that which remains phenomenologically invariant under all possible consistent theories of reality, (in a Quinean sense). It enables a conception of "consciousness" *within* a purely materialistic perspective!

> My third and final —but still *intentional* axiom of realism is that of the "interface". I maintain that any realist perspective *must* assume the specifically *ontic* existence of some connection between externality and experience. But this "interface", because of the relativistic implications of Cassirer's "Theory of Symbolic Forms" as discussed above, must be taken in its broadest, "*heterophenomenological*" interpretation. It must be taken as the "mathematical ideal"[8] of all realist conceptions of the sensory boundary. It is the specifically *ontic* posit of the actual existence of this interface which, I conclude, supplies the existence and the "substance" of mind itself.[9] I am, (we are), conscious: i.e. *consciousness ontologically exists*!

> The perspective above supplies my realist answer to the mind-brain problem and I believe it holds credibility from my initial beginnings in Chapter 1 all the way through my conclusions of Chapter 13, (d'Espagnat), where I believe it provides the beginnings of a viable solution to Einstein's realist dilemma for Physics itself. I believe it supplies a much simpler and more direct answer to the problems of "decoherence", "complementarity", and "entanglement" for instance than does the quantum formalism itself. This, yet again, is a profound teleological simplification.
>
> But remember that my solution involves a relativization of "materialism" itself, ("materialism is pretty much equivalent to "multitudinism" in d'Espagnat's terminology —and which he

---

[8] See Chapter 9
[9] See Chapter 10

> decisively refutes from the standpoint of modern experimental Physics)!

⇕

> In conclusion, ultimately mine is a thesis of hope. It restores the "mind" to the machine and allows a pathway to the whole of our humanity in all its aspects. If man were *just* a machine, it wouldn't really matter what we did to him, it would be just a question of which noises he, ("it"), subsequently would make – "happy noises" or "painful noises". (Note: This is a statement of Functionalism "in the raw", but I think it is an accurate characterization.)
>
> Without an *actual* mind there can be no ethics, no humanity, no compassion –no matter what the adherents of that view might –and *have*– argued contrarily.
>
> This book defines a pathway to more work than you ever dreamed of, but I think the destination of that path makes the work worthwhile for all of us –*and for our children*!

Bibliography

Barrow, John D. . *Pi in the Sky: Counting, Thinking and Being*. Little, Brown & Company 1992

Bell, J.L Toposes and Local Set Theories, an Introduction. Dover Publications, 1988

Benacerraf, P., and H. Putnam, eds. []: *Philosophy of Mathematics*. 2nd Ed Cambridge University Press. 1983

Birkhoff, Garrett and Mac Lane, Saunders *A Survey of Modern Algebra* Macmillan Company   1955

Bohr, Niels. Transcript of Tape-Recorded Interview of Niels Bohr by Thomas S. Kuhn, Leon Rosenfeld, Aage Petersen, and Erik Rudinger on November 17, 1962. Niels Bohr Library & Archives, American Institute of Physics, College Park, MD USA. Session V, www.aip.org/history/ohilist/http://www.aip.org/history/ohilist/4517_5.html

Cassidy, David. Uncertainty: the Life and Science of Werner *Heisenberg*. New York. W.H. Freeman. 1992.

Cassirer, Ernst. *The Theory of Symbolic Forms* Yale University Press 1957, (original publication in German Volume 1, 1925, Volume 2 1927, Volume 3 1929,   Berlin. Bruno Cassirer.

Cassirer, Ernst. "*Substance and Function & Einstein's Theory of Relativity* (bound as one). Dover Publications   1923

Cassirer, Ernst. *Spirit and Life* (An essay included in *The Philosophy of Ernst Cassirer*). New York. Tudor 1949

Cassirer, Ernst. *Kant's Life and Thought* (Hayden translation) Yale 1981

Cassirer, Ernst. *The Problem of Knowledge* (Woglom and Hendel translation) Yale 1950

Cassirer, Ernst. Determinism and Indeterminism in Modern Physics Yale University Press 1956 (Benfey Translation of original , Goeteborg, 1937)

Churchland, Patricia. *Neurophilosophy* Bradford Books 1986

Churchland, Paul. *Matter and Consciousness* MIT Press 1990

D'Espagnat, Bernard. *On Physics and Philosophy* Princeton University Press 2006

D'Espagnat, Bernard. Veiled Reality: An Analysis of Present-Day Quantum Mechanical Concepts. Westview Press 2003

Dennett, Daniel. *Consciousness Explained.* Little, Brown, and Company 1991

Dreyfus, Hubert. *What Computers Still Can't Do* MIT Press. 1992

Durant, Will. *The Story of Philosophy* Pocket Books 1926

Edelman, G. M. "*Bright Air, Brilliant Fire.*" BasicBooks 1992

Einstein, Albert. *Essays in Science* Philosophical Library 1934

Fine, Arthur. *The Shaky Game* University of Chicago Press 1986

Freeman, W.J. Chaotic oscillations and the Genesis of Meaning in Cerebral Cortex Springer-Verlag 1994

Freeman, Walter J. *How Brains Make Up Their Minds* Phoenix Press 1999, (also see his extensive repertoire of publications on his webpage: http://sulcus.berkeley.edu/ (all free to download)

Gleick, James. *Chaos Theory* Penguin 1988

Goldblatt, Robert. Topoi: The Categorial Analysis of Logic Dover 1984

Gould, Stephen Jay. 1994. *The Evolution of Life on the Earth* Scientific American. October, 1994. Volume 271, Number 4

Hilbert, David. . *The Foundations of Geometry* (Translation by E.J. Townsend) Open Court 1910

Hofstadter, Douglas. *Gödel, Escher, Bach* Vintage 1979

Iglowitz, Jerome. *Virtual Reality: Consciousness Really Explained* (First Edition), Lulu Press. (Originally published online) http://www.foothill.net/~jerryi/COMPILED9-2-00.pdf 1995/1998)

Iglowitz, Jerome. *Mind: The Argument from Evolutionary Biology, (A Working Model)* 2004 http://www.foothill.net/~jerryi/BIOLOGY11.pdf

Iglowitz, Jerome. Consciousness: a Simpler Approach to the Mind-Brain Problem, (Implicit Definition, Virtual Reality

and the Mind) 2005
http://www.foothill.net/~jerryi/HILBERT1.pdf
Iredale, Mathew. TPM, The Philosophers's Magazine, Issue 54
Kant, Immanuel. *Prolegomena to Any Future Metaphysics* (Muhaffy translation) Liberal Arts Press 1950
Kant, Immanuel. *Critique of Pure Reason* (Norman Kemp Smith translation) St Martin's Press 1961
Kuhn, Thomas. *The Copernican Revolution* Harvard College 1957
Kumar, Manjit. Quantum: Einstein, Bohr and the Great Debate about the Nature of Reality. W.W. Norton & Co. 2008
Lakoff, G. Women, Fire and Dangerous Things Chicago-London. University of Chicago Press 1987
Lara, Blaise. Semantics and Factor Analysis: an approach to the interpretation of factors semantics and factor analysis: an approach to the interpretation of factors
http://www.hec.unil.ch/people/blara/lara/factors0.htm
Mac Lane, Saunders. *Mathematics: Form and Function* Springer-Verlag, New York, 1986
Marchal, Bruno. Online paper:
http://iridia.ulb.ac.be/~marchal/publications/SANE2004 MARCHALAbstract.html
Maturana, Humberto and Varela, Francisco. *The Tree of Knowledge* Shambala Press, 1987
Moore, Walter. *Schroedinger: Life and Thought* Cambridge University Press 1989 *Note: I really don't like this book* but the Schroedinger quotes in themselves are priceless. The psychologism is atavistic in the extreme and caused me to put it aside!
Minsky, Marvin. *The Society of Mind* Touchstone 1985
Nagel, Ernest. *The Structure of Science.* Harcourt, Brace & World, 1961
Nagel, Ernest & Newman, James R., *Goedel's Proof* NYU Press, 2001 [Note: I read the original of this book as a very young man.]

Penrose, Roger. *The Emperor's New Mind"* Oxford University Press, 1989

Quine, Willard Van Orman: *Word and Object* MIT Press, 1960

Quine, Willard Van Orman *"From a Logical Point of View"* Harper Torchbooks, 1953

Reid, Constance. 1970. *Hilbert* Springer-Verlag.

Resnick, Michael 1992. *Mind, Vol. 101/401"* January 1992

Russell, Bertrand. *The Autobiography of Bertrand Russell* Little, Brown, & Co. 1967

Russell, Bertrand. An Essay on the Foundations of Geometry Dover 1956

Schlick, Moritz. *General Theory of Knowledge"* (Translation by Albert E. Blumberg). New York: Springer-Verlag, 1974/1917

Shapiro, Stewart. Categories, Structures, and the Frege-Hilbert Controversy: The Status of Meta-mathematics *Philosophia Mathematica* (III) **13**, 61–77. doi:10.1093/philmat/nki007 2005

Smart, H. Cassirer's Theory of Mathematical Concepts in The Philosophy of Ernst Cassirer. Tudor Publishing 1949

Stefanik, Richard. Structuralism, *Category Theory and Philosophy of Mathematics* Washington: MSG Press 1994

Stewart, Ian. *Fibonacci Forgeries* Scientific American, May,. Volume 252, Number 5 1995

Van Fraassen, Bas C. *Laws and Symmetry* Clarendon Press 1989

Van Fraassen, Bas C. *Quantum Mechanics, an Empiricist View"* Clarendon Press. 1991

Weyl, Hermann. . *David Hilbert and His Mathematical Work* Bulletin of the American Mathematical Society 50 1944

Wilder, Raymond. . Introduction to the Foundations of Mathematics John Wiley & Sons 1952

ENDNOTES

Extended Dedication

[1] Dedication to the First, (original), Edition: "This dedication will be different from what you are used to. If you choose to skip it therefore, that is your decision. What I choose to put into it, however, is mine. I have lived long as a relative hermit and as a fanatic to the cause of these ideas. Many people dear to me have been forced to pay the price. I dedicate this book to all these compassionate and forgiving souls who have had the tolerance to put up with, and some even to love me:

To my (few) intellectual friends: to Ruelle Denney, whose kindly, (and genuinely aristocratic), response to my youthful naïveté and arrogance I will forever remember, to Tom Owens who, in the kindness of his heart was the first willing to risk apoplexy from my initial two and three-page quotations and quivers of "!"'s, to Dr. Arnold Leiman who was the first comprehending being to tell me I was not a raving megalomaniac, to Dr. Hubert Dreyfus who caused me to read Maturana and Varela, to Dr. David Elliott who, over the last year and a half, through his generosity of spirit and kindness has helped me to endure the unendurable. And lastly, mostly, to my dear friend, David Casacuberta who, though he remains an unrecalcitrant Naturalist, (:-)), in his largeness of spirit and innate decency, has helped me to perfect what is, from his point of view, an enemy's plan of battle. I can never thank him enough.

To my family: I could never give back what you gave to me. I lacked the normal background of human interaction, (because of the circumstances of my childhood), to communicate to you the real love I have always felt for you. And beyond that, my fanaticism and almost total distraction towards the resolution of the problem set for me have robbed you of precious time and attention. But my purpose, beyond the duties of my own spiritual obligation, was to do *you* honor! I hope that happens. But, if my answer is right, it is important for you as well as for me -I hope it will make life better for you, and, if not for you, then for your grandchildren and theirs.

Endnotes

To "Pops", to "Momma Jung", to Doug, to Rich, to "Bee", (Burbank Jr.), and to Matt, who unselfishly gave me the real family I never had, I am truly and forever grateful.

To my mother and father -I wish I could have made your lives better, and to my brother Ron -I wish we could have been closer. It was probably my fault.

To my wonderful daughters, Chenin-blanc Yic-mun-fuung Iglowitz and Mook-lan Sauvignon Iglowitz. In you, God has truly blessed me, and I know it every day. I love you guys.

And finally and especially, to my wife of 24 years, Christina Teresa Sun-Jung Iglowitz, I could never have done it without you. This is the "holy crusade" we talked about on our first date high in the Berkeley hills, (Chinese girls don't kiss?!) I guess it's how I "conned" you into marrying that strange creature. Well, here it is. I have learned, (so far as I am capable of learning it), decency and compassion from you who, I *still* think, embodies these traits more fully than any other human being I have ever met, and I will be forever in awe of you. I love you now, and, whatever happens, will love you till the day I die.

Jerome Iglowitz

October 22, 1998"

Preface

[I] W.V.O. Quine, 1953, pps.42-43
[II] W. V.O. Quine 1960, pps.3-4, my emphasis
[III] Boorstin quoting Kepler in "The Discoverers", Random House.

Chapter 1
[I] see Chapter 2 for citation
[II] my emphases
[III] Sometimes an image works better than an explanation. There was a wonderful episode on "The Outer Limits" television show, I think, that made the point graphically. Let me describe and summarize it:

A spaceship has been detected approaching Earth, and one of the earthly technicians has been assigned to stay in touch with one of its occupants over the years before it can actually land. It will take years because of its distance from the Earth. The Earth-based technician begins to fall in love with "her" and they develop a romance through their communication over this time. Finally the magic day arrives and he goes to the spaceport to finally meet and hold his new love. The spaceship door opens and she emerges. But "she" has the form of an Octopus! Think about that graphic image relative to my claim! Suppose we were all "blind"!

Chapter 2

[I] [Dennett 1991]

[II] Birkhoff & Mac Lane 1955

[III] Resnik 1992

[IV] Stefanik, 1994

[V] Schlick, 1974 (translation)/1917 (original), my emphasis

[VI] I feel I have completed most of the ancillary and exploratory work already, but I definitely need help on the level of foundations and of mechanics. This book explores that level itself. Ultimately I hope to encourage some inspired mathematical genius to develop the actual mathematical "calculus" of the brain. He will stand with Newton in history.

[VII] I will clarify this transition into "axioms" shortly. Mac Lane's book dealt exclusively with axiom systems.

[VIII] Hilbert –from Shapiro, 2005, my emphasis

[IX] ibid

[X] This is one aspect of what I have termed "the static problem". (Iglowitz, 1995, Dennett Appendix) I am not totally happy with that answer –I think the current paper addresses it more honestly.

[XI] I fell in love with mathematics way back then but was horrified when I glimpsed her concealed ugliness reflected in her "makeup mirror", (of set theory) and, shocked, sadly abandoned her! I turned instead to pursue the biological and philosophical implications of my original insight –and this has been the driving force and focus of most of the rest of my life.

[XII] The question of the desired structure of these sections of the book arises immediately. Frankly I hope you will turn to the source material for a more-than-amateur exposition. I specifically recommend Richard Stefanik's "Structuralism, Category Theory and Philosophy of Mathematics" for a very deep and lucid introduction to the subject and further references. But if you're starting here you will at least hear an introduction to the subject, and a linkage with my own ideas —which linkage is quite deep.

[XIII] Resnik 1992

[XIV] Benacerraf, 1983

[XV] Benacerraf is considered the founder of mathematical structuralism

[XVI] Saunders Mac Lane is widely regarded as one of the most significant mathematicians of the 20th century.

[XVII] Mac Lane, 385, my emphasis

[XVIII] Stefanik 1994, my emphasis

[XIX] Stefanik 1994

[XX] I think this argument would not be viable under intuitionist logic for instance.

[XXI] Expanding Resnik, 530

[XXII] which is exactly the sense of "the objects" of mathematical structures

[XXIII] please refer back to the prior Hilbert quote —the objects are defined by the whole of the axiom system

[XXIV] Iglowitz, 1995

[XXV] This does not necessarily lead to epiphenomenalism, at least not in its ordinary sense. My reflections in the Freeman appendix, [Iglowitz, 2005] suggest another usage. The feedback, incorporating intentional perspectives, (axioms), through the primitive, (and fixed), evolutionary objects opens a possibility. Of course, even this usage could be interpreted on the level of primitive axioms. It is a problem of prediction and organization. —see my letter to Rosen quoted later. I think they serve as operational metaphors.

[XXVI] See Bell 1988 for word usage

[XXVII] Kant didn't particularly like this name himself either, but he was forced into it. See the later citations from Kant himself which explains his reasoning. I think it was his greatest mistake. I have termed it "ontic indeterminism" which I think expresses his conception far better.

[XXVIII] Cassirer, 1957, p. 76

[XXIX] I tried to synopsize Cassirer's "Symbolic Forms" in Chapter 5 of my MS. I truly love Cassirer's mind, but his writing style is oblique in the extreme.

[XXX] See my presentation of Cassirer's alternative logic to follow.

[XXXI] Bell, 1988, 245, my emphasis

[XXXII] See Chapter 4 and Iglowitz, 2005 for a specific rationale and a close parallel in W.J. Freeman's non-hierarchical brain map.

[XXXIII] Maturana is another crucial and brilliant source necessary to the problem at this point.

[XXXIV] Iglowitz, 2005, and especially its Freeman Appendix

[XXXV] See "Afterward: Lakoff/Edelman" [Iglowitz, 1995] for a discussion of mathematical "ideals" which bears on this discussion. Afterthought: I appended a graphical rendition of the discussion of "ideals" to Chapter 9 which might make it easier to follow my conclusions.

[XXXVI] this relates to the issues of "hierarchy" which I will discuss in Chapter 3.

[XXXVII] Iglowitz, 1995

[XXXVIII] Which, of course, harks back to Kant

[XXXIX] This is precisely the question that structuralism addresses.

[XL] E.g. Maturana, Edelman, W.J. Freeman, etc.

[XLI] Maturana, 1987

[XLII] Goldblatt: "Topoi: The Categorial Analysis of Logic", Goldblatt, Robert, Dover 1984, p.3

[XLIII] I once again strongly encourage you to turn to the sources themselves.

[XLIV] Iglowitz, 2005

[XLV] Iglowitz 2005

[XLVI] Shapiro's "contentful" seems to equate pretty much with "ontological"

[XLVII] I disagree, and so, I think, would the "young Hilbert".

[XLVIII] This is precisely my point –I think it is precisely the issue. I think it is not vicious at all but is instead perfectly "consistent" (sic) with the whole of Hilbert's *early* perspective!

[XLIX] "assertatory" = "contentful" = "ontological"????

[L] ibid] (my emphasis)

[LI] and why not, Platonism aside? Saunders Mac Lane, [Private correspondence], expressed a view equivalent to Shapiro's to me which I will address presently.

Chapter 3

[I] I would dearly love to hear input from real, (but open-minded), mathematicians of a philosophical bent on this phase of my argument. This is about as far as I can take it. I truly need and would sincerely value their input.

[II] Compare also Lakoff: 1987, p.353. "Most of the subject matter of classical logic is categorization."

[III] Cassirer 1923 pps.3-4   He continues: "The Aristotelian logic, in its general principles, is a true expression and mirror of the Aristotelian metaphysics. Only in connection with the belief upon which the latter rests, can it be understood in its peculiar motives. The conception of the nature and divisions of being predetermines the conception of the fundamental forms of thought. In the further development of logic, however, its connections with the Aristotelian ontology in its special form begin to loosen; still its connection with the basic doctrine of the latter persists, and clearly reappears at definite turning points of historical evolution. Indeed, the basic significance, which is ascribed to the theory of the *concept* in the structure of logic, points to this connection. ..."

[But] "... The work of centuries in the formulation of fundamental doctrines seems more and more to crumble away; while on the other hand, great new groups of problems, resulting from the general mathematical theory of the manifold, now press to the foreground. This theory appears increasingly as the common goal toward which the various logical problems, that were formerly investigated separately, tend and through which they receive their ideal unity."

It is just this "general mathematical theory of the manifold" to which he refers at the end which, I will argue, forces an even further extension of Cassirer's own arguments.

[IV] See Iglowitz, 1995, Chapter 2 for a full discussion

[V] This passage, (mirroring, incidentally, the mathematical "power set"), suggests also the absolute hierarchy of concepts, (and theories), already implicit in the classical conception. Cassirer's alternative, (which I will discuss shortly), reveals a new possibility, developing into his theory of "symbolic forms" which I have elaborated and tried to simplify in Chapter 7 of this book.

[VI] Please forgive the Capital, but the problem lies in talking about "the concept of the concept" and the fact that this is not in any sense trivial to the issue.

[VII] Note: This is a usage of the word entirely distinct from my later usage of the word!

[VIII] Cassirer still saw perceptual objects as the basis of his functional rule, however.

[IX] ibid, P.23. Rosch and Lakoff have argued in more recent times, (based in hard empirical data), that the categories of actual human beings, actual human cultures, actual human languages are not, in fact, grounded in the classical Aristotelian "Concept" but are based, instead, in prototype, metaphor, metonymy, association, radial categories, etc. But what are these, (in their anthropological totality), but the *free posit* of rules of category formation? Cassirer has provided a more classical and rigorous conceptualization. It incorporates the possibility of *all* (consistent) rules in a classical formulation.

Clearly this *does* better correspond with ordinary and scientific usage than does the classical concept. It is the functionality of our definitions which specifies the concept. The mathematical "subset" is the limiting rather than the typical case.

[X] ibid P.16

[XI] See prior footnote: Stewart, "Fibonacci Forgeries"

[XII] ibid p.26

[XIII] ibid

[XIV] In my opinion Cantor is a perfect illustration of the case.

[XV] For the first time I understood the gaps between the conceptual lucidity of the opening few days of any given mathematics course to the "therefore…" it had so invariably falsely claimed.

[XVI] See Wilder, 1952

[XVII] Goldblatt, Robert, Dover 1984, p.1

[XVIII] Which I also read that summer.

[XIX] rather than its ontic references

[XX] This is related to W.J. Freeman 1994 which connection I will pursue shortly

[XXI] Edelman, 1992

[XXII] Cassirer, 1923, pp. 288-289

[XXIII] See Iglowitz, 1995, Chapter 2
[XXIV] I think it is *the most complex*
[XXV] As Edelman's noted: "*certain symbols do not match categories in the world.*"
[XXVI] and the brain is *surely* an operative organ
[XXVII] W. V.O. Quine 1960, pps.3-4, my emphasis
[XXVIII] Freeman, 1995, my emphasis. I will repeat this citation in reference to an argument by Shapiro as well shortly.
[XXIX] Alternatively, as combined with the mathematical conception of the "Ideal".
[XXX] Which was never responded to
[XXXI] identical
[XXXII] Iglowitz, private correspondence
[XXXIII] Kuhn, 1957
[XXXIV] ibid
[XXXV] ibid
[XXXVI] Penrose, 1989
[XXXVII] Iglowitz 2005
[XXXVIII] W.V.O. Quine, 1953, pps.42-43
[XXXIX] See my "Rosen" discussion later.
[XL] Cassirer's "Substance and Function" is an excellent reference to this "abstractive logic". See especially the first few chapters.
[XLI] What *kind* of conditionality is another issue. Material implication, for instance, is not a direct gift from God. See Quine, 1953
[XLII] After Quine's usage.
[XLIII] W.J. Freeman has objected to this usage, but I think if I qualify it to be: "the (*not necessarily hierarchically*) reactive", I think he might approve.
[XLIV] An idea discussed with a correspondent who suggested it. D.E., ~2005
[XLV] See my illustration "Bounds and Limits". [Iglowitz, 2005]

Chapter 4

[I] Maturana and Varela, 1987
[II] Is this not the *usual* case between conflicting theories and perspectives?
[III] Edelman, 1992, pps.236-237, his emphasis.
[IV] Iglowitz, 1995, especially Chapter 4
[V] together: *all* the possible conceptual contexts
[VI] See the later discussion of mathematical "ideals" which bears on this discussion.
[VII] this relates to the issues of "hierarchy" which I will discuss shortly
[VIII] [A recent reference 2009: See Durant on Kant where the same kind of arguments are made.]
[IX] Cf Lakoff, 1987. Also see Iglowitz, 1995, "Afterward: Lakoff, Edelman..."
[X] Edelman, 1992, the problem is that he does not really explore this dimension, but W.J.Freeman —as presented shortly —does so explicitly!
[XI] Freeman, 1994, my emphasis
[XII] Freeman, 1995, my emphasis.
[XIII] Freeman, 1994
[XIV] See prior footnote about his ambivalent use of the word "spatial".
[XV] Freeman, 1994
[XVI] Freeman, 1995
[XVII] Freeman, 1994
[XVIII] Maturana, 1987, pps 163-4
[XIX] Edelman, 1992, p.27
[XX] Freeman, 1994, my emphasis
[XXI] Freeman, 1994, my emphasis
[XXII] See Maturana, 1987 and Edelman, 1992
[XXIII] My function, however, is to introduce a mechanics —which I have done. Merleau-Ponty is not "my philosopher", but the concept seems pregnant.
[XXIV] W.J. Freeman, 1997
[XXV] "Of the virtually unlimited information available in the world around us, the equivalent of 10 billion bits per second arrives on the retina at the back of the eye. Because the optic nerve attached to the retina has only a million

output connections, just six million bits per second can leave the retina, and only 10,000 bits per second make it to the visual cortex.
...After further processing, visual information feeds into the brain regions responsible for forming our conscious perception. Surprisingly, the amount of information constituting that conscious perception is less than 100 bits per second. Such a thin stream of data probably could not produce a perception if that were all the brain took into account; the intrinsic activity must play a role.
...Yet another indication of the brain's intrinsic processing power comes from counting the number of synapses, the contact points between neurons. In the visual cortex, the number of synapses devoted to incoming visual information is less than 10 percent of those present. Thus, the vast majority must represent internal connections among neurons in that brain region." (This is very much in accord with both Maturana's and W.J. Freeman's conceptions.)
.... Although six million bits are transmitted through the optic nerve, for instance, only 10,000 bits make it to the brain's visual processing area, and only a few hundred are involved in formulating a conscious perception –too little to generate a meaningful perception on their own. *The finding suggested that the brain probably makes constant predictions about the outside environment in anticipation of paltry sensory inputs reaching it from the outside world.*" (My emphasis)
From Scientific American March 2010 "The Brain's Dark Energy"
Marcus Reichle, Washington University School of Medicine in Saint Louis

But Reichle does not draw the obvious conclusions, as indeed, nobody else does. His conclusions are confounded by the epistemological paradox of his own arguments –*his* is a brain also, and subject to the same limitations. Those answers lie in the relativism of epistemology I propose.

[XXVI] cf Dennett, Dreyfus on the "large database problem"

[XXVII] This is *typically* the case. A project manager, for instance, must deal with all, (and often conflicting), aspects of his task -from actual operation to acquisition, to personnel problems, to assuring even that there are meals and functional bathrooms! Any one of these factors, (or some combination of them), -even the most trivial- could cause failure of his project. A more poignant example might involve a U.N. military commander in Bosnia. He would necessarily need to correlate many conflicting imperatives -from the geopolitical to the humanitarian to the military to the purely mundane! Or, in a metaphor on the earlier discussion, he might need to take a "Marxist" perspective for one aspect of his task, and a "royalist" perspective for another!

[XXVIII] Simple adequacy is quite distinct from information or parallelism however.

[XXIX] See Iglowitz, 1995: Lakoff/Edelman appendix for a discussion of abstraction and hierarchy

[XXX] See Birkhoff & Mac Lane, 1955, p.350, discussion of the "duality principle" which vindicates this move. More simply put, and using Edelman's vision, it is a question of which end of the "global mapping" we look from!

[XXXI] The "anthropic principle" as usually interpreted, on the other hand, is clearly self-serving and tautological. There is another deeper sense of the principle I discussed in the section: The JCS Review which I think is more pertinent. (Chapter 12)

[XXXII] Freeman has objected to my characterization of the human brain as an "organ of response". I understand his objection, as it seems to imply acceptance of "stimulus-response" causality" —which is clearly not my intention. At this level of discussion, I think the characterization is warranted however.

[XXXIII] Maturana and Varela, 1987

[XXXIV] See Dreyfus on the "large database problem". Also see Appendix A of Iglowitz, 1995 for a "combinatory" counterargument.

[XXXV] See Cassirer, 1923

[XXXVI] See prior note about Freeman's objection to "response"

[XXXVII] "HIKE" (:-) A very tiny bit of humor.

[XXXVIII] see P.S. Churchland re: Hume

Chapter 5

[I] This is the subject of the beginnings of this paper which is itself the best beginning reference.

[II] My function, however, is to introduce a mechanics —which I have done. Merleau-Ponty is not "my philosopher", but the concept seems pregnant.

[III] ibid p.25

[IV] op. cit P.25

[V] ibid p.26

[VI] ibid P.24
[VII] ibid p.25, my emphasis
[VIII] Wilder, 1967, P.18
[IX] my emphasis
[X] Cassirer, 1923, P.26

Chapter 6

[I] Kant, Prolegomena, p.10
[II] "Prolegomena", P. 11
[III] cf Chapter 7
[IV] ibid
[V] Maturana and Varela, 1987
[VI] See Chapter 7 re: Quine
[VII] Maturana and Varela, 1987
[VIII] afterwards "Maturana"
[IX] ibid P.48, my emphasis
[X] ibid Pps. 39-40
[XI] ibid P.51
[XII] ibid P.63
[XIII] ibid Pps.74-75
[XIV] ibid Pps.63-64
[XV] ibid P.96
[XVI] ibid P.74
[XVII] ibid Pps.80-81
[XVIII] ibid Pps.95-102, (my emphasis)
[XIX] ibid Pps.147-148
[XX] ibid Pps.157-159
[XXI] ibid p.159
[XXII] ibid Pps.163,164
[XXIII] ibid P.124, my emphasis
[XXIV] ibid Pps.129-133, my emphasis

[XXV] op.cit p.133
[XXVI] ibid Pps.133-134
[XXVII] cf Dennett, 1991
[XXVIII] cf P.S. Churchland, 1986, Dennett, 1991
[XXIX] Dennett, 1991, P.382, my emphasis
[XXX] An Inquiry into Meaning and Truth", Bertrand Russell, Pp. 14-15
[XXXI] cf Fine, 1986. p.97
[XXXII] op.cit Pps.234-244, my emphasis
[XXXIII] cf Penrose
[XXXIV] cf Chapter 5
[XXXV] This is also, obviously, a reiteration of Maturana's "razor's edge".
[XXXVI] Kant, "Prolegomena" pps.36-37
[XXXVII] Kant, "Critique of Pure Reason", 2nd edition, 333, translated by Woglom and Hendel, and cited in Cassirer: "The Problem of Knowledge", 1950, Pps. 101-102 I prefer this to Smith's rendering.
[XXXVIII] cf Chapter 5
[XXXIX] Maturana & Varela, 1987
[XL] Afterwards "Maturana"
[XLI] Kant, "Critiqueof Pure Reason"
[XLII] ibid P.96
[XLIII] ibid Pps.63-64
[XLIV] H. Hertz, "Die Prinzipien der Mechanik", p.1 ff, my emphasis

Chapter 7

[I] Quine, 1953, pps.42-43
[II] cf heading above!
[III] Penrose 1989
[IV] Using Dreyfus' term again
[V] Cassirer, 1953, p. 75
[VI] my emphasis
[VII] ibid, p.75

[VIII] my emphases
[IX] ibid
[X] ibid
[XI] H. Hertz, "Die Prinzipien der Mechanik", p.1 ff, my emphasis
[XII] Cassirer, op cit p.76
[XIII] ibid
[XIV] Cassirer, 1954, p.76
[XV] ibid
[XVI] ibid
[XVII] see Chapter 3
[XVIII] ibid
[XIX] ibid
[XX] ibid
[XXI] ibid p.77, my emphasis
[XXII] ibid
[XXIII] Cassirer 1923
[XXIV] ibid, P.446, my emphasis
[XXV] ibid, p.447
[XXVI] ibid, p.446
[XXVII] ibid, p.447
[XXVIII] my emphasis
[XXIX] Cassirer, 1923, pps.374-379, my emphasis

Chapter 8

[I] Van Fraassen, 1991, pps.4-5
[II] my emphasis
[III] ibid
[IV] ibid
[V] ibid p.94
[VI] Cassirer, 1953, p.77
[VII] ibid. pps. 77-78, my emphasis

[VIII] ibid, my emphasis
[IX] ibid
[X] ibid
[XI] Cassirer, 1923, p.446
[XII] ibid, my emphasis
[XIII] ibid, p.446
[XIV] ibid, p.447
[XV] ibid, p.447
[XVI] Edelman, 1992
[XVII] See Birkhoff & Mac Lane, 1955, p.350, discussion of the "duality principle" which vindicates this move. More simply put, and using Edelman's vision, it is a question of which end of the "global mapping" we look from!
[XVIII] ibid Pps.147-148
[XIX] ibid Pps.157-159
[XX] op.cit Pps.234-244, my emphasis
[XXI] my emphases
[XXII] W. V.O. Quine 1960, pps.3-4

Chapter 10

[I] cf Iglowitz, 1995

Chapter 12

[I] Durant, 1926
[II] My emphasis
[III] Durant, 1926
[IV] See Iglowitz, 1995, Chapter 3
[V] My emphasis
[VI] See Chapter 3, this paper
[VII] My emphasis
[VIII] Durant, Will. 1926, my emphasis

[IX] Here at this point, I am sad to say, he broaches his integrity by attributing a dishonest motive to Kant. "The truth is that Kant was too anxious to prove the subjectivity of space as a refuge from materialism; he feared the argument that if space is objective and universal, God must exist in space, and therefore be spatial and material. This, (Durant's), is an *ad hominem* argument, pure and simple and does not do justice to Durant's own intrinsic intellectual integrity.

[X] i.e. William James

[XI] ibid, p.447

[XII] Cassirer, 1923, pps.374-379, my emphasis

[XIII] See Smart, 1949. Smart, though not in agreement, does an excellent job of elucidating the essential perspectives and is well worth reading.

[XIV] H. Hertz, "Die Prinzipien der Mechanik", p.1 ff, my emphasis

[XV] ibid, P.446, my emphasis

[XVI] ibid, p.447

[XVII] ibid, p.446

[XVIII] End repeat of Chapter 5 section.

[XIX] Iglowitz, 1995

[XX] Russell, Bertrand. 1967

Chapter 13

[I] D'Espagnat, 2002

[II] D'Espagnat, 2006, P.45

[III] Ibid, P. 39

[IV] Wlodek Duch, private correspondence –circa 1991-5

[V] d'Espagnat, 2006

[VI] Penrose, 1989, pps. 226-227

[VII] D'Espagnat, 2002

[VIII] ibid, Page 47

[IX] ibid, page 48

Chapter 14

[I] W.V.O. Quine, 1953, pps.42-43
[II] W. V.O. Quine 1960, pps.3-4, my emphasis
[III] Penrose, 1989. P. 243, my emphasis
[IV] Bas Van Fraassen, Quantum Mechanics, p.17
[V] "beneficial" is itself a synthetic a priori perspective
[VI] i.e. at the "fine-grained" level of mind
[VII] or to gain reflective insights on them

Chapter 15

[I] P.S. Churchland, 1988, P.260
[II] cf Appendix D, (Penrose)
[III] And their damnable and blatant arrogance about it!
[IV] Van Doren, 1938

Appendix A

[I] Dennett, 1991
[II] i.e. -relative to Dennett's problem
[III] See Cassirer 1923, and Iglowitz, 1995
[IV] i.e. vis-à-vis current process
[V] See Iglowitz 2005
[VI] Consider the world-views implicit in paranoia or schizophrenia, for instance, or in bipolar orientations
[VII] Or "objectivist logic" after Lakoff's usage
[VIII] Dennett, 1991, P.382

Appendix B

[I] cf Wlodek Duch, (footnote #74, P.566), for instance

[I] Hertz, "Die Prinzipien der Mechanik"

[II] ibid, p.75
[III] ibid
[IV] ibid

Appendix E

[I] Goldblatt, 1984
[II] ibid, p.93
[III] ibid, p.94
[IV] Ibid, p.107

Index

(Note: Page numbers followed by "n" and another number—either Arabic or Roman—lead to a footnote or endnote.)

abstract domains, 345–46
abstract frame of experience, 268–71, 297n25
abstraction, challenges to, 92, 165–72, 202
academic philosophy, author's perspective on, 443–44n1, 467–69, 551
A/D converters. *See* hierarchical/non-hierarchical converters
aesthetics consideration, 121–23
algebraic principle of duality, 128n41
analytic vs. synthetic statements (Quine), 19–20
Anglo/American vs. Continental philosophical traditions, 551
anthropic principle, 110, 380
Aristotelian logic, 88–89, 199–200n12, 319, 331–32, 488–90, 494–98, 552, 584nIII
artificial intelligence, 27–28
Aspect, Alain, 395, 398–99, 416, 422
assertive statements, necessarily (Shapiro), 79, 82–83, 84, 117, 129, 133
attention, theory of, 203–4
automorphism, 181–86, 347, 369, 372, 375–77, 379
autopoiesis/autopoietic unity (Maturana), 133, 227–29, 231–32, 310
axioms
    as definitions of concepts, 334–35
    of experience, 266–67, 268–71, 278–79, 297n25
    of externality, 243–52, 259–60, 352, 385n34, 415–16, 430–31
    as fundamental operative units of the brain, 62
    of interface, 355–57
    knowledge as axiomatic, 188
    as rules, 169
axiom systems
    biological conception of, 37, 38–39, 111–12, 169–71
    in C.I.D., 106–8, 112–13, 196
    as embodying spatiality, 74–77
    meaning in, 35–37, 102, 334–35
    non-hierarchical functional alternatives, 321
    ordering's foundation in, 102–3, 107
    relationality, 118–20, 125–28, 208–9
    relative consistency, 82–83, 118
    Shapiro and author on, 80–82
    as source of ultimate concepts, 207–8
    "static problem," 473–86
    virtual objects in, 57, 59–60, 61
    See also Hilbert, David

Banach/Tarski theorem, 104
beauty, scientific, 122–23
behavior
    brain as processor for efficient response, 46–47, 68, 166, 178, 313
    naïve perceptual objects as virtual artifacts of, 38, 43, 46, 68, 72–73, 181–82, 210, 251, 273, 409
    organizational function of mind for, 42–43, 116, 192–95, 375
    as source of mind, 49, 178–80, 195, 238–39
    structural coupling and, 232–34
    trigger/boundary condition relationship to objects, 145, 224
beliefs, mechanistic brain's ability to have, 72–73. *See also* strategic brain
Bell, John L., 70, 71–72, 115, 346, 417
Bell, John Stewart, 394–95, 397–98
Bell's Theorem, 413–15
Benacerraf, P., 64, 70–71, 96, 103, 346
Berkelian vs. critical idealism, 281n38, 300, 331, 443n1
biological evolutionary perspective (Iglowitz)
    Cassirer's concept vs., 93, 552
    cognitive closure in, 206–7, 221–24, 234–38, 256, 347, 371–72
    concordance as biology's proper conclusion, 173–76
    Einsteinian realism and, 448–50
    evolutionary niches, 180
    formal and abstract problem, 166, 167–71
    information issue, 138
    logic as biologically based, 48, 75, 76n51, 128–33, 171–72, 175, 197–200, 310–13, 526–27
    objective existence of reality and, 177–80
    ontic indeterminism and, 48–49, 253
    operative and dynamic functions, 241
    organizing principle of evolution, 42–43, 242, 253, 409–11, 422–23, 450
    quantum physics application to, 389
    realism's imperative, 185–87, 188–89
    symbolic forms and, 310–15, 505–7
    See also brain, human; epistemological relativized biology; schematic operative model; structural coupling
biological phenomenology (Maturana and Varela), 227
Bohr, Niels
    Cassirer and, 326–27, 328–29, 532–36, 541–42
    complementarity principle, 326, 328, 329, 531, 536–41, 545
    on epistemology and realism, 323–31, 531–33, 536–50
    on lack of ontological realism, 67, 459
    multivalued functionalism, 323–30, 532–33, 536–41, 546–47
    scientific contribution of, 391–92
    on scientists and philosophers, 543–44

brain, human
    axiom systems as applied to, 37, 111–12
    as "blind" mechanism, 40–41, 42, 131, 150–51, 161, 167, 372
    calculus for, 526–27, 581nVI
    embodiment in epistemological relativism, 506–7
    logic as embodied in, 129
    mechanistic view of, 27–28, 30–38, 70
    modeling of objects in, 62, 107, 113–14, 116, 151–65
    neuronal structure and, 233–35
    other minds issue, 418
    processing power of, 130–33, 587–88nxxv
    as proposed focus of study for mind-body solution, 460–61
    quantum mechanics as structural model for, 378, 432–33
    reduction of consciousness to, 476
    self-organization of rules in, 38, 160, 161, 412
    self-referentiality of, 433–34
    substance of, 367n11
    See also strategic brain; substance of mind

calculus, 94–95, 104–5, 139n4
Cantor, Georg, 96–101
cardinality of objects, arbitrariness of, 405
Cartesian theatre of the mind. *See* homunculus argument
Cassirer, Ernst
    on Aristotelian logic, 89, 584nIII
    Bohr and, 326–27, 328–29, 532–36, 541–42
    color phi phenomenon and, 479–81
    complementarity principle, 284n46, 286–87, 329–30
    complexity of work, 561–64
    critical idealism in, 69, 281n38, 331
    epistemological relativism, 14, 69–70, 72, 124, 276–78, 286–89
    on epistemology's role in physics, 51–52
    on experience, 265–70, 276–78
    on Hilbert's implicit definition, 204n27, 294n13, 298
    influence on author, 87, 94, 553, 561, 563–64
    Kant vs., 379
    on knowledge as axiomatic, 188
    Lakoff and, 492–500
    limitations of perspective, 215–17, 377–78, 525–26, 527–28
    logical concept reformulation, 88–94, 103, 201–7
    mathematical structuralism of, 56, 69–70, 72, 78, 553–60
    on naïve realism, 383–84
    on naturalism, 276–78, 301, 499
    ontic indeterminism contribution, 330, 560–61
    on ordering within concepts, 87, 93

    on percepts, 112–14
    on presentation, 92, 93, 201–3, 205–7
    on series, 208n39
    Shapiro and Mac Lane vs., 552–60
    significance of, 315
    on specialized concepts, 115
    See also concept as functional rule; symbolic forms
category, traditional philosophical, 444n1, 489–90
category cue validity, 203n24
category theory, 55, 63–74, 77–86, 114–16, 556, 585nIX
causal dependence, 333
causation vs. triggering, 230–31
cellular process and logic, 129, 171–72, 310–13
centrality principle for theories, 122
Chalmers, David, 137
chaos theory, 160, 241, 516
Churchland, Patricia S., 28, 76n51, 165, 447, 467
C.I.D. *See* concept of implicit definition (C.I.D.)
circular causality, 43, 46
classical logical concept, 88–89, 199–200n12, 319, 331–32, 488–90, 494–98, 552, 584nIII
closure, cognitive, 206–7, 221–24, 234–38, 256, 347, 371–72. *See also* Maturana, Humberto
cognition
    biological/evolutionary basis for, 311–12, 313–15
    Cassirer's culture extension, 306n46
    C.I.D. and, 197–98
    as continual test of intentional arc vs. perceptual objects, 32–33, 43, 411–12, 417–18
    as coordination of atomic primitives, 229n23
    defined, 223n10
    as informationally indeterminate, 168
    Lakoff's idealized cognitive models, 491n17, 492, 502, 504
    metaphysical externality issue for, 253, 292–93, 309–10, 348–49
    mind as constitutive concept of the brain, 174
    other minds' existence and, 368
    particular-to-universal process of, 299–300
    problem of perceptual objects as, 223–24n11
    as process vs. representation, 44
    quantum physics and "normal" cognition, 428–29
    relativity of, 330
    See also interface, cognator/object; strategic brain
color phi phenomenon (Dennett), 187, 473–86
complementarity
    Bohr's principle of, 326, 328, 329, 531, 536–41, 545
    Cassirer's principle of, 284n46, 286–87, 329–30

comprehensiveness, theoretical, 304, 305–7, 309n53, 315–22
computer language as machine language at its source, 23n2
concept as functional rule (Cassirer)
    biological evolutionary perspective vs., 93, 552
    C.I.D. as generalization of, 106–8, 112–13, 170, 196, 204–8, 210–11, 321–22n73, 527–28
    classical concept vs., 169–70, 332, 584nIII
    continuous generation, 209
    derivation through free act of unification, 205, 206
    as immersed in own perspective, 331–33, 335
    importance of, 34–35, 453–54
    Lakoff on logical concepts and, 493–500
    logic reformulation by, 88–94, 103, 201–7
    nature of precept and, 112
    as new form of consciousness, 91, 93–94, 169–70, 205, 206
concept of implicit definition (C.I.D.) (Iglowitz)
    Cassirer's concept and, 106–8, 112–13, 170, 196, 204–8, 210–11, 321–22n73, 527–28
    definition, 208–9n39
    Hilbert as basis for, 106–8, 112–13, 196
    knowing virtual objects within biology, 176
    logic as biologically based, 197–200
    ontic indeterminism and, 193
    origins of, 191–97
    outline of, 205–11
    presentation and, 201–3, 205–7, 211–13, 217–18
    relevance to mind-body problem, 213–16
    schematic operative model and, 173–76, 214n59
    scientific implications of, 182–83
    theory of attention, 203–4
concepts
    axiom systems as source of ultimate, 207–8
    classes and sets as specialized, 115
    consciousness in, 107
    Hilbert on, 36, 61–62, 334–35
    Kant's constitutive concept, 174
    Lakoff on, 489–500
    ordering within, and Cantor's diagonal proof, 96–101
    See also virtual objects
concordance as biology's proper conclusion, 173–76
congruence, Maturana and Varela's, 225, 260
connective invariants. *See* invariant relationality
consciousness
    Cassirer's concept as, 91, 93–94, 169–70, 205, 206
    C.I.D as basis of, 113–14, 205, 215
    denial of, 239

    Edelman's neglect of, 515–16
    intentionality and, 482–84
    materialist explanation's flaw, 24n2, 29, 34
    naturalism in theories of, 24n3
    in operative ordering of concepts in brain and, 34–38, 107
    as primary problem in mind-body question, 137
    reduction to brain, 476
    as relations system (James), 364
    schematic operative model's applicability to, 76–77, 177
conservation of autopoiesis, 231–32
consistency, Shapiro on Hilbert's axioms, 82–83
container schema (Lakoff), 489–90, 492, 502
Continental vs. Anglo/American philosophical traditions, 551
continuity of sentiency, 465–66n48
cortical vs. sensory mapping, 156–59
critical idealism
    in Cassirer, 69, 281n38, 331
    in Kant, 14, 223n7, 253–56, 385
cross categorization (Lakoff), 489n9
cross-reduction in symbolic forms, 305, 307, 308–10, 331
cultural forms (Cassirer), 298–304, 306n46, 314

decoherence, 408
deep ordering, 58
defined, mathematical version of, 196n4
Dennett, Daniel
    author's critique of, 469–70
    color phi argument, 187, 473–86
    heterophenomenological understanding, 109, 140–42, 253, 475n5
    mechanistic view of mind/body problem, 55–56
    naturalism of, 472–73
    on organisms' lack of "epistemic engines," 242
    "static problem," 401, 473–86
d'Espagnat, Bernard
    application to author's thesis, 439–40
    author's discovery of, 387–88n5, 436–37
    background on, 396–98
    on Bell's Theorem, 413–15
    on consciousness, 24n3
    epistemological vs. physics perspective, 389
    on local reality, 111, 395, 398–401
    overview, 393–94
    pante aporeton, 430–31
    parallelism with self-organized evolutionary optimization, 409–11, 422–23, 450

on predictivity of quantum mechanics, 406
quantum formalism, 425–28
realism and, 67, 395–96
"realism of the accidents" argument, 186–87, 421–23
"static problem" and, 401–3
"Veiled Reality" interpretation of quantum mechanics, 428–29, 449
diagonal proof on rational vs. real numbers (Cantor), 96–101
dimensional logic, inapplicability to mind-body problem, 517–18
Diophantine logic, 94
dualism, 214–17
Durant, Will, 359–68, 384, 418, 594nIX

Edelman, Gerald M., 147–50, 157–58, 180, 235n27, 310, 507–18
Einstein, Albert, 67, 390, 391–92, 393, 477
Einsteinian realism, 277n33, 401, 415–17, 434–36, 448–50
embodied logic, 129
embodiment, 225, 250, 506–7
emergence, 23n2, 28–29
engineering argument for schematic model of "mind," 145–46
environment. *See* externality; naïve perceptual objects
epiphenomenalism, 582nXXV
epistemologically relativized biology (Iglowitz)
    Bohr and Cassirer parallels, 330–31, 535–36
    connectivity map of brain connections, 526–27
    interface, cognator/object, 351–52
    Kant and, 222–24, 360–61
    Lakoff and "hierarchy" vs., 518–23
    Maturana and, 316–22
    necessity of, 252–53
    validity of, 335–36
epistemological relativism
    Bohr and, 323–31, 531–33, 536–50
    Cassirer's, 14, 69–70, 72, 124, 276–78, 286–89
    closure, 221–24
    conclusion, 262–64
    embodiment in, 506–7
    experience and, 266
    Hilbert and, 333–34
    interface thesis and, 354
    intrinsic processing power of the brain and, 132–33
    introduction, 219–21
    Kant and, 222–26, 253–56, 291–92n3
    Lakoff's problems with, 501–3
    naturalism and, 266, 276–78, 336
    See also symbolic forms

epistemology
  automorphism, 181–86, 347, 369, 372, 375–77, 379
  Bohr on, 323–31, 531–33, 536–50
  as fundamental mind-body question, 55–56, 220
  importance of complexity to effectiveness of, 514n53
  Kant on, 275
  language's function, 316–17
  logical dilemma vis-a-vis experience, 271–75
  Quine on, 19–20, 265–66
  role in physics, 51–52, 132
  See also knowing
equivalence classes in neuronic structures, 159–60, 161
evolutionary congruence, critique of, 225–26. *See also* biological evolutionary perspective
existence problems (Shapiro and Mac Lane), 118, 129
exotic mathematics
  mathematic structuralism and category theory, 55, 63–74, 77–86
  sources and perspectives, 55–63
  space and axiom systems, 74–77
experience
  as axiom of realist reason, 266–67, 268–71, 278–79, 297n25
  definition, 278
  equipotent theories of reality for, 270–71
  as expression of process, 276n29
  interface as synthesis of externality and, 256, 348
  invariant primitive of, 249–50
  as invariant relationality across all orientations, 223–24n11, 270–71, 278, 307, 376–77
  logical dilemma of epistemology, 271–75
  matching of theory to, 297n25
  naïve objects as prototypes of, 521
  naturalism relativized, and Cassirer, 276–78
  preconceptual, 492
  quantum physics as tool to describe, 399–400, 521–23
  Quine on, 19–20, 265–66, 267, 268–69, 270
  reference and, 322n74
  as understood through mathematical ideals, 520
  as uninterpreted primitive, 223, 252–53
  See also symbolic forms
extensionality, 89, 92
externality (external reality)
  author's goal to preserve, 24
  as axiom of realist reason, 243–52, 259–60, 352, 385n34, 415–16, 430–31
  Bohr on, 67, 459
  as a bound to evolutionary possibility, 177–80
  Cassirer on, 69–70

closed cognitive systems and, 221–24, 256
cognition's challenge in dealing with, 253, 292–93, 309–10, 348–49
denial of in Maturana's paradox, 237–38
emergence thesis, 28–29
existence question as one of faith, 461–62
foundations of mathematics as, 55
impossibility of actual knowledge of, 355–56
as interface, cognator/object, 49–50, 256, 351–52
Kant on, 24–25, 222–23, 348–49, 368
localized objects as lacking (physics), 67
as modulator of organism's internal activity, 234
naturalism's mistake in its claims for, 285–86
as necessity for realism, 292–93
operational objects correlating with, 108–10
Quine on, 20
See also ontic indeterminism; realism; reference, metaphysical

feature detection (Freeman), 156–57, 158
feedback loop in mind/brain mechanism, 32, 43–44, 411–12, 417–18
Feynman, Richard, 425–26
Fine, Arthur, 247
finitary proof theory, 84
formal and abstract problem, 165–76
Freeman, Walter J.
    automorphism, 181–86
    biological model for object relationship, 116, 151–65
    on circular causality and intentional arc, 32–33
    equivalence classes, 310
    non-hierarchical mapping of sensory input onto brain, 43, 45, 46, 410, 412
    spatial integration, 356n7
free will, Bohr's use of, 326n85
Frege, Gottlob, 35, 117, 128
functionalism, 214n60, 283–84, 288. *See also* concept as functional rule

general group/integral domain, existence of, 551–52
Gleick, James, 372
global mapping (Edelman), 147–50
"God's eye view"
    Durant and JCS reviewer, 381–84
    Edelman and, 148–50, 511, 513
    impossibility of, 165, 185, 256, 263, 381, 446–47
    Lakoff, 506–7
Goldblatt, Robert, 78, 556
Goodman, Nelson, 474

Gould, Stephen, 170–71
G.U.I. metaphor for cognator/object interface, 46–47, 50, 146–51, 161, 166

Halmos, Paul, 94
Heisenberg, Werner, 407n48
Helmholtz, H.L.F. von, 480
Hertz, Heinrich, 142, 204–5n28, 263, 280, 382, 533
heterophenomenological understanding (Dennett), 109, 140–42, 253, 475n5
hidden variable, defined, 413n59
hierarchical/non-hierarchical converters
    intentionality and, 404, 425, 528–29
    virtual objects as, 73, 111, 116, 374
    See also non-hierarchical structures
hierarchy
    author's epistemological perspective, 518–23
    biology as symbolic form, 505–7
    Cassirer and, 493–500
    classical concept and, 488–90
    conclusion, 523–24
    Edelman, 507–18
    emergence theory, 23n2
    introduction, 487
    Lakoff, 488, 493, 497–98, 499–504
    naturalism's dependence upon, 245–46, 319–20
    Putnam's internal realism requirements, 500–501
    quantum mechanics' challenge to, 406
Hilbert, David
    Cassirer on, 558
    on concepts, 36, 61–62, 334–35
    Heisenberg and, 407n48
    influence on author, 47, 105–8, 110–13, 125–28, 196
    on invariance in axiom systems, 120n35, 125–28
    mathematical structuralism contribution, 59, 70–71, 78–86, 128, 211–12n50
    naïve perceptual objects in axiom systems, 61–62, 81
    ordering's foundation in axiom system as a whole, 102
    relative consistency proofs, 84, 117–18, 129, 133
    Schlick on, 116
    Shapiro vs. author on, 79–80, 82–83, 118
    Van Fraassen on, 296
    See also implicit definition
homunculus argument
    author's solution to, 37, 47, 58, 60, 62, 217, 517
    Edelman's neglect of, 515–16
    as sourced in presentation vs. attention in logic, 203n23, 217

hormonal level in reactivity of organism, 313–14

idealism, 239, 281n38, 300, 331, 443n1. *See also* critical idealism
idealized cognitive models (ICMs) (Lakoff), 129, 491n17, 492, 502, 504
implicit definition (Hilbert)
    application to mathematics, 57
    Cassirer on, 204n27, 294n13, 298
    development of, 35–37
    distinction between earlier and later conceptions, 38–39
    influence on author, 105–8, 110–12
    as logical/mathematical objects, 75, 107–8, 176, 178–80, 207n36
    operative knowledge of internal functioning, 47
    as reinterpretation, 127–28
    Schlick on, 59
    scope of beyond Cassirer's concept, 333–34
    Shapiro on, 79
    virtual objects of as knowable, 35, 176, 212, 214
infinity and Cantor's diagonal proof, 95, 98–100, 103–4
information
    Edelman's competence vs., 509–10, 511n51, 514
    evolution's process method without, 77n53
    Freeman on, 152–55, 158–59
    intentionality in strategic brain and, 170
    mind as function of organization vs., 42–43, 116, 192–95, 375
    realism without, 213
    schematic operative model and, 168, 177, 180–81, 225, 258, 312
    structural coupling argument against, 138
informational model (Raichle), 131
intentionality
    in author's theses, 361–62
    brain's strategic function and, 170, 347
    consciousness and, 482–84
    in constitution of virtual objects, 32–33, 43, 166, 308, 409, 411–12, 417–18, 481–82
    emotional aspect of, 314
    evolutionary origins of, 404–6, 425, 528–29
    interface axiom and, 355
    as key to mind/brain partnership, 44, 195
    language as part of, 248
    Merleau-Ponty/Freeman and, 32–33
    realistic imperative and, 189
    vs. referents as substance of the mind, 115
    as "static problem" solution, 473–86

interface, cognator/object (Iglowitz)
    author's conclusions, 461–63
    as axiom of realist reason, 355–57
    Bohr and, 330, 542, 546–47
    G.U.I. example, 46–47, 50, 146–51, 161, 166
    and interpretation of experience, 249
    materialism and, 49–50
    mathematical ideals and, 345
    as metaphysical entity, 49–50, 256, 351–52
    overview, 353–57
    and scientific realism, 189–90
    as synthesis of externality and experience, 256, 348
    See also structural coupling
internal realism requirements (Putnam), 500–501
invariant relationality
    in axiom systems, 118–20
    biology's lack of (Freeman), 155
    Cassirer's, 288–89, 305, 307, 354
    in Einsteinian theory, 277n33
    as epistemological relativism basis, 72
    experience and, 223–24n11, 249–50, 266, 270–71, 278, 307, 376–77
    in mathematical ideal, 344–45
    naturalism and, 277, 286
    organizational processing applicability of, 210, 378
    predictive organization vs., 294
    preserving, 123–28, 264, 436
irenic relativism, 277n33

James, William, 131, 359, 362–66, 545–46, 547–48
Journal of Consciousness Studies (JCS) reviewer, 359, 369–81

Kant, Immanuel
    Cassirer vs., 379
    constitutive concept, 174
    critical idealism, 14, 223n7, 253–56, 385
    Durant's realist critique of, 359–62, 366–68, 594nIX
    epistemological relativism of, 222–26, 253–56, 291–92n3, 360–61
    on epistemology, 275
    on externality, 24–25, 222–23, 348–49, 368
    mind-body perspective, 50–51, 219–20, 222–23, 384–86
    ontological sympathy with author, 24–25
    the reader's challenge, 15
    as realist, 443n1
Kepler, Johannes, 21

Kline, Morris, 104
knowing
    as axiomatic, 47, 188
    Cassirer on impossibility of, 69–70
    of external reality as indeterminate, 255
    Maturana and Varela on goal of project, 226n19
    operative, 49
    virtual objects of implicit definition as knowable, 35, 176, 212, 214
    See also entries starting with epistemolog...
Kuhn, Thomas, 544–45
Kumar, Manjit, 388n5, 394–95, 413n59

Lakoff, George
    category cue validity, 203n24
    epistemological relativism of, 497–98, 499–504
    ICMs, 129, 491n17, 492, 502, 504
    on internal realism, 500–501
    on logic, 90n6, 488, 492–94, 497–98, 499–500
    radial categories, 490–91, 497
language, function of, 248–49, 256, 316–17
Laplacean ideal, 318–19n69
Leibniz, Gottfried, 47, 60, 75, 553–54, 555, 556
linguistic idealism, 239
locality of objects
    Cassirer on impossibility of knowing, 69–70
    quantum challenge to, 67, 111, 394, 395, 398–401, 413–15
logic
    as biological/evolutionary faculty, 48, 75, 76n51, 128–33, 171–72, 175, 197–200, 310–13, 526–27
    Cassirer's conceptual reformulation, 88–94, 103, 201–7
    classical concept, 88–89, 199–200n12, 319, 331–32, 488–90, 494–98, 552, 584nIII
    dimensional, 517–18
    Diophantine, 94
    embodiment of, 129
    epistemological dilemma vis-a-vis experience, 271–75
    idealistic (as in mathematical ideals) basis for, 519–21
    in implicit definition, 75, 107–8, 176, 178–80, 207n36
    Lakoff on, 90n6, 488, 492–94, 497–98, 499–500
    of operative knowing, 49
    ordering as logical relationship, 107, 112–13
    relativism in, 19–20
    schematic operative model and, 174–75
    substance of mind as, 485
    See also concept of implicit definition (C.I.D.); set theory

looping and circular probability machine, mind as, 32, 43–44, 411–12, 417–18

Mac Lane, Saunders
    on author's conception, 110–11
    category theory, 114–16
    existence problems, 118, 129
    influence on author, 56, 95–96
    mathematical objects as virtual positions in axiom systems, 57, 59–60
    mathematical structuralism, 56, 64–66, 114
    rule of reference issue, 552, 556
materialism
    author's perspective, 29–34, 48–49
    author's reversal of Freeman's diagram, 45
    axioms as physically conceived, 38–39
    brain perspective in, 41–47
    conclusion, 50–53
    consciousness explanations, problems with, 24n2, 29, 34–38
    Edelman's reductions, 507–10
    emergence and, 28–29
    epistemologically relativized, 526–27
    hierarchy in, 23n2, 319–20, 406
    interface, 49–50
    introduction, 27–28
    Maturana and Varela's relativized, 252–53, 263
    organization of experience in, 278–79
    other minds issue, 39–41
    symbolic forms and, 48
    See also naturalism
mathematical ideals, 306n45, 339–49, 351, 503, 518–21
mathematical objects, necessity of existence problem, 118
mathematical Platonism, 20–21
mathematical structuralism
    author's discovery of, 56–57, 59–60, 96–101
    Bell, J.L., 71–72
    Benacerraf's perspective, 64
    Cassirer's, 56, 69–70, 72, 78, 553–60
    Hilbert's contribution to, 59, 70–71, 78–86, 128, 211–12n50
    infinity and Cantor's diagonal proof, 95, 98–100, 103–4
    introduction, 55
    Mac Lane and, 56, 64–66, 114
    meaning in, 296–97
    Resnik's, 57, 59, 63–64, 67
    schematic operative model and, 57–58, 68–69, 73–74, 101–2
    set theory vs., 66–67, 105
    Shapiro on, 78–86, 117

Stefanik, 70–71
survey of, 56–59
mathematics, author's influences in, 94–96, 581nXI
mathematics, functional concept of. *See* concept as functional rule
Maturana, Humberto
    autopoiesis/autopoietic unity, 133, 227–29, 231–32, 310
    on biological organism as closed system, 223, 234–36
    causation vs. triggering, 230–31
    congruence problem, 225, 260
    conservation of autopoiesis, 231–32
    epistemological relativized biology contribution, 316–22
    externality axiom, 243–52
    introduction/summary, 223–28
    Kant and, 256
    naturalist argument, 239–47, 249, 316–17
    on nervous system structure, 157
    relativized materialism, 252–53, 263
    representational paradox, 137–38, 237–39
    on structural determination of systems, 27
    Tree of Knowledge, 137–38
    on unicellular organization of response, 313
    See also structural coupling
meaning
    autonomous theory of, 213
    in axiom systems, 35–37, 102, 334–35
    mathematical sense of, 296–97
    as sourced in Hilbert's implicit definition, 334
    variation in non-hierarchical systems, 83n62
mechanics, reducing structure of brain to, 27–28, 30–38, 70. *See also* schematic operative model
Merleau-Ponty, Maurice, 32, 46, 161–62
metamathematics, 79–83
metaphysical form, abstract limits of, 348. *See also* externality
mind-body/brain problem
    author's challenge to tradition, 443–44n1, 467–69
    category theory and Mac Lane, 114–16
    conclusions on author's solution, 445, 455–59
    outline of author's argument, 565–73
    overview, 87
    quantum mechanics' applicability to, 388–90
    realism's inability to solve, 24
    ultimate social importance of, 463–65, 468–69
    See also concept of implicit definition (C.I.D.); interface, cognator/object; schematic operative model; substance of mind
multitudinism conception of reality. *See* locality of objects

Nagel, Ernest, 28n4, 211n48
naïve perceptual objects
    in axiom systems, 61–62, 81
    Cassirer on, 69–70, 92, 215–17
    as cognition (problem of), 223–24n11
    cognition's continual test of intentional arc vs., 32–33, 43, 411–12, 417–18
    as in correspondence to biological process, 246
    decoherence, 408
    as free mental creations (acts of unification), 206, 212, 214–15
    hierarchical vs. non-hierarchal structure, 319–22, 374
    as implicitly defined logical objects, 35, 107–8, 176, 178–80, 210, 212, 214
    in intentionality solution to "static problem," 473–86
    mathematical ideals as relativized example, 339–49
    as organizational and virtual artifacts of behavior, 38, 43, 46, 68, 72–73, 181–82, 210, 251, 273, 409
    as triggers/boundary conditions vs. motivators to behavior, 145, 224, 228–29
    See also schematic operative model
naturalism
    anti-realism and, 276n26
    author's perspective, 251, 274–76, 291–93, 321–22, 336, 447–48, 459–62, 463n44
    biological extension of, 274–75, 459–60
    Cassirer on, 276–78, 301, 499
    comprehensiveness claim of, 306n46, 315–22
    confusion of organization with phenomena, 276n26, 285–86, 382–83
    in consciousness theories, 24n3
    Dennett's, 472–73
    epistemological relativism and, 266, 276–78, 336
    hierarchy of scale in, 245–46, 319–21
    Lakoff's argument against, 499
    Maturana and, 239–47, 249, 316–17
    schematic operative model as extension of, 481–82
    substance under author's definition, 278–79
    See also materialism
nervous system, 156–60, 234–35, 313
neuronal group selections (TNGS) (Edelman), 507–12
neuronal structure, 157–59, 233–35
neurophilosophy, 52, 60, 63, 388
Newton, Isaac, 554
Newtonian physics, 164–65, 393
non-hierarchical structures
    in axiom systems, 321
    mapping of sensory input onto brain, 43, 45, 46, 410, 412
    meaning variation in, 83n62

naïve perceptual objects and, 319–22, 374
in schematic operative model, 246
See also hierarchical/non-hierarchical converters
non-local entanglement, 399, 403, 418, 422
non-locality principle in quantum theory, 413–15
non-representational models. *See* schematic operative model
non-topological mapping by the brain, 152–58
"normal" cognition and quantum physics, 428–29

objective existence of reality. *See* externality
objects. *See* naïve perceptual objects; schematic operative model; virtual objects
objects-Platonists, 57, 64, 65
observation in quantum physics, 401
Occam's razor, 100, 123
ontic indeterminism (Iglowitz)
    author's biologically based argument for, 48–49, 253
    author's conclusions, 463–66
    behavior without information and, 116, 375
    Bohr's contributions to, 330
    Cassirer's contributions to, 330, 560–61
    diagrammatic illustration, 193
    Kant's, 223, 255–56, 385
    machine analogy to human brain, 30–34
    other minds issue, 39–41, 368, 420–21
    overview, 188
    strategic brain and, 33
    value of, 463–64
    virtual nature of perceptual objects, 37–38
ontogeny, 228–29, 230
ontological existence postulates, 49–50. *See also* externality
operational closure of organism, 234–35, 240–41
operational concept. *See* concept of implicit definition (C.I.D.)
operational objects. *See* virtual objects
operative knowing. *See* schematic operative model
ordering
    Cassirer, 87, 93
    within a concept and Cantor's diagonal proof, 96–101
    consciousness in process of, 34–38, 107
    as foundation in axiom system, 102–3, 107
    as logical relationship, 107, 112–13
organization
    as basis for aesthetic principle, 122
    comprehensive vs. categorical and prediction, 318–19
    as evolutionary principle, 42–43, 242, 253, 409–11, 422–23, 450

invariant relationality, 210, 294, 378
of mind for efficient response, 42–43, 116, 192–95, 375
modeling of virtual objects, 138–42
naïve perceptual objects' role, 38, 43, 46, 181–82, 273, 409
naturalism's confusion of with phenomena, 276n26, 285–86, 382–83
objects as servants of, 143
vs. reference in theories of experience, 322
in schematic operative model, 46–47, 68, 73–74, 166, 178, 198, 313
self-organization of rules of the brain, 38, 160, 161, 409–12, 422–23, 450
theoretical principles, 124–25
other minds issue, 39–41, 185, 352n3, 366–68, 418–21, 465–66

pante aporeton, 430–31
parallelism
    critique of Maturana and Varela and, 225
    d'Espagnat on, 409–11, 422–23, 450
    in structural coupling, 256–61
particular phenomena vs. metaphysical reality, 282–83
Penrose, Roger
    on accuracy of scientific theories, 14, 392–93n15
    on aesthetic criteria in mathematics and sciences, 122–23
    anthropic principle, 110, 380
    on quantum physics, 43, 108–9, 320, 405, 411, 522
perception
    algorithmic relation to nameless reality, 110
    as internal mental process based in biology, 206–7
    Merleau-Ponty and Freeman on, 32–33, 46
    See also naïve perceptual objects
percepts, function of, 73, 108, 112–14, 162–63, 169–70
peripheral vs. cortical neurons, connection type, 157–59
personality principle (Schrödinger), 420n79
"The Philosophy of Ernst Cassirer" (Cassirer), 562
physics, laws of
    challenges to science, 273
    epistemology and, 132
    as mathematical translations, 120–21
    Newtonian, 164–65, 393
    social interface with, 390–93
    See also quantum physics
Piaget, Jean, 20–21
planetary system example of invariance, 119–20
Poincaré, Henri, 103–4
preconceptual experience (Lakoff), 492
predictivity of quantum mechanics, 400, 406–8
present, structural, 235

presentation
- Cassirer's challenge to as logic foundation, 92, 93, 201–3, 205–7
- C.I.D's relationship to, 211–13, 217–18
- Edelman on, 511–12
- paradoxes created by, 206
- as problem of consciousness, 137
- *See also* naïve perceptual objects

primitive objects of experience. *See* naïve perceptual objects

Ptolemean physics, modern, 120–21

Putnam, H., 500–501

quantum formalism, 124, 425–28, 452–53
quantum physics
- applicability to mind-body problem, 388–90
- circular causality, 43
- decoherence, 408
- Einsteinian realism and, 415–17, 434–36, 449–50
- as experience defining tool, 399–400, 521–23
- introduction, 387–88
- locality of objects, 67, 111, 394, 395, 398–401, 413–15
- non-local entanglement, 399, 403, 418, 422
- "normal" cognition and, 428–29
- observation in, 401
- other minds issue, 418–21
- Penrose on, 43, 108–9, 320, 405, 411, 522
- perspective-shifting nature of, 394–95
- predictivity of, 400, 406–8
- reduction of material to subatomic particles, 23–24
- scientific realism and, 438–40
- social interface with, 390–93
- strategic brain and, 427–29, 432
- as structural model for human brain, 378, 432–33
- twin-slit experiment, 404–5
- validation of theory, 393n16

Quine, W. V. O.
- on knowledge and belief system as closed, 223
- on knowledge and experience, 19–20, 265–66, 267, 268–69, 270
- on non-existence of objects, 114, 115, 336–37
- on science as an experience-bounded field of force, 126–27

radial categories (Lakoff), 490–91, 497
radical empiricism (James), 364
Raichle, Marcus, 130–33, 588nxxv
re-afferent model, 109

realism
- Bohr on, 67, 323–31, 459, 531–33, 536–50
- Cassirer on naïve realism, 383–84
- d'Espagnat and, 67, 395–96
- Einsteinian, 247, 277n33, 401, 415–17, 434–36, 448–50
- imperative of, 185–87, 188–89
- Kant's ontic indeterminism and, 255–56
- necessary assumptions of, 351–52
- preamble/overview, 23–25
- Putnam's internal realism requirements, 500–501
- Russell on naïve realism, 243
- utilitarian approach to, 293, 314, 321, 332–33, 364–65
- without information and presentation, 213
- *See also* externality; naturalism; scientific realism

"realism of the accidents" argument (d'Espagnat), 186–87, 421–23

realist reason, axioms of
- experience, 266–67, 268–71, 278–79, 297n25
- externality, 243–52, 259–60, 352, 385n34, 415–16, 430–31
- interface, 355–57

real number continuum, 525

reduction, 301, 304–5, 310n54, 319, 321, 476. *See also* materialism

re-entrant topobiological maps (Edelman), 507, 512–13

reference, metaphysical
- author's opposition to, 321
- closed cognitive systems, organisms as, 347
- experience and, 322n74
- lack of need for, 346–47
- naturalism's grounding in absolute, 292–93
- reevaluation of, 297–98
- Shapiro and Mac Lane, 552, 556
- *See also* representation

relationality
- axiom systems, 118–20, 125–28, 208–9
- Cassirer's focus on general-to-specific, 91–92
- as locus of reality, 142
- schematic models and naïve perceptual objects, 57–58, 107, 112–13, 138–46, 556, 559–60
- of scientific realism, 251–52
- *See also* invariant relationality

relative consistency proofs (Hilbert), 84, 117–18, 129, 133

relativism, mathematical, 339–49. *See also* epistemological relativism; epistemological relativized biology

relativized materialism, 252–53, 263

representation
- lack of for perceptual objects, 38, 47, 58, 75–76
- lack of in biological systems, 236–38

Maturana and Varela's argument against, 224, 237–39
moving beyond, 154–55, 158, 160, 162–63, 167, 181–83
schematic operative model as counter to, 57–58, 76, 137–39, 176–77
Resnick, Michael, 57, 59, 63–64, 67
response. *See* behavior
Rosch, Eleanor, 488
Rosen, Robert, 118–19
rules, brain's use of, 38, 160–61, 174, 379, 412, 422–23, 450. *See also* concept as functional rule
Russell, Bertrand, 242–43, 384, 525

schematic operative model (Iglowitz)
    in author's response to *JCS* reviewer, 374–76
    automorphism and Freeman, 181–86
    biological/evolutionary case, 49, 150–65, 167–72, 364
    C.I.D.'s relationship to, 173–76, 214n59
    color phi phenomenon and, 479–82
    conclusions, 187–90
    concordance as biology's proper conclusion, 173–76
    consciousness and, 76–77, 177
    d'Espagnat and realism of the accidents, 186–87, 421–23
    engineering argument, 145–46
    externality correlation challenge, 108–9
    formal and abstract problem, 165–72
    G.U.I. example, 46–47, 50, 146–51, 161, 166
    information and, 168, 177, 180–81, 225, 258, 312
    mathematical structuralism and, 57–58, 68–69, 73–74, 101–2
    Maturana's paradox and, 237n29
    meaning and, 296–97
    naïve perceptual objects as virtual servants of, 143
    non-hierarchical nature of, 246
    object/model relationship, 57–58, 138–46, 556, 559–60
    objects as organizational process, 46–47, 68, 73–74, 166, 178, 198, 313
    representation vs., 57–58, 76, 137–39, 176–77
    shape principle for theories and, 454
    structural coupling and, 246
    summary, 176–80
    workability of, 447–48
Schlick, Moritz, 36–37, 102, 116, 209, 334
Schrödinger, Erwin, 43, 407–8, 420n79
science
    challenges of modern physics to, 273
    constructive role of philosophy for, 52
    as experience-bounded field of force, 126–27
    hierarchy's breakdown at quantum level, 320

    as organizer of phenomena, 291
    Penrose's aesthetic criteria for, 122–23
    questioning of foundation of realism, 23
    symbolic forms and, 354, 533–35, 550
    See also physics; scientific realism
science of mind. *See* mind-body/brain problem
scientific epistemological relativism, 565–73. *See also* epistemological relativized biology
scientific realism
    author's perspective, 291–94, 448–54
    confusion of experience with organization of experience, 354–55
    d'Espagnat on, 67
    importance in author's thesis, 415
    interface and, 189–90
    quantum physics and, 320, 438–40
    reducing mind to biology, 175
    strategic brain and, 443–59
    tenets of, 189
    Van Fraassen on, 295–96n18
    variations on depending on theory, 392–93
    viable relationality of, 251–52
    See also biological evolutionary perspective; naturalism
self-referentiality of the human brain, 433–34
semantic view of theories (Van Fraassen), 295–96
sensory surface, 233–34
sensory vs. cortical mapping, 156–59
series as logical and conceptual organization, 208n39, 552
set theory
    category theory as supplanting, 78
    in classical logical concept, 89n4, 331–32, 488–90
    vs. concepts as basis for mathematics, 87n2
    mathematical structuralism vs., 66–67, 105
    mathematicians critical of, 104
    problem as logical primitive of thought, 64–65
shape principle for theories, 122, 123, 124, 454
Shapiro, Stewart, 78–86, 117–18, 129, 133, 552, 556
simplicity, triumph of, in biology, 170–71
Smart, H., 562–63
social interface with physics, 390–93
space and axiom systems, 74–77
spatial distribution of function in the brain, 158–60
spatial integration (Freeman), 356n7
specialized concepts, classes and sets as, 115
specific phenomena vs. metaphysical reality, 282–83
"spiritual" relativism (Cassirer), 303
"static problem," the, 401–3, 473–86

Stefanik, Richard, 59, 65, 67, 70–71
strategic brain
    applying to d'Espagnat's views, 401
    intrinsic processing power of, 131–33
    lack of direct connection to external reality, 423
    as looping, circular probability machine, 32, 43–44, 411–12, 417–18
    meaning as organizational in, 297
    ontic indeterminism and, 33
    quantum physics and, 427–29, 432
    scientific realism and, 443–59
    See also intentionality; schematic operative model
structural coupling (Maturana and Varela)
    behavior and, 232–34
    as connection between organism and externality, 349, 380
    conservation of autopoiesis, 231–32
    as ground for biology, 250–51, 370
    parallelism in, 256–61
    quantum physics and, 416
    in schematic operative model, 246
    structural present in, 235–37
    as substance of mind, 379
    triggering and, 224, 228–31, 236–37, 258–59
    weaknesses of, 225, 261–62
structuralism. *See* mathematical structuralism
subjectivity of reality as perceived, 37–38. *See also* schematic operative model
"Substance and Function" (Cassirer), 561
substance of mind
    author's thesis, 347–49
    challenge of finding, 221–22
    as constitutive concept/rule of the brain, 174, 379
    intentionality and, 115, 347
    as logic, 485
    See also interface, cognator/object
substance vs. function as key to unity (Cassirer), 284
SUPERB scientific theories (Penrose), 14, 123
Swabey, W. C., 354
symbolic forms (Cassirer)
    author's perspective, 14, 48–49, 279–89, 298–310, 330–37, 450–51
    biology as, 310–15, 505–7
    Bohr's multivalued functionalism and, 323–31, 541–42
    Cassirer's thesis, 294–98
    cross-reduction in, 305, 307, 308–10, 331
    experience as invariant relationality across all orientations, 376–77
    influence on author, 123–25
    mathematical ideals and, 306n45, 339–49
    mathematical structuralism and, 69–70

science and, 354, 533–35, 550
symmetry assertion (Cassirer), 281
synthetic vs. analytic statements (Quine), 19–20

theory, matching of experience to, 297n25
theory of attention, 203–4
"Theory of Symbolic Forms" (Cassirer), 561–62. *See also* symbolic forms
Thurston, William P., 104
topological vs. non-topological aspects in nervous system, 156–60
train argument (Einstein), 477–78
translation vs. reduction, 321
translatory invariants. *See* invariant relationality
triggering in structural coupling, 224, 228–31, 236–37, 258–59
twin-slit quantum mechanics experiment, 404–5

ultimate reality. *See* externality
unicellular perspective, 170–171, 310–13
utilitarian approach to realism, 293, 314, 321, 332–33, 364–65

Van Fraassen, Bas, 294–96
Varela, Francisco. *See* Maturana, Humberto
"Veiled Reality" interpretation of quantum mechanics (d'Espagnat), 428–29, 449
virtual embodiment, 250. *See also* schematic operative model
virtual objects (of the mind)
   in axiom systems, 57, 59–60, 61
   externality correlation, 108–10
   as hierarchical/non-hierarchical converters, 73, 111, 116, 374
   as implicitly defined, 35, 107–8, 176, 178–80, 210, 212, 214
   intentionality in, 32–33, 43, 166, 308, 409, 411–12, 417–18, 481–82
   naïve perceptual objects as, 38, 43, 46, 68, 72–73, 181–82, 210, 251, 273, 409
   ontic indeterminism and, 37–38
   organizational modeling of, 138–42
   as perceptual and conceptual, 62, 63, 70–71, 126, 174–75
   See also concept of implicit definition (C.I.D.); schematic operative model

Weyl, Hermann, 104
Wilder, Raymond, 81–82, 95, 210, 557
Wirzcup, Isaac, 94–95
Wittgenstein, Ludwig, 488

Zermelo-Frankel set-theoretic foundation for numbers, 66

www.ingramcontent.com/pod-product-compliance
Lightning Source LLC
Chambersburg PA
CBHW021129230426
43667CB00005B/69